Calcium signalling

Lars Kaestner

Calcium signalling

Approaches and Findings in the Heart and Blood

With 96 figures

 Springer

Dr. Lars Kaestner
Universität des Saarlandes
Institut für Molekulare Zellbiologie

ISBN-13 978-3-642-34616-3   ISBN 978-3-642-34617-0 (eBook)
DOI 10.1007/978-3-642-34617-0

Die Deutsche Nationalbibliothek verzeichnet diese Publikation in der Deutschen Nationalbibliografie;
detaillierte bibliografische Daten sind im Internet über http://dnb.d-nb.de abrufbar.

Springer Medizin
© Springer-Verlag Berlin Heidelberg 2013

Cover Design: deblik Berlin

Printed on acid-free paper

Springer Medizin is brand of Springer
Springer is part of Springer Science+Business Media (www.springer.com)

Foreword

This book is a habilitation treatise. I aimed to both fulfil the formal requirements of the habilitation treatise as defined in the habilitation regulation of the Medical Faculty at Saarland University and provide a resource for colleagues, students and all that are interested in calcium signalling and/or optical imaging methods to record it. As such, I hope this booklet is more than a file in academic qualification, and that it might also serve as a useful tool whenever the universal messenger calcium crosses the track.

The biological examples are recruited from two major fields, cellular cardiology and red cell research. I am grateful to the people who not only introduced me to these subjects, but accompanied my research for many years, who are Prof. Peter Lipp and Prof. Ingolf Bernhardt, respectively. Although both fields seem to be independent, future research, systems biology approaches and translational methods will reveal interconnections going far beyond the obvious ones described in the Introduction. Therefore, the current compilation can only be seen as a snapshot, and I am looking forward to the future gain of knowledge.

Finally, I would like to thank all of my colleagues who accompanied my work and have contributed to the scientific outcome so far. I am grateful to my family, who suffer from restrictions caused by the intensity of the scientific passion. Last, but not least, I thank Springer for publishing this script.

Homburg/Saar, September 2012   Lars Kaestner

Contributors

Parts II-IV contain contributions of the following people:

Bennekou, Poul
Department of Biology
University of Copenhagen
Denmark

Bernhardt, Ingolf
Laboratory of Biophysics
Saarland University
Saarbrücken
Germany

Bloch, Wilhelm
Molecular and Cellular Sports Medicine
German Sport University Cologne
Germany

Bogdanova, Anna
Veterenary Physiology
University Zürich
Switzerland

Boldyrev, Alexander A.
International Biotechnilogy Centre
Moscow State University
Russia

Chien, Kenneth R.
Massachusetts General Hospital
Boston
USA

Christophersen, Palle
Neurosearch A/S
Copenhagen
Denmark

Edelmann, Ludwig
Molecular Cell Biology
Saarland University
Homburg/Saar
Germany

Fleischmann, Bernd K.
Life & Brain Center
University of Bonn
Germany

Flockerzi, Veit
General & Clinical Pharmacology
Saarland University
Homburg/Saar
Germany

Freichel, Marc
Institute of Pharmacology
University Heidelberg
Germany

Gassmann, Max
Veterinary Physiology
University Zürich
Switzerland

Hammer, Karin
Department of Pharmacology
UC Davis
USA

Hanske, Gabriela
Aesculap Pharmacy
Dortmund
Germany

Held, Brigitte
GRADE BioMed First
Goethe University
Frankfurt/Main
Germany

Jones, Larry R.
Krannert Institute of Cardiology
Indiana University
Indianapolis
USA

Jung, Achim
SECTOR Cert GfZmbH
Cologne
Germany

Jung, Jennifer
Life Science
University of Applied Science
Zweibrücken
Germany

Justus, Isabel
Pharmacy Board
Bremen
Germany

Kahl, Valentin
Ibidi GmbH
Munich
Germany

Kirchhefer, Uwe
Pharmacology & Toxikology
Wilhelms University
Münster
Germany

Kraegeloh, Annette
Leibnitz Institute for New Materials
Saarbrücken
Germany

Kraushaar, Udo
Natural and Medical Science Institute
Reutlingen
Germany

Lipp, Peter
Molecular Cell Biology
Saarland University
Homburg/Saar
Germany

Maia, Sara
Laboratory of Biophysics
Saarland University
Saarbrücken
Germany

Makhro, Asya
Veterenary Physiology
University Zürich
Switzerland

Müller, Oliver
Institute for Genetics
University of Cologne
Germany

Müller, Torsten
JPK Instruments
Berlin
Germany

Neumann, Joachim
Pharmacology & Toxikology
Martin-Luther University
Halle/Saale
Germany

Bach Nguyen, Duc
Molecular Biology
Hanoi University of Agriculture
Vietnam

Oberhofer, Martin
Molecular Cell Biology
Saarland University
Homburg/Saar
Germany

Oleinikow, Katharina
Life Science
University of Applied Science
Zweibrücken
Germany

Pahlavan, Sara
Molecular Cell Biology
Saarland University
Homburg/Saar
Germany

Ruppenthal, Sandra
Molecular Cell Biology
Saarland University
Homburg/Saar
Germany

Schmitz, Wilhelm
Pharmacology & Toxikology
Wilhelms University
Münster
Germany

Scholz, Anke
Molecular Cell Biology
Saarland University
Homburg/Saar
Germany

Schumann, Christian
Leica Microsystems CMS GmbH
Wetzlar
Germany

Schwarz, Sarah
Ursapharm Arzneimittel GmbH
Saarbrücken
Germany

Steffen, Patrick
Experimental Physics
Saarland University
Saarbrücken
Germany

Tabellion, Wiebke
Roche Diagnostics GmbH
Mannheim
Germany

Tian, Qinghai
Molecular Cell Biology
Saarland University
Homburg/Saar
Germany

Viero, Cedric
Molecular and Experimental
Medicine
Cardiff University
United Kingdom

Vogel, Johannes
Veterenary Physiology
University Zürich
Switzerland

Wagner, Christian
Experimental Physics
Saarland University
Saarbrücken
Germany

Wagner-Britz, Lisa
Laboratory of Biophysics
Saarland University
Saarbrücken
Germany

Wang, Jue
Molecular Cell Biology
Saarland University
Homburg/Saar
Germany

Weiss, Erwin
Veterinary Medicine
University of Cambridge
United Kingdom

Weissgerber, Petra
Experimental Pharmacology
Saarland University
Homburg/Saar
Germany

Zantl, Roman
Ibidi GmbH
Munich
Germany

Content

I Summary

II Calcium Signalling Methodology

III   Calcium Signalling in Cardiac Myocytes

IV   Calcium Signalling in Red Blood Cells

Summary

Introduction

The importance of calcium in the vital functions of cells was first recognised in 1883 by Sidney Ringer, who performed experiments demonstrating that minute amounts of calcium were necessary to maintain heart muscle contractility[1].

Currently, approximately 130 years later, our knowledge of calcium as a versatile signalling molecule has increased tremendously, and we have a good understanding of its general principles[2-4]. As described in these reviews, calcium signalling is a toolkit, and the spatio-temporal modulation of calcium concentration translates into a plethora of signals that trigger processes ranging from cell proliferation to cell suicide[5]. Life is based on millions of such processes, which can be competitive with, independent of or synergistic with other processes, and builds a network that results in physiological function.

In the physiology of vertebrates, one of the most illustrative synergistic functions is the relationship between the heart and blood. Obviously, the primary function of the heart is to pump the blood through the body, and the primary function of the blood is to carry oxygen and nutrients. The heart could not pump without the oxygen and nutrients delivered by the blood, the blood could not collect the oxygen and nutrients without being pumped by the heart, and oxygen could not be delivered to any part of the body without the cardiomyocytes and red blood cells working »hand in hand«. Although this relationship sounds straightforward, most of the physiological details are much more complex. Calcium signalling it is vital for cardiac myocytes because calcium mediates the electrical stimulation of mechanical contraction, whereas in red blood cells, calcium is predominantly a suicidal signal.

The contrast can hardly be greater. A cardiac myocyte contracts 2.5 billion (10^9) times in an 80 year human life and undergoes as many calcium transients, whereas in red blood cells, a significant calcium increase is the signal of death, which occurs in each red blood cell just one time. However, in every hour, the human body produces as many red blood cells as the heart contains cardiomyocytes (approximately 9 billion). Because the numbers of produced and dying red blood cells are balanced, the total number of calcium signals in red blood cells in the human body during an 80 year lifetime amounts to 6 x 10^{15}. Hence, in a human body, the number of calcium signals in red blood cells is only a few thousand-fold less than the total number of calcium signal cycles in cardiac myocytes.

In the following seven ► chapters (2-8), a general overview of strategies for studying calcium signalling is presented, whereas in part II more concrete problems and approaches are detailed. ► Chapters 9 and 10 and the corresponding parts III and IV provide evidence and examples of calcium signalling in cardiac myocytes and red blood cells, respectively. An outlook to future research on cellular cardiology and red blood cells is provided in ► chapter 11.

Direct evidence - the digital approach

Historically, the importance of calcium was discovered when experiments were performed in the presence and absence of calcium. Thus, Sidney Ringer discovered the essential role of calcium in the process that is now named excitation–contraction coupling (ECC)[1]. Ringer performed experiments with tap water and distilled water and identified the presence of calcium as a necessary requirement for ECC. This early experiments hinted that the physiologically relevant calcium concentration is rather small and is in the μM range. For Ringer, the »contamination« of the tap water with calcium was sufficient for his discovery. Because the contamination of all types of chemicals with small amounts of calcium is quite common, a nominally calcium-free solution cannot be regarded as being completely calcium-free; the solution requires an appropriate amount of a calcium-sequestering buffer, such as ethylene glycol tetraacetic acid (EGTA), ethylenediaminetetraacetic acid (EDTA) or 1,2-bis(o-aminophenoxy)ethane-N,N,N',N'-tetraacetic acid (BAPTA).

In red blood cells, George Gardos observed a breakdown of the potassium gradient in the presence of calcium under certain conditions[6,7]. This equilibration of the potassium gradient could be inhibited by calcium chelators[7]. Aside from the fact that the question of whether red blood cells contain calcium has a long history[8-10], this initial publication by Gardos inspired membrane function-related calcium investigations in red blood cells.

Although this digital approach is the most direct way to probe for the involvement of calcium, it generally lacks quantitative information. Nevertheless, experiments in the presence and absence of calcium are still popular and valid for performing negative or positive control experiments[11-14].

Fluorescence-based visualisation

Fluorescence-based visualisation is a very powerful method for exploring calcium signalling because it is a method where living cells can be investigated with relatively minimal disturbance. Fluorescence-based methods can deliver the following information:

- a high temporal resolution, down to the order of milliseconds[15].
- spatial resolution down to diffraction-limited spots[16,17], and in the future, potentially even lower (cp. references [18,19] and section 3.4 - Further optical sectioning techniques). The sensor itself can improve the spatial resolution by being targeted to particular organelles, cells or organs.
- quantitative information once the sensor (preferentially, but not exclusively, a ratiometric sensor) is calibrated.

Due to the universality of fluorescence-based calcium measurements, this chapter is by far the most comprehensive chapter of part I.

All fluorescence-based visualisation methods rely on a combination of molecular sensors and sophisticated technology to read out the modulated signal, which is usually fluorescence. Both topics, the sensors and detection technology, will be separately discussed in the following sections. Parts of the sections are modified from Kaestner and Lipp, 2007[16].

3.1   Calcium sensors

We are familiar with two major light-molecule interactions, absorption and emission. The absorption of light by molecules is a well-known process. However, to resolve a molecular entity by absorption,

at least a 1% attenuation of the light is necessary. According to the Lambert-Beer law, this attenuation would require a concentration of at least 10^7 M, which is equivalent to 10^{16} molecules. In contrast, in fluorescence recordings, a single molecule can be detected, especially when the background fluorescence is very low. Thus, fluorescent techniques are almost exclusively used for calcium imaging. On the sensor side, small molecule dyes and genetically encoded calcium sensors are available.

Small molecule dyes are still the most popular sensors for measuring cellular calcium signalling due to their wide range of available indicators and relatively simple application protocols. The most comprehensive overview of commercially available calcium sensors in the ultraviolet (uv) and visible light range is presented in the latest edition of the Handbook of Fluorescent Probes[20]. In addition to the protocols and advice in this book, Lipp *et al.* provided a useful guide for the selection and use of fluorescent calcium indicators approximately 10 years ago[21], which is still relevant because very little has changed in the small molecule calcium sensor field in the last decade. The only recent extension in the calcium sensor range was into the red/far-red spectral region using dyes such as Asante Calcium Red or Asante Calcium NearIR, which were brought to the market by Teflabs Inc. (Austin, USA). Practical reports on these new dyes are still sparse[22]; therefore, it is difficult to evaluate their utility. However, Asante Calcium Red might be an interesting sensor for the red blood cells, because it is an ratiometric dye, and other popular ratiometric calcium sensors, such as Indo-1 and Fura-2, can not be used in red blood cells[13] (► chapter 16).

Apart from aequorin, which is separately discussed (► chapter 4), genetically encoded calcium sensors were introduced to the scientific community 15 years ago[23]. Initially based on calmodulin as the conformation-changing element to allow Förster Resonance Energy Transfer (FRET) between two coupled fluorescent proteins, the selection of sensors has steadily increased and has introduced new features that are based on different calcium-binding domains, such as troponin[24,25] or on circularly permuted fluorescent proteins[26-28]. The common advantages of genetically encoded calcium sensors include the possibility of specific sub-cellular localisation[23], expression *in vivo*, potentially in a selected cell type of an entire organ, such as the heart[29], and high biocompatibility, which allows for long-term measurements[30]. The disadvantages can include a possibly complicated transfection or transduction into the target cells (section 6.2 - Genetic manipulation) and a potentially complicated read-out and interpretation of the signals, due to the complex photophysical properties of fluorescent proteins[31]. An overview of the current genetically encoded calcium sensors was presented by Gensch and Kaschuba[32].

3.2 Spatio-temporal considerations

Before discussing the details of fluorescence detection techniques, note that a high acquisition speed is necessary to resolve the highly dynamic calcium transients that are associated with every contraction, especially for following calcium signalling in cardiomyocytes. At the same time, it is important to maintain a high spatial resolution.

There are a number of highly developed, so-called diffraction-limited imaging methods to achieve optical sectioning (necessary for high-resolution image acquisition), such as two-photon imaging, selective plane imaging (SPIM) or structured illumination (for details, see below). However, the most popular and versatile imaging method that offers high resolution and is used in living cells is laser scanning confocal microscopy. Because high-speed acquisition is essential when imaging cell/protein signalling, the technological and methodological approaches available today to achieve that goal are discussed.

The basic principle of laser scanning confocal microscopy is an optical sectioning of the specimen

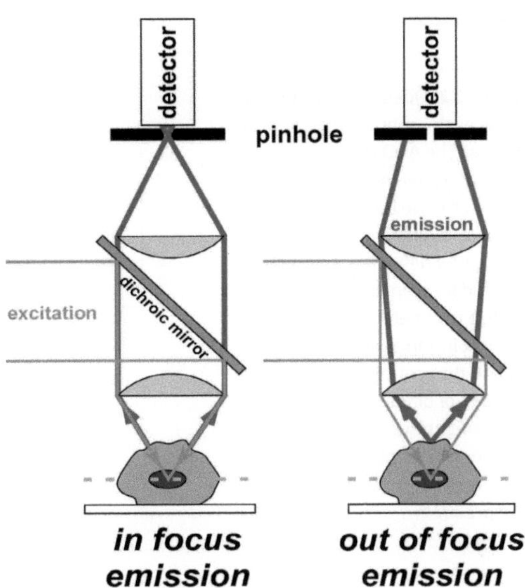

◨ Fig. 3.1 The confocal principle. The excitation light (green) is focussed onto a specimen with a microscope objective, and the emitted fluorescence (red) is returned, separated from the excitation light by a dichroic mirror and focussed onto a detector (e.g., a photon multiplier tube). If the fluorescence is collected from within the plane of focus (left panel) then the emitted fluorescence will pass through the pinhole, which is a spatial filter. If the fluorescence originates from out-of-focus planes (right panel), the emitted light will not be focussed though the pinhole, and most of the light will be blocked by the pinhole. This figure is a reprint from Lipp and Kaestner, 2006[33]. (For interpretation of the references to color in this figure legend, the reader is referred to the web version of the original article.)

along the optical axis. In single photon excitation, this goal is achieved by excluding the light that originates from above or below the plane of focus. This exclusion may be attained by fixed pinholes or variable irises, depending on the construction of the confocal microscope. For an illustration of the confocal principle, refer to ◨ Fig. 3.1.

The trade-off for the optical sectioning capabilities offered with this approach is that the measurements can only be obtained from a single point at one time (however, see the multi-point scanner below). Because of this limitation, it becomes obvious that a true confocal image cannot be seen by the human eye; instead, a computer is required to construct the resulting image from the consecutive points recorded during the scanning process.

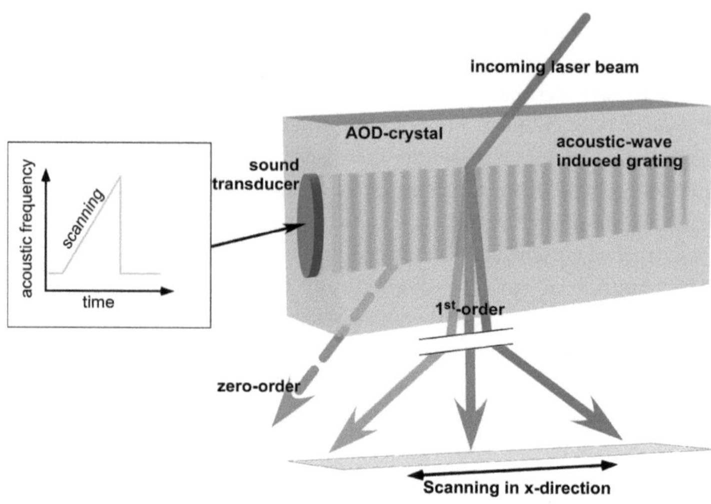

Fig. 3.2 Operating principle of an acousto-optical deflector (AOD) crystal. The AOD crystal is an optical crystal through which an acoustic wave is transmitted that induces a diffraction grating. The first-order diffracted laser beam is used for scanning. Because the lattice parameter can be changed almost instantaneously by modulating the acoustic frequency, the degree of diffraction and hence movement (scanning) of the laser beam is inherently fast. This figure is a reprint from Kaestner and Lipp, 2007[16].

Scanning an entire high-resolution image by such a mechanism is a time-consuming process; therefore the image acquisition rates have traditionally been rather low.

3.3   Confocal scanning techniques

Below is an enumeration of the two basic scanning processes that are currently available in various incarnations:

- Scanning the image by moving the specimen (microscope stage). This technique is the historical approach of the first confocal device patented by Marvin Minsky[34]. This technology is still used in single molecule applications and is commercially available from PicoQuant GmbH (Berlin, Germany).
- Scanning a single laser beam across the specimen. This method is much faster than stage scanning and is currently the standard approach. Single beam scanners are usually equipped with galvanometer-based scanning mirrors for the x- and y-direction. Although this arrangement preserves the simplicity of the scanning electronics and acquisition algorithms because the recording is usually performed during the linear phase of the mirror movements, it is inherently slow because the mirrors must be physically moved. Image

acquisition rates with a reasonable resolution are usually limited to single-figure numbers of frames per second. This rate can be significantly increased by switching from »normal« mirrors to resonating mirrors (particulary in the more demanding x-direction). Using this technique, the frame rates can easily reach 100 frames per second or more depending on the image size. Another approach to gain almost an order of magnitude in scanning speed is to replace the x-scanning mirror with an a̲cousto-o̲ptical d̲eflector (AOD) crystal. The operating principle of an AOD crystal is illustrated in **Fig. 3.2**. This mass-free scanning raises the frame-rate to several hundred Hertz. The price for this speed is a reduced spatial resolution in the x- and z-directions because AOD crystal operation is wavelength dependent and cannot be used for the de-scanning of the image (Stokes' shift). As a result, the emitted fluorescence is not a stationary point; however, it is linearly moving point at the level of the detector. Thus, the pinhole must be replaced by a slit. Such a scanner is commercially available from Visi-Tech International Ltd. (Sunderland, UK). For this type of device, the restriction in scanning speed in experiments is no longer the scanning mechanism itself, but mostly the amount of fluorescence light available in conjunction with the detection efficiency and viability of the specimen.

In addition, specialised devices are commercially available (Molecular Devices Corp., Sunnyvale, USA) that combine stage scanning and beam scanning: the x-scan is performed with a mirror, and the microscope stage is moved for the y-scan.

A compleatly different approach for increasing the frame-rate of confocal scanning is the idea of simultaneously exciting with more than one beam. This concept can be realised either by a line of points, such as the swept-field microscope by Prairie Technologies Inc. (Madison, USA), or by a 2-dimensional array of points. Because the current implementations of the latter approach use several thousand parallel scanning beams. These machines are referred to as kilo-beam array scanners. Such scanners have many advantageous properties that are essential in calcium imaging. These advantages include high acquisition speeds (as fast as the attached camera can capture images), high efficiency in terms of the simultaneous imaging of thousands of beams, low bleaching and low photo toxicity[35]. These kilo-beam scanners are available in two versions:

- The Nipkow-disk system is based on a rotating disk with a specific pattern of pinholes, which was originally invented, designed and constructed to code and transmit television images[36]. This scanning principle was made popular approximately 15 years ago by Yokogawa Electric Corp. (Tokyo, Japan) in its confocal scanning unit (CSU-10). This device overcame the original drawback of single disk scanners, namely the extremely low excitation throughput, by including a second disk with matching micro-lenses that served as a light collector (focussing the excitation light into each of the pinholes). This approach massively increased the excitation throughput and represents one of the major advancements in confocal microscopy, because it allowed long-term live-cell imaging with high acquisition rates[37]. The operating principle is displayed in ◘ Fig. 3.3. A second version of the Nipkow disk was recently implemented by TILL Photonics GmbH (Munich, Germany). This design uses

a single disk that is completely covered by hexagonal micro-optics; however, in contrast to the tandem disk, the micro-optics are not lenses, but curved mirrors. The mirrors reflect the light onto a fixed reflecting optics and simultaneously focus the light through the 35 µm pinhole in the centre of the curved mirrors, unifying the two functions of a tandem disk in a single disk design.

- 2D-array scanners actively move the array of parallel laser beams generated by a micro-lens array across the specimen. In this type of scan-head, in contrast to the Nipkow disc system, the micro-lens-array and pinhole-array are stationary. The only moving part is a single mirror, which is responsible for scanning and de-scanning at the front surface and for rescanning the image across the detection camera on its back surface. A schematic sketch of a multi-beam array scanner is presented in ◘ Fig. 3.4. The novel concept of the 2D-array scanner by VisiTech International Ltd. (Sunderland, UK) also includes changeable pinhole sizes for variable resolutions in the z-direction of the optical axis, a feature that is not found in the current versions of Nipkow-disk based systems. For high-speed imaging (video-rate and above), the linear movement of the scanning mirror allows for an easy synchronisation between image generation by the confocal head and image detection by the attached camera.

One of the most frequent concerns regarding kilo-beam scanners is the possible crosstalk between the pinholes, which is mostly prominent in thicker specimens. There is little doubt that single beam scanners might be more appropriate when deep penetration with single photon confocal microscopy is the prime goal of the experiment; however, in situations where one could compromise on the penetration depth (less than 20 µm) such experiments still benefit from the higher quantum yield of back-thinned camera chips (compared with photon multiplier tubes) and the resulting minimised impact on the tissue viability. The situation is strictly different when considering multiphoton applications, as outlined below.

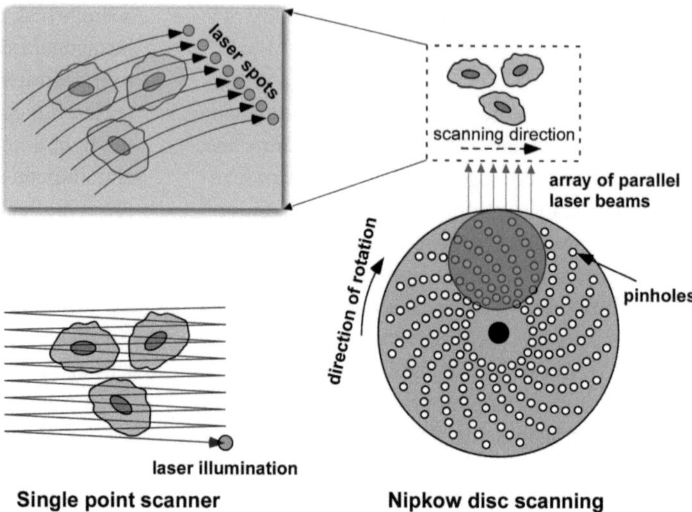

Fig. 3.3 Left: the scanning principle of popular multibeam confocal microscopes with Nipkow disks. Right: the setup of a tandem disk that consists of aligned micro-lens and pinhole disks. The light path across the tandem disk is also indicated. This figure is modified from Lipp and Kaestner, 2006[33].

3.4   Further optical sectioning techniques

In addition to the confocal point and multipoint scanners there are a number of other optical sectioning techniques:

- In contrast to single point scanners, slit-scanners do not use an individual point; however, they use an entire line for excitation. Consequently, the pinhole is replaced by a slit. The idea is to gain acquisition speed (matching the AOD-driven scanner) by sacrificing a portion of the resolution (see below). A few years ago, this technique underwent a revival with the introduction of the LSM 5 Live by Carl Zeiss Jena GmbH (Germany).

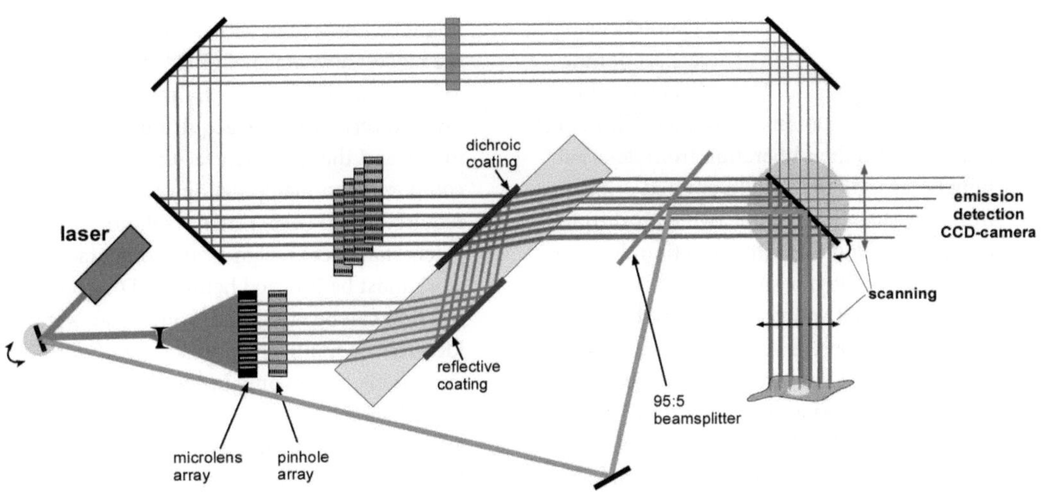

◘ **Fig. 3.4** A schematic design of the kilo-beam array scanner. A laser beam is widened, and a complementary stationary system consisting of a micro-lens array and a pinhole array generate a set of 50x50 beams. These beams pass through a dichroic mirror design that is insensitive to slight position changes and can be quickly changed by a motorised filter wheel. The beam bundle hits the major scanning mirror and serves three functions, scanning, de-scanning and, on its backside, re-scanning. Between the latter two functions, the emitted light passes the dichroic mirror and stationary pinhole array that consists of a set of 5 different pinhole sizes as well as an emission filter (upmost beam bundle). An additional beam path (lower beam path in scetch) illustrates the possibility of performing manipulations in a region of interest within the image, such as fluorescence redistribution after photobleaching (FRAP) - cp. section 7.2 - Optical manipulation.

Multi-photon microscopy is based on non-linear effects during the sample excitation. By condensing the laser energy in time (femtosecond pulses) and space (focus) the energy density in the focus becomes so high that a molecule in the sample can simultaneously absorb two (or more) photons. Because the underlying molecular excitation process remains the same, both photons must deliver approximately half (or other fractions depending on the multiplicity of excitation) of the energy. Half of the energy translates into a doubling of the wavelength, which explains why the far-red and infrared light-emitting Titanium-Sapphire lasers are predominantly used. Thus, if the chromophore requires single photon excitation at 480 nm, the equivalent 2-photon excitation wavelength would be 960 nm. For other multi-photon processes, this shift is multiplied by higher factors. One should be aware that the absorption cross-sections of fluorescent dyes for single photon and multi-photon excitation could be quite different. The

basic advantage compared with single photon confocal imaging is the reduction of photobleaching in the out-of-focus planes because multi-photon excitation is restricted to the focal plane. For a thorough discussion of multi-photon microscopy, see Denk *et al.*[38] In this section, two additional significant advantages of multi-photon excitation are outlined. (a) Deeper penetration depth. Because excitation can be performed with red, far-red or even infrared light, the penetration of the excitation light in living tissue is considerably increased in comparison with shorter wavelengths. The maximal penetration depth is tissue dependent and emission light scattering becomes more prominent with increasing penetration depth. (b) Intrinsic sectioning. As described above, the »multi-photon effect« is restricted to the core of the excitation light focus in the specimen due to the non-linear excitation probability. From this situation, in contrast to single photon confocal scanners that employ optical sectioning on the »emission« side (spatial filter, the pinhole(s)), scanners using multi-photon excitation generate sectioning on the excitation side. Thus, this method is referred to as »excitation sectioning«. This effect translates into their liberation from de-scanning because all of the light emitted originates from the excitation volume, which is diffraction limited, and the light collection does not require the ability for spatial discrimination. As a result, multi-photon scanners do not need pinholes on the detection side. Similar to single beam confocal scanners, the construction of a 2D image is also realised by scanning processes in multi-photon microscopy. In the simplest cases, the scanning is performed by two galvanometer-controlled mirrors with all of the limitations discussed above. The application of fast AOD crystals for x-scanning is more complex when using pulsed femtosecond light sources because diffraction in AOD crystals is wavelength dependent and femtosecond-pulsed lasers produce a spectral band (with 100 fs pulses the bandwidth of the resulting spectrum can reach approximately 10 nm). This effect means that the degree of diffraction of the excitation light will be different for the »red« and »blue« components of the excitation spectrum. Nevertheless, multiple approaches exist that offer possible solutions[39,40]. In addition, LaVision BioTec GmbH (Bielefeld, Germany) has introduced a commercial multi-photon, multi-point, multiplexed scanner (up to 64 parallel excitation points arranged in a line), which allows for up to 64 times faster image acquisition in principle.

– Selective plane imaging (SPIM), also called light sheet microscopy, relies on the illumination of the sample by a light sheet that is perpendicular to the optical axis of the microscope. The revival of this technique[41] is based on a sample holder with a rotational axis parallel to the gravitation field, which enables rotation while maintaining the sample distortion-free. The serial optical sectioning is achieved by moving the sample through the light sheet and collecting an image for every position. Although the lateral resolution might be lower than with confocal techniques, the biggest advantage of SPIM is its ability to approach the diffraction limit in all three axes. Nevertheless, the acquisition speed is limited, and the specimen set-up is much more complex when manipulation of the cells is required for the experiment (solution changes, patch-clamp, etc.). Significant computational work must be invested before a 3D reconstruction can be visualised. Similar, but not identical, to multi-photon excitation, »SPIM excitation« occurs only in the focal plane (the »light sheet«).

– Structured illumination is another technique that allows for optical sectioning at the level of axial resolution offered by a classical, pinhole-based confocal arrangement. The principle of structured illumination is an optical grid, placed in the illumination beam in such a way that the image of the grid is projected exactly into the focal plane of the microscope objective. For the data acquisition, the representation of the grid in the focal plane is moved laterally, and images are acquired in each position. The lateral movement of the grid-image can be obtained by a piezo (if the grid itself is moved as in the initial setup[42]) or by moving a glass block in the excitation light path (as incorporated in the commercialised version (Apotome) by the Carl Zeiss Jena GmbH (Germany)). Exposures are taken from three defined grid-image positions. From three such consecutive exposures, an image of the optical section is calculated. In contrast to confocal laser scanning, structured illumination imposes fewer requirements on the illumination source (no laser is necessary) and is less expensive. In terms of acquisition speed, structured illumination recordings can be faster than classical confocal recordings (see above). However, the acquisition speed is limited by the read-out of the camera used, and because there three exposures are necessary for the calculation of one image, it is inherently slower than other camera-based confocal systems, such as kilo-beam systems or slit scanners (see above). This »delay« can be partially overcome by using the principle of the »running aver-

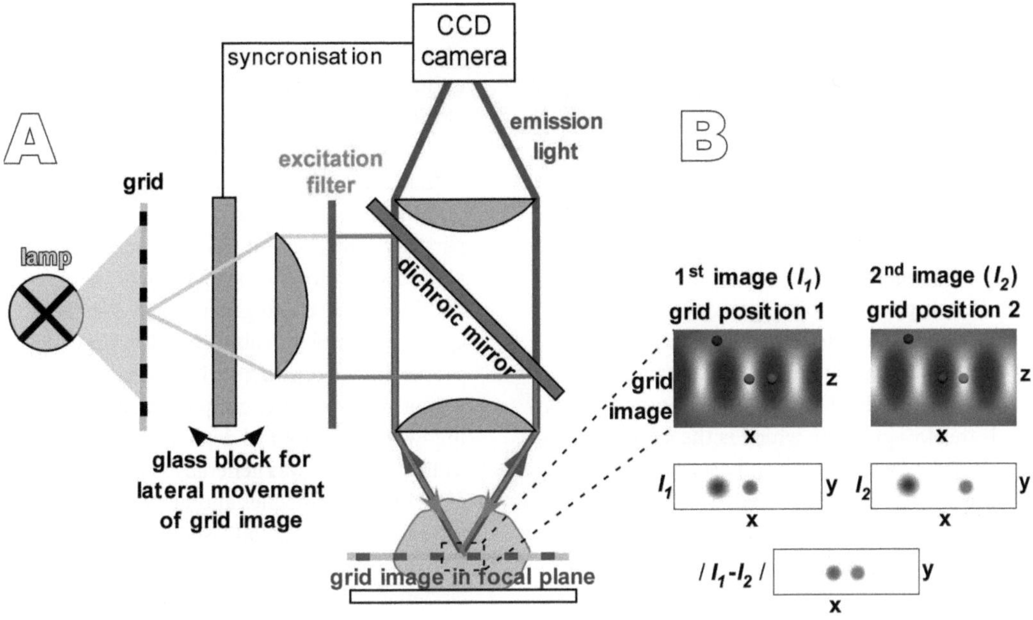

■ **Fig. 3.5** Simplified scheme of optical sectioning by structured illumination. Part A shows the modified optical arrangement. An optical grid is placed in the illumination beam in such a way that the image of the grid is exactly in the focal plane of the microscope. The grid image is moved laterally in the focal plane by swinging a glass block in the excitation beam path as shown in the figure. Exposures are taken in three fixed grid image positions. Based on these three exposures, an image of an optical section is calculated. Part B depicts the principle of image calculation out of two exposures instead of three (for simplification). The upper row shows sections of the x/z-plane with the grid image (dark ovals): the red and green beads are in the focal plane, and the blue bead is out of the focal plane. Below, there are diagrams of the x/y-plane that the camera would be exposed to in this experiment. In the left column, the green bead is shaded by the grid image, whereas the red bead is perfectly illuminated and produces a sharp image on the camera. The blue bead is out of focus and results in a blurred image on the camera. In the right column, the grid position has changed; therefore, the red bead is no longer visible, and the green bead gives a sharp image. For the blue bead, the situation is virtually unchanged, and it is mapped exactly as in the first grid position. If the two x/y-exposures, as shown in the middle row, are now subtracted from each other, the structures that have been out of focus vanish, whereas the absolute value of the structures that are in focus provide the required optical section (bottom equation). This figure is a reprint from Lipp and Kaestner, 2006[33]. (For interpretation of the references to color in this figure legend, the reader is referred to the web version of the original article.)

age«; that is, exposures 1-3 provide the first image, exposures 2-4 the second, and so on. However, if the process one is viewing is very fast, there is not only a smear in the x/y-plane but also a resolution decrease in the z-axis because the algorithm relies on the assumption that there are no sample changes within the three exposures needed to calculate the image. When comparing structured illumination with confocal imaging, one must take into account that the structured illumination image is already a processed image. For a detailed comparison see Weigel *et al.*[43]

Two additional subjects should be discussed with respect to (diffraction limited) imaging technologies:

- The time resolution can almost always be increased by sacrificing the spatial resolution. The first step in this direction is the reduction of the number of pixels per image. This reduction can be realised either by limiting the overall size of the image, by binning adjacent pixels or, at its extreme, by reducing image sizes to individual lines (line-scans). Going further, if the entire fluorescence signal of the microscope is detected in a point detector, confocal spot measurements can be utilised[44]. In this case, the sampling rates can be as high as several MHz. Nonetheless, even without any optical spatial resolution, the functional spatial resolution (even beyond the diffraction limit) can originate from the experimental design by making use of Förster Resonance Energy Transfer (FRET).

— The spatial resolution can be increased to an almost spherical point spread function by illuminating and detecting the specimen from more than one angle. One of these incarnations is the SPIM (see above). Another version is the 4π-approach, which is the simultaneous recording with two objectives from opposing directions, commercialised (but presently discontinued) by Leica Microsystems GmbH (Wetzlar, Germany).

3.5   Super-resolution imaging techniques

Super-resolution usually refers to imaging technologies that reach resolution limits below the size of a diffraction-limited spot. Although these technologies have been rarely used for calcium imaging thus far, the most common methods are mentioned below:

— Total internal reflection fluorescence (TIRF) microscopy, which is based on evanescent wave imaging, is offered as an add-on by virtually all of the major microscope manufacturers. At the angle of total internal reflection, excitation around the phase boundary is induced by the energy transfer of the evanescent wave. In most cases, these are cells growing on a glass substrate (a glass cover slip). This evanescent wave travels parallel to the cover slip, illuminating a layer less than 100 nm above the cover slip. Therefore, this technique is the method of choice for membrane-related imaging and can be applied for calcium transients[45].

— Another technique is stimulated emission depletion (STED). This methodology requires a high technical complexity; currently, a simplified version has been implemented in a commercial product line of Leica confocal scanners (Wetzlar, Germany). Briefly, an already excited fluorophore is illuminated with light of the same wavelength as the emission of the fluorophore. This additional illumination will induce the depletion of the excited state. The point spread function of the excitation spot is partially »switched off« by a donut-shaped depletion structure that »cancels out« (depletes) the fluorescence and generates an effective emission spot with sub-diffraction-limited properties. This principle is realised by providing an excitation light pulse that is directly followed by a 20-40 ps depletion pulse[46]. The STED principle was generalised to all types of switching principles and named reversible saturable (or switchable) optical fluorescence transition (RESOLFT)-microscopy[47] and includes approaches that are based on switchable proteins and small molecular dyes (see below). Further variants are ground state depletion (GSD)-microscopy or saturated pattern excitation microscopy (SPEM).

— Photo-activated localisation microscopy (PALM)[48] and stochastic optical reconstruction microscopy (STORM)[49] are basically the same method. The principle is to switch fluorescent molecules in such a way that, in a certain surrounding, only one molecule emits light. The position of this molecule can then be determined with (theoretically) unlimited precision. This process will be repeated until the positions of all of the molecules of interest have been obtained, and out of these data, an image can be reconstructed with a resolution of the precision limit achieved for the molecular position. These types of microscopy systems are commercially available from Carl-Zeiss Jena GmbH (Germany) or Nikon Corp. (Tokyo, Japan). This technique can be combined with TIRF (cp. above) to obtain a high resolution in the z-direction as well.

— Structured illumination microscopy (SIM) was introduced previously (section 3.4 - Further optical sectioning techniques, and ◘ Fig. 3.5). This principle can also be extended to theoretically unlimited resolution. The extension of SIM towards super-resolution was pursued by Mats Gustafsson[50,51]. Currently, structured illumination microscopes are available from Applied Precision (Issaquah, USA), Nikon or Zeiss.

- A completely different approach is based on the scanning near-field optical microscopy (SNOM). In this method, one does not find an optical image plane. Instead, a probe with a diameter of 100 nm or less is used. This sensor is mechanically scanned across the specimen and the probe head needs to stay within a 20 nm distance of the surface of interest.

- Fluorescence, once generated, exhibits an exponential decay, which is termed the fluorescence lifetime and is a defined characteristic for every fluorophore. Because the fluorescence lifetime is a material constant and concentration independent, it has been demonstrated to be a potentially useful parameter for microscopic images through fluorescence lifetime microscopy (FLIM). Interestingly, the lifetime of a fluorophore can be extremely sensitive to the surrounding in its proximity (the nano-environment). This interaction between the lifetime and its environment can be used for calcium imaging with FLIM[52]. Another extremely interesting approach is the combination of FLIM and FRET (discussed below) and is called FIFA (FRET-induced fluorescence lifetime alterations).

Depending on the particular implementation technology, most of the latter approaches are inherently slow, and in their practical use, they are far away from video-rate imaging. This technical limitation restricts their significance predominantly to the determination of »long lasting« calcium concentrations, such as resting calcium and other experiments that are not time-critical.

Aequorin-based measurements

There are special proteins that do produce light by themselves (bioluminescence; no excitation light is necessary). An example is the aequorin-GFP-complex (GFP, green fluorescent protein) that naturally occurs in the jellyfish *Aequorea victoria*. Interestingly, the famous paper by Shimomura[53] that identified GFP as a protein only mentioned GFP in one sentence because the paper was primarily dealing with luminescent aequorin. Because the operating principles and measurement techniques are different from those of fluorescent probes, they are discussed in their own chapter.

Aequorin became a popular calcium sensor after its first use as such in 1968[54] and before the age of the current generation of small molecule dyes, which started in 1980[55] (cp. section 3.1 - Calcium sensors). This protein shares the advantages of the genetically encoded calcium sensors as outlined in section 3.1 and does not require a light source for illumination, thereby circumventing the autofluorescence of the cells. It can report calcium in a wide concentration range, between 0.1 μM and 100 μM[56,57]. However, when expressed in cells (cp. section 6.3 - Genetic manipulation), the transfected or transduced DNA codes for apoaequorin, which requires the small organic molecule coelenterazine to form aequorin. In the bioluminescent reaction, aequorin reacts with 3 calcium ions to produce apoaequorin coelenteramide, carbon-dioxide and light at 466 nm. For the reconstitution of aequorin, the availability of coelenterazine was shown to be the rate-limiting step[58].

In addition to the expression of apoaequorin, purified aequorin is commercially available (Molecular Probes, Eugene, USA) and can be injected into cells as a calcium sensor. Although the protocols for the use of aequorin are demanding, a number of applications, such as *in vivo* imaging of calcium[59] or calcium assays for high-throughput screens[60], benefit from the properties of aequorin. The use of aequorin to visualise calcium dynamics was recently reviewed.[61]

Measurement of calcium transport across membranes

5.1   Flux measurements

Investigations measuring transport across a membrane were initially facilitated by the discovery of radioactivity, which allowed for a precise and quantitative measurement of the movement of radioisotopes across cell membranes. Radioactive isotopes were added to the cells, and after a given time, transmembrane transport can be stopped using low temperature; the cells were washed and lysed, and the radioactivity was measured using an appropriate device, such as a Coulter counter. Thus, the influx can be determined.

Similarly, cells were »loaded« with a radioactive isotope, washed and the experiment could be initiated. After a given experimental time, the supernatant was measured for radioactive components to determine the cellular efflux. In contrast to the patch-clamp technique (see below), tracer flux measurements allow for the probing of electrogenic and electroneutral transport processes.

Although the first measurements were performed with radioactive sodium isotopes[62], the technique was extended to radioactive calcium, which is still a state-of-the-art technique particularly for the investigation of mammalian red blood cells[63] due to their lack of cell organelles and internal calcium stores.

Other than radioactive tracer flux measurements, there are additional means of measuring ion concentrations in a reservoir, such as ion-selective electrodes, flame photometry or atomic absorption spectroscopy. All of these methods have advantages and disadvantages but share the shortcoming of measuring only bulk solutions of large cell populations without spatial and with very limited temporal resolution.

5.2   The patch-clamp technique

The patch-clamp technique, as it was introduced by Neher, Sakmann and colleagues[64,65], revolutionised the investigation and understanding of membrane transport. This methodology allows for the detection of ions crossing the membrane in channels by the detection of the current that is created when ions move through permeable protein structures in the otherwise isolating lipid bilayer. A glass pipette, with a tip opening that is typically 1-3 μm, is brought into contact with the cell. Inside the pipette and in the bath are the electrodes of an electrical circuit. The contact between the pipette and cell needs to be tight enough to produce an electrical resistance in the range of Gigaohms between the electrodes. Such a pipette-cell connection is called a »gigaseal«. The intention is that the currents caused by channel openings are greater than the leak current of the seal. The described configuration is called the »cell-attached« mode. However, for a direct measurement, one prefers to have a single membrane between the two electrodes. Therefore, the membrane patch below the pipette either needs to be destroyed or needs to be the only maintained piece of membrane to reach the »whole-cell« or »inside-out« configuration, respectively. The whole-cell technique is achieved either by applying a negative pressure to the pipette or by disruption using an electrical pulse. The inside-out configuration can be reached by moving the pipette relative to the cell or, alternatively, if the cells are not attached to a surface, by moving the cell across the liquid-air interface. A more detailed technical and conceptual description can be found elsewhere[66,67].

However, once set up, the challenge is to discriminate calcium ions from other ions. The easiest and most straightforward method is to restrict the ions in the solutions to calcium[11]. Other approaches include the pharmacological inhibition of disturbing channel activity or inhibition by appropriate voltage protocols. Alternatively, ion channels can be selectively activated according to their activation properties, such as ligand binding[68], pressure activation[69] or potential induction[70].

Molecular biology based approaches

6.1 Determination of molecular identities

If there are indications for the involvement of a molecule or protein in the process of calcium signalling (e.g., from patch-clamp recordings as stated above), there are many techniques that can be used to identify an unknown player. This biochemical and molecular biology-based methodology is reviewed in other sources[71,72]. Examples include high-pressure liquid chromatography (HPLC), 2D-gel electrophoresis, mass spectrometry, matrix-assisted laser desorption/ionisation in combination with time-of-flight mass spectrometer (MALDI-TOF) and expression analysis with DNA chips or next-generation sequencing.

6.2 Antibody-based techniques

To test for a known signalling protein, antibody-based techniques are very popular. These techniques can be performed on extracted protein mixtures in an enzyme-linked immunosorbent assay (ELISA), a Western blot[73,74], or inside cells. The latter approach is called immunochemistry, or more precisely immunocytochemistry[63] or immunohistochemistry[70] depending on whether the staining is performed on single cells or in tissue, respectively. Except for the rare cases where the cells are accessible for immunocytochemistry and the antibody can bind to the protein of interest from outside the cell, antibody-based techniques are primarily restricted to dead cells or tissues and lack any dynamic information. However, there are technically advanced options, such as microinjection or infusion by a patch-clamp pipette, which allow for the insertion of antibodies into a living cell.

The most widely used immunochemistry practise is a two-step process characterised by two antibodies. The primary antibody identifies the molecule of interest; it may itself contain a fluorescent tag or it may contain binding sites for secondary antibodies that are fused to fluorescent tags. The latter method has the advantage of containing an amplification step but is inherently non-quantitative because the ratio of primary to secondary antibodies is difficult to determine quantitatively. Antibodies are available in virtually all colours throughout the entire range of the visible spectrum. There are kits available that allow the relatively simple labelling of one's own antibodies. The chromophore is usually an inorganic group. However, another recently established approach to labelling proteins in fluorescent imaging uses the so-called quantum dots (Q-dots), which have several interesting properties. These nanostructures (5-15 nm in diameter) contain crystalline cores (usually cadmium mixed with selenium or tellurium). This core is the »heart« of the quantum dot, as it provides the structure and mechanism for the generation of fluorescence. Further coatings protect the core, make it biocompatible and provide binding sites for the attachment of the Q-dot to the target protein. Q-dots are advantageous because they have an extremely broad excitation spectrum while maintaining a symmetrical and narrow emission band. Furthermore, Q-dots exhibit extremely low rates of photobleaching, especially when compared with small molecule dyes, but they also have drawbacks, such as an unpredictable fluorescence blinking.

6.3 Genetic manipulation

Once a protein has been genetically identified, it can be manipulated *in vivo* or *in vitro*. Genetic manipulation is also required if genetically encoded calcium sensors (see section 3.1 - Calcium Sensors, and ▶ chapter 4 - Aequorin-based measurements) are to be expressed.

In vivo manipulation can be performed by breeding transgenic animals. Without going into detail, this transgenic manipulation can occur totally (in all of the cells of the organism), or in specific organs by making use of organ-specific promoters or can be induced using a drug-induced determination process that can be either total or organ specific. Transgenic manipulations are most commonly performed in mice due to their relatively short generation cycle. Such manipulations can involve either protein over expression[70,73] or protein knock-out[74].

The concept of genetic manipulation *in vitro* is less complex; however, due to the lack of a translational machinery, this method is not applicable to mammalian red blood cells. The genetic information that is needed either to over-express or to delete/downregulate the protein of interest by RNA interference (RNAi) or associated methods must be incorporated into the cell of interest. There are a number of chemical transfection methods available, such as catalysing DNA cross-membrane transport using of calcium phosphate, polycations or dendrimers, in addition to physical transfection approaches, such as electroporation, microinjection or particle guns. Even better success rates have been achieved with biological transfection methods, such as lipofection, which leads to a 5- to 100-fold increased transfection rate compared with chemical methods[75]. These methods are successful in many model cells and cell lines but provide only very low transfection rates in terminally differentiated cells, such as cardiac myocytes.

An alternative that, under certain conditions, can even be applied *in vivo* is viral gene transfer. Using cardiac myocytes we tested Semliki Forest virus[76], lentivirus[77], adenovirus[78] and adeno-associated virus (AAV)[79]. In our hands in cardiac myocytes, Semliki Forest virus transduction led to a transduction rate lower than 5%. In neurons the expression of the transduced gene was achieved rapidly (within 6 hours); however, it killed all of the cells within 24 hours[80]. In contrast, lentiviral gene transfer required approximately a week for protein expression. This delay is too long for established culture systems of cardiac myocytes[17,30] but can be sufficiently rapid for cell lines. Furthermore, lentiviruses may serve a role in gene transfer in animals[75]. Using an adenoviral gene transfer and AAV gene-transfer approaches, protein expression reached sufficient levels within 24 hours for most proteins tested[30,80]. Viral expression is stable over at least one week, and we could not find adverse effects of virus transduction and induced gene expression. Although there are reports of adenovirus-mediated gene silencing in the heart[81], in our hands, the expression of the sensor could not be achieved in the heart after injection into either neonatal rat hearts or in adult mouse hearts. This result is likely due to the immune responses of the animals[75]. In contrast, AAVs yielded a sufficient level of expression in mouse hearts to allow for cellular investigations. Because AAV infection begins with a receptor-mediated endocytosis, the different serotypes, in combination with distinct intracellular processing, give rise to specific tissue tropisms[82]. In addition, tissue-specific expression was fostered by cloning the appropriate tissue-specific promoter into the AAV genome.

Manipulation of calcium

7.1 Pharmacological manipulation

Calcium can be manipulated by the pharmacological disturbance of equilibria that are maintained by the cell. Such a disturbance can be permanent using the application of a calcium ionophore, such as A23187[14,83] or ionomycin, to provoke an intracellular situation similar to the digital approach discussed in ▶ chapter 2 - Direct evidence - the digital approach.

An even more elegant approach is to induce a transient disturbance in cardiomyocytes, the application of 20 mM caffeine leads to the opening of the internal calcium stores, namely the sarcoplasmic reticulum (SR), by opening the ryanodine receptors[74,84] (cp. ▶ chapter 9 - Calcium signalling in cardiac myocytes). Caffeine is easily washed out, and the effect on the ryanodine receptors is reversible. Therefore, after the initial caffeine exposure, one can study the reestablishment of the cellular equilibrium. In detail, this approach requires a local perfusion system, and as soon as the perfusion is switched to caffeine, one observes a calcium transient. The amplitude of this transient corresponds to the SR filling, and the exponential decay corresponds to the activity of the sodium-calcium exchanger. If the caffeine is replaced before the calcium reaches the baseline, a faster exponential decay overwhelms the initial one, which represents the activity of the sarcoplasmic/endoplasmic reticulum calcium ATPase. If the SR is »emptied« by caffeine, one can probe the calcium refill of the SR by electrical field stimulation[74].

7.2 Optical manipulation

The first example of optical manipulation is a rather »virtual« manipulation, the so-called fluorescence redistribution after photobleaching (FRAP) technique. In this approach, the visibility of molecules, calcium or calcium-handling organelles[84] is manipulated. The redistribution process (sometimes referred to as »recovery«) can be visualised by optical imaging methods (cp. ▶ chapter 3 - Fluorescence based visualisation) and dynamic behaviour can be determined.

A more direct optical manipulation is the uncaging of calcium, which can be performed as a global uncaging using an ultraviolet (uv) flash lamp[85] or as 2-photon photolysis using a femtosecond pulsed laser[86]. With this method, calcium can be elevated or buffered without changing any other parameter in the cell. The 2-photon photolysis allows a highly localised calcium release at the size of a diffraction-limited spot and even allows the mapping of receptors in a cell. This feature is unique to 2-photon-based uncaging in contrast to one-photon uv-laser uncaging, which allows a similar focus in the x/y plane but leaves a double cone of photoconversion in the z-direction.

7.3 Combined approaches

The approaches described in 7.1 - Pharmacological manipulation, and 7.2 - Optical manipulation, can be directly combined by the photolysis of caged A23187. This principle was improved based on the light-sensitive protein channelrhodopsin2, which was introduced as an optogenetic tool in 2005[87]. Channelrhodopsin2 is a light-activated cation channel from green algae (Chlamydomonas reinhardtii) that can be used to induce action potentials in cardiac myocytes[88]. Channelrhodopsin is also calcium permeable[89], and this property has been used to investigate calcium signalling[90].

In addition, channelrhodopsins have been engineered to improve their function and/or to induce new properties[91]. Therefore, one can expect a calcium-selective channelrhodopsin to be available in the future.

Calcium-induced function

8.1 The concept of calcium-induced function

The concept of calcium-induced function has certain aspects in common with ▸ chapter 2 - Direct evidence - the digital approach. However, the latter concept was used as a digital approach, whereas the concept introduced in this section, describes a modulation in function as an indirect read-out of a modulation in calcium. This concept is a very complex approach because function can be modulated by many parameters that are directly or indirectly influenced by calcium. Therefore, primary and secondary effects are difficult or impossible to discriminate. In addition, the cellular function varies among different cell types, and in contrast to the previous chapters, no general assays or procedures are available; however, only specific readouts reflect the function of the cell. Because such readouts can be complicated in terms of the technical set-up and are only applicable to a particular cell type, the next sections describe singular examples.

8.2 Cardiac contraction

Contraction is the functional parameter of the heart that is mediated by calcium; however, the actual readout can be diverse. The readout can be based on measurements of the entire organ *in vivo* by means of sonography or magnet resonance imaging, which is depicted in ◼ Fig. 8.1.

Alternatively, it can be measured on isolated hearts by a Millar tip catheter[70]. As imaging is based on geometric changes and other functional parameters, such as ejection fraction or fractional shortening, the Millar tip catheter is a

pressure measurement with direct correlation to a force.

These two methods of measurement have also been translated to the cellular level. The measurement of cell length is a popular readout for cellular contraction[30,45]. Because the amount of cellular contraction depends on cell size, sarcomere length measurements are a more universal alternative. A general problem with such measurements in culture is the artificial contraction condition: on one side, the cells adhere to virtually inelastic glass or plastic surfaces, and the rest of the cell has no matrix at all. An approach that at least partly addresses this issue was introduced by Müller *et al.*[45] (cp. ▸ chapter 13), where adult cardiomyocytes were cultured on elastic surfaces.

Alternatively, carbon fibres can be attached to cardiomyocytes and force transducers can directly read out the contraction force[93].

8.3 Red blood cell aggregation

The formation of clots is a crucial function of the blood as a reaction to the injury of a vital body, and our present knowledge of coagulation cascades is extensive[94]. In contrast, thrombosis is a life-threatening pathophysiological event. In both situations the role of the red blood cells is believed to be passive. However, there is evidence for the active involvement of red blood cells in the clot-forming process, as discussed in ▸ chapter 10 - Calcium signalling in red blood cells.

With respect to the red blood cells, there are three different types of aggregation:

1. The aggregation of red blood cells induced by the cross-blood group binding of antigens

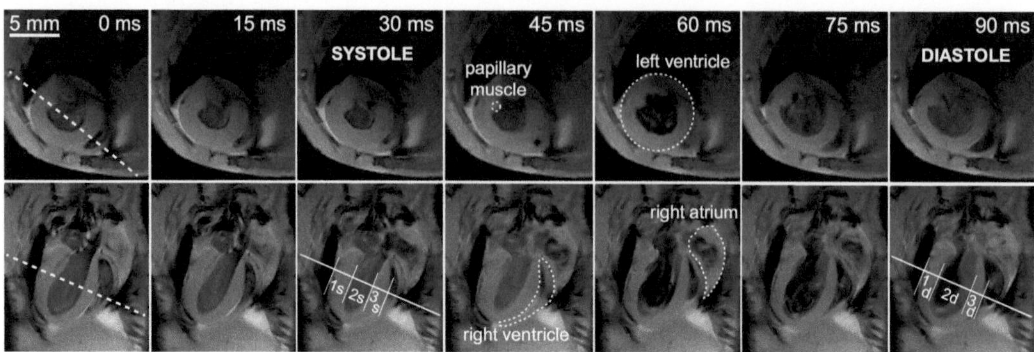

☐ **Fig. 8.1** Magnet resonance imaging of small rodents. The image series depicts a beating heart in a 15 weeks old mouse over-expressing Rac-1[70,92]. The upper row shows cross sections of the short axis as indicated by the dashed line in the lower leftmost image. The lower row shows the corresponding cross sections of the long axis as indicated by the dashed line in the upper leftmost image. Images represent a slice thickness of 1 mm. Typical morphological parameters to extract are the left ventricular posterior wall thickness (1) the left ventricular inner diameter (2) and the inter-ventricular septal thickness (3), whereas the »s« denotes systolic and the »d« diastolic. Note the pathologically dilated right atrium. Imaging was performed on a BioSpec Avance III (Bruker BioSpin MRI GmbH, Ettlingen, Germany) comprising of magnetic field strength of 9.4 Tesla using a cine-FLASH blackblood technique, respiratory gated and ECG triggered. For the saturation of blood signal intensity, a global inversion pulse (flip angle of 180°) and an inversion time of 150 ms was applied.

and antibodies of the AB0-blood group system is termed agglutination. Such processes are artificial but play an important role in transfusion medicine. Because the agglutination is very easy to detect and is visible to the naked eye, this process can be utilised to determine blood groups, such as the AB0 bedside test. Agglutination is not considered for functional tests, as intended in this chapter.

2. Red blood cells have a tendency to spontaneously form aggregates that are also known as rouleaux formation. This aggregation can be modulated by external properties or by properties of the cells (aggregability). However, this type of aggregation is defined as a reversible process and can easily be dispersed by fluid flow. The current knowledge in this field was recently summarised in a dedicated textbook[95].

3. Aggregation of red blood cells that are presumably induced by specific binding, which is irreversible up to a separation force of at least 25 pN[96]. This aggregation behaviour is induced by an increase in intracellular calcium in red blood cells[14] (cp. ▶ chapters 25 and 26). This finding establishes aggregation

as the intended functional readout of calcium signalling.

Aggregation is traditionally measured either by microscopic investigation by quantifying a microscopic aggregation index[97] or by indirect methods that measure properties from red blood cell suspensions as a liquid. Such techniques include sedimentation-associated procedures, light-interacting readouts, measurements of viscosity and other rheometric methods[95]. However, when it comes to adhesion force measurements rheometric techniques have widely been used[98,99]. These methods are all of an indirect nature and suffer from a limited overall picture, such as not knowing how many cells are involved or the influence of changes in red blood cell deformability. An alternative is provided by force measurements on the single cell level, which is depicted in ▶ chapters 25 and 26. For these quantitative measurements, two different techniques, holographic optical tweezers and atomic force microscope-based single cell force spectroscopy were utilised to measure the forces between two red blood cells.

When exerting forces on cells with optical tweezers, one suffers from a limited force regime due to cell damage with increasing laser power, which indicates that an upper limit for measuring the adhesion forces between cells exists[96]. In contrast, a lower limit to the measurable adhesion forces for single cell force spectroscopy that is due to both, the limited force resolution of the system and squeezing of the cells during the measurements, which can possibly induce adhesion force artefacts. However, the different types of aggregation (especially (2.) and (3.)) require very careful control experiments.

Calcium signalling in cardiac myocytes

Calcium signalling in cardiac myocytes consists of a basic scaffold of four molecular entities: the voltage-activated calcium channel and sodium-calcium exchanger (NCX) in the plasma membrane and the ryanodine receptor (RyR) and the sarcoplasmic/endoplasmic reticulum calcium ATPase (SERCA) in the membrane of the sarcoplasmic reticulum. The interplay among these four proteins is presented in ◘ Fig. 9.1. These four players can be modulated by their abundance (expression), regulation through other proteins, phosphorylation, geometrical arrangement or by other means. Furthermore, the different calcium buffers play crucial roles, and other calcium transport proteins, such as inositol triphosphate (IP_3)-receptors[101,102], transient receptor potential (TRP)-channels[103,104] and *N-methyl-D-aspartate*-(NMDA) receptors[105], just to name a few, have a further modulatory function. The entirety of this modulation of the calcium signals in the heart is so complex that a comprehensive description would go beyond the scope of this summary. Therefore, the reader is referred to textbooks[106,107] and recent reviews[108-113]. Instead, in this section, a summary is given of the results that are detailed in part III.

The first chapter in part III (▶ chapter 17)[74] deals with an important aspect of developmental cardiology. As mentioned above, the voltage-activated calcium channel is the initial calcium input entity. This channel consists of four subunits. The α_1 subunit is believed to contain the ion channel pore and most of the functional properties[114], and the β_2 subunit is an intracellular ancillary unit that binds to the α_1 subunit and was believed to cause only modulations of the channel properties[115]. The α_2/δ complex is an auxiliary unit that modulates the channel; therefore, this complex is a drug target[116]. In ▶ chapter 17[74], a $Ca_V\beta_2$-deficient knock-

out mouse was designed and did not survive past day E11. Investigation of the E9.5 embryos and characterisation of the corresponding embryonic cardiac myocytes revealed diminished L-type Ca^{2+} currents, which led to a functionally compromised heart. This effect caused the secondary effect of defective remodelling of intra- and extra-embryonic blood vessels as a cause of the embryonic death in the $Ca_V\beta_2$ knockout mice.

In ▶ chapter 18[73], calcium modulation, namely the modulation of the RyR by junctin, which is a protein that complexes with the RyR and two other proteins (calsequestrin and triadin 1), is probed by the cardiac-specific overexpression of junctin in transgenic mice. As expected, all of the tested SR-associated parameters were altered in the transgenic mice: the sarcoplasmic reticulum calcium load was reduced, the calcium-spark frequency was reduced, the decay of the calcium-sparks was prolonged and the basal RyR phosphorylation was increased. However, as in all types of complex systems, changes in a single parameter will result in adaptation or compensation mechanisms of other parameters. For the junctin-overexpressing mice, down regulation of the NCX was observed. Because of these changes, the cardiomyocytes of the transgenic mice, in contrast to wild type mice, could not or could only slightly adapt their calcium transients and cellular contraction to the frequency changes of the electrical stimulation.

▶ Chapters 19 and 20, by Hammer *et al.*[84] and Tian *et al.*[17], respectively, investigate cardiac myocytes and their preservation in culture. As such, these studies are a continuation of the investigation by Viero *et al.*[30] (▶ chapter 15) and contain methodological aspects. Nevertheless, one of the major remodelling processes of cardiomyocytes in conventional culture is the loss of the T-tubular

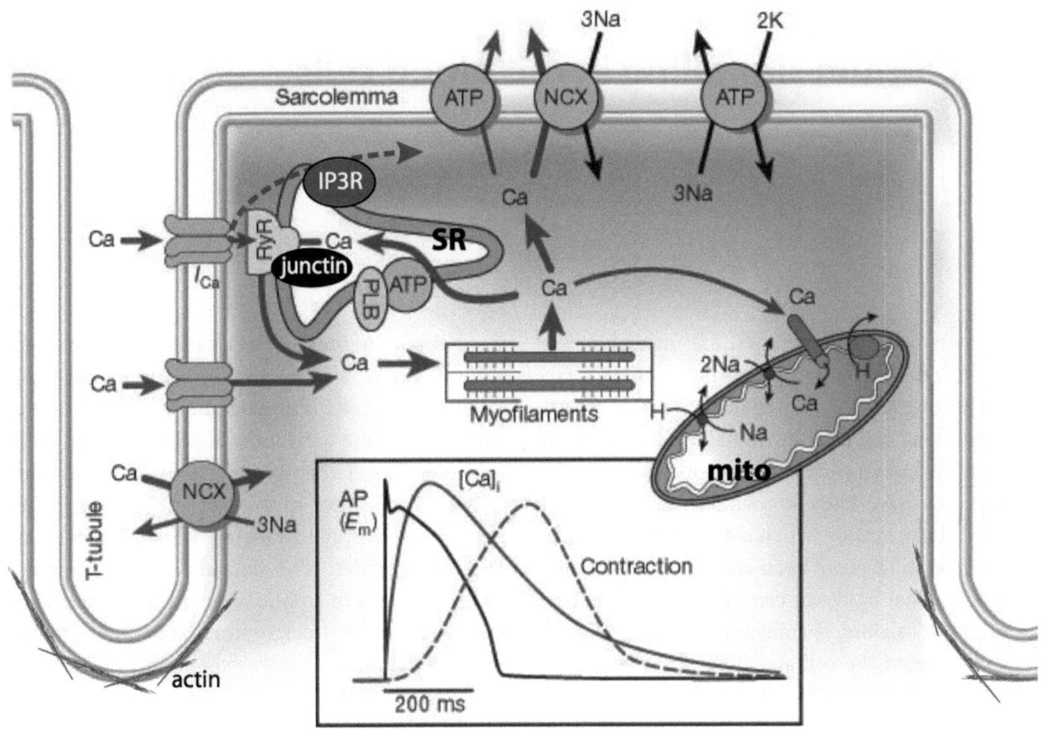

☐ Fig. 9.1 Calcium signalling in cardiac myocytes. The red arrows depict calcium entry into the cytoplasm, whereas green arrows represent calcium extrusion into the extracellular space and the intracellular calcium stores. The following abbreviations were used: SR, sarcoplasmic reticulum; mito, mitochondrium; I_{Ca}, voltage-activated calcium channel; NCX, Na$^+$/Ca^{2+} exchange; ATP, adenosine-5′-triphosphate; RyR, ryanodine receptor; IP3R, inositol triphosphate receptors; PLB, phospholamban. The inset shows the time course of an action potential (black line), Ca^{2+} transient (blue line) and contraction (red dotted line) measured in a rabbit ventricular myocyte at 37 °C. This figure is modified from Bers 2002[100]. (For interpretation of the references to color in this figure legend, the reader is referred to the web version of the original article.)

membrane invaginations. This loss of T-tubules becomes increasingly evident as a symptom in cardiac disorders, such as following myocardial infarction[117], in diabetic cardiomyopathy[118,119], during heart failure in general[120-122] or, particularly, during the transition from cardiac hypertrophy to heart failure[123]. It is still not known whether the detubulation is a cause or a consequence of the mentioned cardiac diseases. Within ▶ chapters 19 and 20, there are two major findings that are discussed in detail: (i) T-tubules sequentially pinching off in an outward direction and (ii) the remodelling of the actin skeleton precedes the T-tubules. In these studies the conservation of the actin resulted in the preservation of the T-tubular structure. However, the presented indications suggest that the spatial distraction of the functional units that mediate calcium-induced calcium release is the major component that causes altered cardiomyocyte functionality.

Calcium signalling in red blood cells

Numerous excellent reviews are available that cover calcium signalling in cardiac myocytes (see above); however, a comprehensive review regarding red blood cell calcium signalling has not been written. Although certain effects of calcium in red blood cells have been known for more than 50 years[6,7], the means by which calcium enters the red blood cells still has not been completely elucidated. However, what has been clear for quite some time is how the calcium is transported out of the cell once it entered the cell and how the internal calcium concentration is maintained at a low concentration (in general, approximately 60 nM[124]) via the calcium pump. The calcium pump was initially described and characterised by Schatzmann[125-127], has been further investigated and reviewed by many others[124,128-132] and is still not completely understood[133].

Based on flux measurements, Halperin *et al.*[134] proposed a voltage-activated cation channel permeable to calcium. Soon after, the existence of a non-selective cation channel was shown using the patch-clamp technique[135]. However, it was only shown in 2000, that this channel is indeed permeable to calcium[11] (cp. ▶ chapter 21). In the following years, a number of further channels permeable to calcium have been described. These channels are reviewed in ▶ chapter 28[69] and an overview of the calcium transporting entities in red blood cells is summarised in ◘ Fig. 10.1. To date, carriers for calcium transport have been suggested[124] but never proven.

In addition to activating the calcium pump, intracellular calcium inhibits flippase (EC$_{50}$ of 400 nM)[143] and activates the Gardos channel (EC$_{50}$ of 4.7 μM)[144], the lipid scramblase (EC$_{50}$ of 29 μM)[145], the protein kinase Cα (EC$_{50}$ of 35 μM)[146] and calpain (EC$_{50}$ of 40 μM)[147].

The sum of these effects leads to cellular symptoms that are similar to the signs of apoptosis in nucleated cells. For example, the inhibition of flippase and the activation of scramblase provoke a breakdown of the asymmetric lipid distribution between the inner and outer membrane leaflet; therefore, phosphatidylserine will be exposed on the outer leaflet. The activation of the Gardos channel results in a potassium efflux that is associated with cell shrinkage. In conjunction with calpain activation and cleavage of actin, cells undergo vesiculation. The activity of protein kinase Cα likely contributes to the apoptosis-like effects; however, the role of protein kinase Cα still needs to be elucidated (cp. ▶ chapter 27). Because red blood cells do not contain mitochondria, they can - per definition - not undergo apoptosis. Therefore, Lang *et al.*[5] defined eryptosis as the red blood cell equivalent of apoptosis. The network of calcium signalling in red blood cells as far as concerned in part IV is summarised in ◘ Fig. 10.2.

It is physiologically interesting that all of the signs of eryptosis appear to support the intercellular adhesion of red blood cells[14,96]. This aggregation is not as obvious as the agglutination that follows an antigen-antibody reaction in blood of non-matching AB0-blood groups (cp. section 8.3 - Red blood cell aggregation). However, there is further experimental support from experimental and clinical investigations of a prolongation of the bleeding time in low red blood cell counts[148-150]. Evidence that the intercellular adhesion process between red blood cells may play a role *in vivo* was recently provided by Noh *et al.*[151]. In this study, an increase in the intracellular calcium concentration of red blood cells, which is associated with phosphatidylserine exposure, could be related to prothrombotic activity *in vivo* in a rat

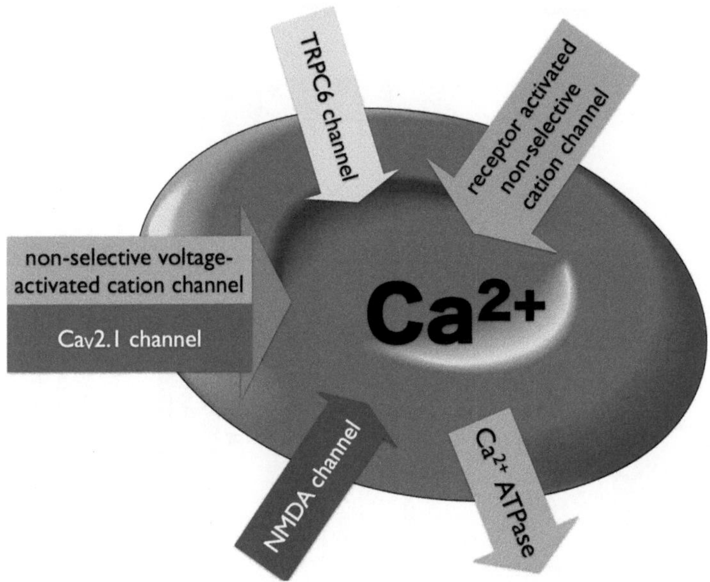

◘ **Fig. 10.1** Overview of calcium transport pathways across the red blood cell membrane. For calcium efflux the situation is clear - in human red blood cells the calcium pump is the only known, but very effective transporter (references see text). For the calcium efflux there is a number of different channels. The TRP C6[136], the Ca$_V$2.1[137] and the NMDA channel[138] were identified by Western blots. The non-selective voltage activated cation channel was functionally described[11,135,139,140]. Since the Ca$_V$2.1 was so far not observed in electrophysiological measurements, it was hypothesised that the non-selective voltage-activated cation channel and the Ca$_V$2.1 have the same identity[69]. There are several reports about chemical or physical induced currents[83,141,142]; however, it still remains to be elucidated if these are separate channels or induced signalling cascades leading to the activation of one or several of the afore mentioned channels.

model of venous thrombosis. Under pathophysiological conditions, intercellular red blood cell adhesion after calcium influx appears to exert a more pronounced effect. An example is the vasco-occlusive crisis of sickle cell disease patients. In this condition, the entry of calcium is proposed to be mediated by the NMDA receptor that was determined to occur in red blood cells[63,138]. The calcium signalling leading to red blood cell adhesion provides the link between the increased prevalence of the NMDA receptor in sickle cell disease patients[138] and the symptoms of the vasco-occlusive crisis. Further examples include disorders in the ion homeostasis of red blood cells that are associated with thrombotic events in malaria[152] and thalassemia[153,154] patients.

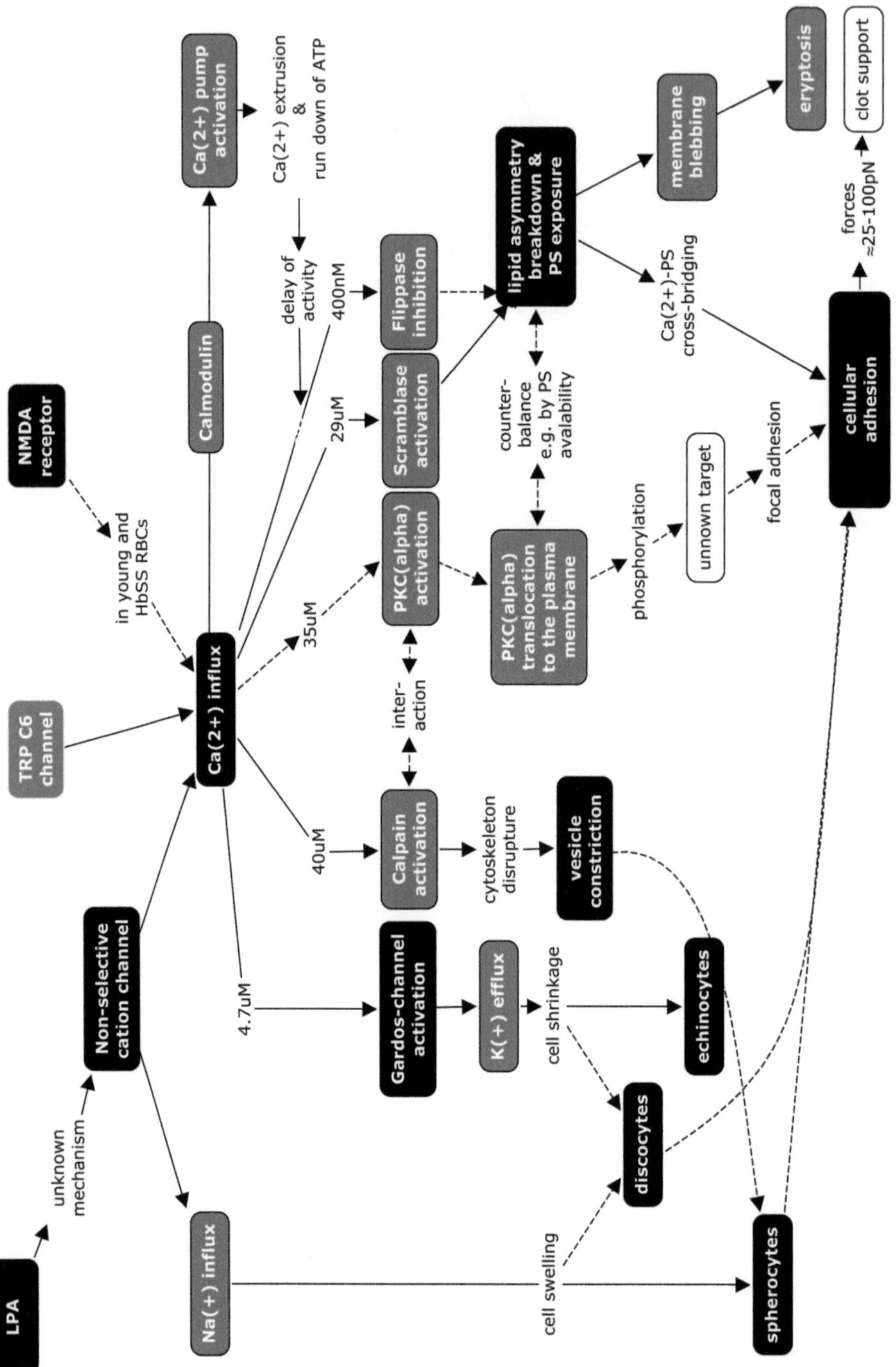

❏ **Fig. 10.2** Calcium signalling in human red blood cells as far as concerned in part IV. The black boxes represent data observed in experiments presented in part IV, the grey boxes contain data from the scientific literature (for references, see text) and the white boxes are hypotheses. The following abbreviations were used: LPA, lyso-phosphatidic acid; TRP, transient receptor potential; NMDA, N-methyl-D-aspartate; HbSS RBCs, sickle red blood cells; PKC, protein kinase C; PS, phosphatydilserine. The signalling can be triggered under pathophysiological conditions or at the end of the life cycle of red blood cells; however, to the current knowledge the calcium signalling (except the calcium pump) seems to have no or little impact on »normal« red blood cell physiology - most calcium entry is believed to occur stochastically.

Perspective

»Ja Kalzium, das ist alles.« is an expression that is attributed to the Nobel-laureate Otto Loewi. This German sentence is often cited in the English literature[155-159] and is, I believe, regularly misinterpreted (»Calcium is everything«), leading to articles titled »Kalzium ist nicht alles«[160]. Instead, a more appropriate translation would be »Calcium is universal.« This concept is also the key to future calcium-related research. The many different signals that are mediated by calcium as a second messenger range from proliferation signals to the induction of apoptosis. Accordingly, the questions of how these signals are modulated, read by the cell and differentiated by the cell will be a general focus in calcium signalling-related research.

The nature of the calcium signal cannot be more different in cardiomyocytes compared with red blood cells as stated in the introduction. However, the same holds true for the history of scientific investigations: the role of calcium in the cardiac contraction was the initial milestone to recognise the importance of calcium as a signalling molecule. In contrast, the physiological importance of calcium as a signalling molecule in red blood cells was only recognised along the timeline given by the original publication summarised in part IV. Therefore, we have a comprehensive knowledge of calcium signalling in cardiac myocytes, whereas in red blood cells, a number of rather basic questions remain to be answered. Nevertheless, both cell types share the feature that a large number of pathologies are associated with impaired calcium signalling as the source, the mediator or the consequence of the disorder. Therefore, the investigation of cellular and sub-cellular calcium signals still is and will remain a major tool in cell biology, biophysics and physiology. The experimental perspective will further shift: for red blood cells, the focus will shift from bulk and population measurements to single cell approaches, as described above. For cardiac myocytes there is already a long-lasting focus on single cell investigations[161]; however, this focus must shift from isolated cells to cells in the context of the tissue.

11.1 References

[1] Ringer, S. A further Contribution regarding the influence of the different Constituents of the Blood on the Contraction of the Heart. J. Physiol. (Lond.) 4, 29–45 (1883).

[2] Bootman, M. D., Berridge, M. J. & Lipp, P. Cooking with calcium: the recipes for composing global signals from elementary events. Cell 91, 367–373 (1997).

[3] Berridge, M. J., Lipp, P. & Bootman, M. D. Signal transduction. The calcium entry pas de deux. Science 287, 1604–1605 (2000).

[4] Berridge, M. J., Lipp, P. & Bootman, M. D. The versatility and universality of calcium signalling. Nature Reviews Molecular Cell Biology 1, 11–21 (2000).

[5] Lang, K. S. et al. Mechanisms of suicidal erythrocyte death. Cellular Physiology and Biochemistry 15, 195–202 (2005).

[6] Gardos, G. The permeability of human erythrocytes to potassium. Acta Physiol Hung 10, 185–189 (1956).

[7] Gardos, G. The function of calcium in the potassium permeability of human erythrocytes. Biochimica et Biophysica Acta (1958).

[8] Abderhalden, E. Über den Blut Kalium Gehalt verschiedener Säugethiere. Hoppe-Seyler's Zeitschrift für Physiologische Chemie 25, 65–115 (1889).

[9] Gulaczy, von, Z. Über den Calciumgehalt der roten Blutkörperchen im menschlichen Blute I. Biochemische Zeitschrift 251, 162–166 (1932).

[10] Schönberger, S. Über den Calciumgehalt der roten Blutkörperchen im menschlichen Blute II. Biochemische Zeitschrift 251, 167–172 (1932).

[11] Kaestner, L., Christophersen, P., Bernhardt, I. & Bennekou, P. The non-selective voltage-activated cation channel in the human red blood cell membrane: reconciliation between two conflicting reports and further characterisation. Bioelectrochemistry 52, 117–125 (2000).

[12] Quintana, A. et al. Sustained activity of CRAC channels requires translocation of mitochondria to the plasma membrane. The Journal of biological chemistry 281, 40302–40309 (2006).

[13] Kaestner, L., Tabellion, W., Weiss, E., Bernhardt, I. & Lipp, P. Calcium imaging of individual erythrocytes: Problems and approaches. Cell Calcium 39, 13–19 (2006).

[14] Steffen, P. et al. Stimulation of human red blood cells leads to Ca^{2+}-mediated intercellular adhesion. Cell Calcium 50, 54–61 (2011).

[15] Kaestner, L. & Lipp, P. Non-linear and ultra high-speed imaging for explorations of the murine and human heart. Optics in Life Science, 66330K–1 – 66330K–10 (2007).

[16] Kaestner, L. & Lipp, P. Towards Imaging the Dynamics of Protein Signalling. Imaging Cellular and Molecular Biological Functions 289–312 (Springer: 2007).

[17] Tian, Q. et al. Functional and morphological preservation of adult ventricular myocytes in culture by sub-micromolar cytochalasin D supplement. Journal of Molecular and Cellular Cardiology 52, 113–124 (2012).

[18] Egner, A. & Hell, S. W. Fluorescence microscopy with super-resolved optical sections. Trends in cell biology 15, 207–215 (2005).

[19] Fishbain, B., Ideses, I. A., Shabat, G., Salomon, B. G. & Yaroslavsky, L. P. Superresolution in color videos acquired through turbulent media. Optics letters 34, 587–589 (2009).

[20] Haugland, R. P. Indicators for Ca^{2+}, Mg^{2+}, Zn^{2+} and Other Metal Ions. Molecular Probes Handbook (2011).

[21] Lipp, P., Bootman, M. D. & Collins, T. Photometry, video imaging, confocal and multi-photon microscopy approaches in calcium signalling studies. Calcium Signalling (2001).

[22] Ljubojević, S. et al. In Situ Calibration of Nucleoplasmic versus Cytoplasmic Ca^{2+} Concentration in Adult Cardiomyocytes. Biophysical Journal 100, 2356–2366 (2011).

[23] Miyawaki, A. et al. Fluorescent indicators for Ca^{2+} based on green fluorescent proteins and calmodulin. Nature 388, 882–887 (1997).

[24] Mank, M. et al. A FRET-based calcium biosensor with fast signal kinetics and high fluorescence change. Biophysical Journal 90, 1790–1796 (2006).

[25] Mank, M. et al. A genetically encoded calcium indicator for chronic in vivo two-photon imaging. Nature Methods (2008).

[26] Baird, G. S., Zacharias, D. A. & Tsien, R. Y. Circular permutation and receptor insertion within green fluorescent proteins. Proceedings of the National Academy of Sciences of the United States of America 96, 11241–11246 (1999).

[27] Nagai, T., Sawano, A., Park, E. S. & Miyawaki, A. Circularly permuted green fluorescent proteins engineered to sense Ca^{2+}. Proceedings of the National Academy of Sciences of the United States of America 98, 3197–3202 (2001).

[28] Nakai, J., Ohkura, M. & Imoto, K. A high signal-to-noise Ca(2+) probe composed of a single green fluorescent protein. Nature biotechnology 19, 137–141 (2001).

[29] Tallini, Y. N. et al. Imaging cellular signals in the heart in vivo: Cardiac expression of the high-signal Ca^{2+} indicator GCaMP2. Proceedings of the National Academy of Sciences of the United States of America 103, 4753–4758 (2006).

[30] Viero, C., Kraushaar, U., Ruppenthal, S., Kaestner, L. & Lipp, P. A primary culture system for sustained expression of a calcium sensor in preserved adult rat ventricular myocytes. Cell Calcium 43, 59–71 (2008).

[31] Fluorescent Proteins I. (Springer: 2012).

[32] Gensch, T. & Kaschuba, D. Fluorescent Genetically Encoded Calcium Indicators and Their In Vivo Application. Fluorescent Proteins II 125–162 (Springer: 2012).

[33] Lipp, P. & Kaestner, L. Image-based High-content Screening - A view from Basic Science. High Throughput-Screening in Drug Discovery 129–149 (WileyVHC: 2006).

[34] Minsky, M. Microscopy Apparatus. (patent 1957).

[35] Hüser, J., High-Throughput Screening in Drug Discovery. (WileyVHC: 2006).

[36] Nipkow, P. Elektrisches Teleskop. (patent 1884).

[37] Kaestner, L., Lipp, P., von Gegerfelt, D., Karlsson, H., Illy, E., Hellström, J. 515 nm all-solid-state laser: the missing link to compact fluorescence excitation in bioanalytical Instrumentation. Photonik International 09/1, 24-26 (2009)

[38] Denk, W., Piston, D. W. & Webb, W. W. Two-photon molecular excitation in laser-scanning microscopy. Handbook of biological confocal microscopy 445–458 (Plenum Press: 1995).

[39] Bouzid, A. & Lechleiter, J. Laser scanning fluorescence microscopy with compensation for spatial dispersion of fast laser pulses. (patent 2002).

[40] Roorda, R. D. & Miesenbock, G. Beam-steering of multi-chromatic light using acousto-optical deflectors and dispersion-compensatory optics. (patent 2002).

[41] Huisken, J., Swoger, J., del Bene, F., Wittbrodt, J. & Stelzer, E. H. Optical sectioning deep inside live embryos by selective plane illumination microscopy. Science 305, 1007–1009 (2004).

[42] Neil, M. A., Juskaitis, R. & Wilson, T. Method of obtaining optical sectioning by using structured light in a conventional microscope. Optics letters 22, 1905–1907 (1997).

[43] Weigel, A., Schild, D. & Zeug, A. Resolution in the ApoTome and the confocal laser scanning microscope: comparison. Journal of Biomedical Optics 14, 014022 (2009).

[44] Parker, I. & Ivorra, I. Confocal microfluorimetry of Ca^{2+} signals evoked in Xenopus oocytes by photoreleased inositol trisphosphate. Journal of Physiology 461, 133–165 (1993).

[45] Müller, O. et al. A system for optical high resolution screening of electrical excitable cells. Cell Calcium 47, 224–233 (2010).

[46] Hell, S. W. & Wichmann, J. Breaking the diffraction resolution limit by stimulated emission: stimulated-emission-depletion fluorescence microscopy. Optics letters 19, 780–782 (1994).

[47] Hofmann, M., Eggeling, C., Jakobs, S. & Hell, S. W. Breaking the diffraction barrier in fluorescence microscopy at low

light intensities by using reversibly photoswitchable proteins. Proceedings of the National Academy of Sciences of the United States of America 102, 17565–17569 (2005).

[48] Betzig, E. et al. Imaging intracellular fluorescent proteins at nanometer resolution. Science 313, 1642–1645 (2006).

[49] Rust, M. J., Bates, M. & Zhuang, X. Sub-diffraction-limit imaging by stochastic optical reconstruction microscopy (STORM). Nature Methods 3, 793–795 (2006).

[50] Gustafsson, M. G. Surpassing the lateral resolution limit by a factor of two using structured illumination microscopy. Journal of Microscopy 198, 82–87 (2000).

[51] Gustafsson, M. G. L. Nonlinear structured-illumination microscopy: wide-field fluorescence imaging with theoretically unlimited resolution. Proceedings of the National Academy of Sciences of the United States of America 102, 13081–13086 (2005).

[52] Lakowicz, J. R., Szmacinski, H., Nowaczyk, K. & Johnson, M. L. Fluorescence lifetime imaging of calcium using Quin-2. Cell Calcium 13, 131–147 (1992).

[53] Shimomura, O., Johnson, F. H. & Saiga, Y. Extraction, purification and properties of aequorin, a bioluminescent protein from the luminous hydromedusan, Aequorea. Journal of cellular and comparative physiology 59, 223–239 (1962).

[54] Ashley, C. C. & Ridgway, E. B. Simultaneous recording of membrane potential, calcium transient and tension in single muscle fibers. Nature 219, 1168–1169 (1968).

[55] Tsien, R. Y. New Calcium Indicators and Buffers with High Selectivity Against Magnesium and Protons - Design, Synthesis, and Properties of Prototype Structures. Biochemistry 19, 2396–2404 (1980).

[56] Miller, A. L., Karplus, E. & Jaffe, L. F. Imaging $[Ca^{2+}]_i$ with aequorin using a photon imaging detector. Methods in cell biology 40, 305–338 (1994).

[57] Blinks, J. R. Use of calcium-regulated photoproteins as intracellular Ca^{2+} indicators. Methods in Enzymology (1989).

[58] Creton, R., Steele, M. & Jaffe, L. Expression of apo-aequorin during embryonic development; how much is needed for calcium imaging? Cell Calcium 22, 439–446 (1997).

[59] Rogers, K. L. et al. Non-invasive in vivo imaging of calcium signaling in mice. PLoS ONE 2, e974 (2007).

[60] Menon, V. et al. Development of an aequorin luminescence calcium assay for high-throughput screening using a plate reader, the LumiLux. Assay Drug Dev Technol 6, 787–793 (2008).

[61] Webb, S. E., Rogers, K. L., Karplus, E. & Miller, A. L. The use of aequorins to record and visualize Ca(2+) dynamics: from subcellular microdomains to whole organisms. Methods in cell biology 99, 263–300 (2010).

[62] Cohn, W. E. & Cohn, E. T. Permeability of red corpuscles of the dog to sodium ion. Proc. Soc. Exp. Biol. Med. 41, 445–449 (1939).

[63] Makhro, A. et al. Functional NMDA receptors in rat erythrocytes. AJP: Cell Physiology 298, C1315–C1325 (2010).

[64] Sigworth, F. J. The patch clamp is more useful than anyone had expected. Federation proceedings 45, 2673–2677 (1986).

[65] Hamill, O. P., Marty, A., Neher, E., Sakmann, B. & Sigworth, F. J. Improved Patch-Clamp Techniques for High-Resolution Current Recording From Cells and Cell-Free Membrane Patches. Pflügers Archive 391, 85–100 (1981).

[66] Sakmann, B., Neher, E. Single-Channel Recording. (Springer: 2009).

[67] Numberger, M. & Draguhn, A. Patch-clamp Technik. (Spektrum Akademischer Verlag: 1996).

[68] Kaestner, L. & Bernhardt, I. Ion channels in the human red blood cell membrane: their further investigation and physiological relevance. Bioelectrochemistry 55, 71–74 (2002).

[69] Kaestner, L. Cation Channels in Erythrocytes - Historical and Future Perspective. The Open Biology Journal 4, 27–34 (2011).

[70] Reil, J. C. et al. Cardiac Rac1 overexpression in mice creates a substrate for atrial arrhythmias characterized by structural remodelling. Cardiovascular research 87, 485–493 (2010).

[71] Holzhauer, M. Biochemische Labormethoden. (Springer: 2009).

[72] Elliott, W. H. & Elliott, D. C. Biochemistry and Molecular Biology. (Oxford University Press: 2009).

[73] Kirchhefer, U. et al. Overexpression of junctin causes adaptive changes in cardiac myocyte Ca^{2+} signaling. Cell Calcium 39, 131–142 (2006).

[74] Weissgerber, P. et al. Reduced cardiac L-type Ca^{2+} current in Ca(V)beta2-/- embryos impairs cardiac development and contraction with secondary defects in vascular maturation. Circulation Research 99, 749–757 (2006).

[75] Kaestner, L., Tian, Q. & Lipp, P. Action potentials in heart cells. Frontiers in Pharmacology 2, 42 (2012).

[76] Lundstrom, K. Semliki Forest virus vectors for gene therapy. Expert opinion on biological therapy 3, 771–777 (2003).

[77] Delenda, C. Lentiviral vectors: optimization of packaging, transduction and gene expression. The journal of gene medicine 6 Suppl 1, S125–38 (2004).

[78] Russell, W. C. Update on adenovirus and its vectors. The Journal of general virology 81, 2573–2604 (2000).

[79] Vig, K. et al. Recombinant adeno-associated virus as vaccine delivery vehicles. Gene Ther Mol Biol 12B, 277–292 (2008).

[80] Kaestner, L., Ruppenthal, S., Schwarz, S., Scholz, A. & Lipp, P. Concepts for optical high content screens of excitable primary isolated cells for molecular imaging. Progress in Biomedical Optics and Imaging II, 737008-1 – 737008-8 (2009).

[81] Christensen, G., Minamisawa, S., Gruber, P. J., Wang, Y. & Chien, K. R. High-efficiency, long-term cardiac expression of foreign genes in living mouse embryos and neonates. Circulation 101, 178–184 (2000).

[82] Büning, H., Perabo, L., Coutelle, O., Quadt-Humme, S. & Hallek, M. Recent developments in adeno-associated virus vector technology. The journal of gene medicine 10, 717–733 (2008).

[83] Kaestner, L., Tabellion, W., Lipp, P. & Bernhardt, I. Prostaglandin E_2 activates channel-mediated calcium entry in human erythrocytes: an indication for a blood clot formation supporting process. Thrombosis and Haemostasis 92, 1269–1272 (2004).

[84] Hammer, K. et al. Remodelling of Ca^{2+} handling organelles in adult rat ventricular myocytes during long term culture. Journal of Molecular and Cellular Cardiology 49, 427–437 (2010).

[85] Reither, G. PKC : a versatile key for decoding the cellular calcium toolkit. The Journal of Cell Biology 174, 521–533 (2006).

[86] Ellis-Davies, G. C. R. Caged compounds: photorelease technology for control of cellular chemistry and physiology. Nature Methods 4, 619–628 (2007).

[87] Boyden, E. S., Zhang, F., Bamberg, E., Nagel, G. & Deisseroth, K. Millisecond-timescale, genetically targeted optical control of neural activity. Nat. Neurosci. 8, 1263–1268 (2005).

[88] Bruegmann, T. et al. Optogenetic control of heart muscle in vitro and in vivo. Nature Methods 7, 897–900 (2010).

[89] Harzheim, D. & Hegemann, P. Rhodopsin-regulated calcium currents in Chlamodymonas. 489–491 (1991).

[90] Caldwell, J. H. et al. Increases in intracellular calcium triggered by channelrhodopsin-2 potentiate the response of metabotropic glutamate receptor mGluR7. The Journal of biological chemistry 283, 24300–24307 (2008).

[91] Lin, J. Y., Lin, M. Z., Steinbach, P. & Tsien, R. Y. Characterization of Engineered Channel rhodopsin Variants with Improved Properties and Kinetics. Biophysical journal 96, 1803–1814 (2009).

[92] Sussman, M. A. et al. Altered focal adhesion regulation correlates with cardiomyopathy in mice expressing constitutively active rac1. Journal of Clinical Investigation 105, 875–886 (2000).

[93] Iribe, G. et al. Axial stretch of rat single ventricular cardiomyocytes causes an acute and transient increase in Ca^{2+} spark rate. Circulation Research 104, 787–795 (2009).

[94] Antovic, J. P. & Blombäck, M. Essential Guide to Blood Coagulation. (Wiley-Blackwell: Oxford, 2010).

[95] Baskurt, O., Neu, B. & Meiselman, H. J. Red blood Cell Aggregation. (CRC Press: 2012).

[96] Kaestner, L. et al. Lysophosphatidic acid induced red blood cell aggregation in vitro. Bioelectrochemistry 87, 89-95 (2012).

[97] Chien, S. & Jan, K.-M. Ultrastructural Basis of the Mechanism of Rouleaux Formation. Microvascular Research 5, 155–166 (1973).

[98] Picart, C., Piau, J. M., Galliard, H. & Carpentier, P. Human blood shear yield stress and its hematocrit dependence. J Rheol 42, 1–12 (1998).

[99] Shin, S., Park, M., Jang, J., Ku, Y. & Suh, J. Measurement of red blood cell aggregation by analysis of light transmission in a pressure-driven slit flow system. Korea-Aust Rheol J 16, 129–134 (2004).

[100] Bers, D. M. Cardiac excitation-contraction coupling. Nature 415, 198–205 (2002).

[101] Lipp, P. et al. Functional InsP3 receptors that may modulate excitation-contraction coupling in the heart. Curr. Biol. 10, 939–942 (2000).

[102] Mackenzie, L. et al. The role of inositol 1,4,5-trisphosphate receptors in Ca(2+) signalling and the generation of arrhythmias in rat atrial myocytes. J. Physiol. (Lond.) 541, 395–409 (2002).

[103] Watanabe, H., Murakami, M., Ohba, T., Ono, K. & Ito, H. The pathological role of transient receptor potential channels in heart disease. Circ. J. 73, 419–427 (2009).

[104] Eder, P. & Molkentin, J. D. TRPC channels as effectors of cardiac hypertrophy. Circulation Research 108, 265–272 (2011).

[105] Seeber, S. et al. Transient expression of NMDA receptor subunit NR2B in the developing rat heart. Journal of neurochemistry 75, 2472–2477 (2000).

[106] Bers, D. M. Excitation-Contraction Coupling and Cardiac Contractile Force. (Kluwer Academic Publishers: Dordrecht, Boston, London, 2001).

[107] Zipes, D. P. & Jalife, J. Cardiac Electrophysiology: From Cell to Bedside. (Saunders: 2009).

[108] Knot, H. J. et al. Twenty years of calcium imaging: cell physiology to dye for. Molecular Intervention 5, 112–127 (2005).

[109] Schaub, M. C., Hefti, M. A. & Zaugg, M. Integration of calcium with the signaling network in cardiac myocytes. Journal of Molecular and Cellular Cardiology 41, 183–214 (2006).

[110] Bootman, M. D., Higazi, D. R., Coombes, S. & Roderick, H. L. Calcium signalling during excitation-contraction coupling in mammalian atrial myocytes. Journal of cell science 119, 3915–3925 (2006).

[111] Laurita, K. R. & Rosenbaum, D. S. Cellular mechanisms of arrhythmogenic cardiac alternans. Prog Biophys Mol Biol 97, 332–347 (2008).

[112] Cheng, H. & Lederer, W. J. Calcium sparks. Physiol. Rev. 88, 1491–1545 (2008).

[113] Fearnley, C. J., Roderick, H. L. & Bootmann, M. D. Calcium Signaling in Cardiac Myocytes. Calcium Signaling 403–422 (Cold Spring Harbor: 2012).

[114] Catterall, W. A. Voltage-Gated Calcium Channels. Calcium Signalling 1–23 (2012).

[115] Richards, M. W., Butcher, A. J. & Dolphin, A. C. Ca^{2+} channel beta-subunits: structural insights AID our understanding. Trends in Pharmacological Sciences 25, 626–632 (2004).

[116] Davies, A. et al. Functional biology of the alpha(2)delta subunits of voltage-gated calcium channels. Trends in Pharmacological Sciences 28, 220–228 (2007).

[117] Louch, W. E. et al. T-tubule disorganization and reduced synchrony of Ca^{2+} release in murine cardiomyocytes following myocardial infarction. J. Physiol. (Lond.) 574, 519–533 (2006).

[118] Stolen, T. O. et al. Interval Training Normalizes Cardiomyocyte Function, Diastolic Ca^{2+} Control, and SR Ca^{2+} Release Synchronicity in a Mouse Model of Diabetic Cardiomyopathy. Circulation Research 105, 527–U47 (2009).

[119] McGrath, K. F. et al. Morphological characteristics of cardiac calcium release units in animals with metabolic and circulatory disorders. Journal of muscle research and cell motility 30, 225–231 (2009).

[120] Cannell, M. B., Crossman, D. J. & Soeller, C. Effect of changes in action potential spike configuration, junctional sarcoplasmic reticulum micro-architecture and altered t-tubule structure in human heart failure. Journal of muscle research and cell motility 27, 297–306 (2006).

[121] Lyon, A. R. et al. Loss of T-tubules and other changes to surface topography in ventricular myocytes from failing human and rat heart. Proceedings of the National Academy of Sciences of the United States of America 106, 6854–6859 (2009).

[122] Song, L. et al. Orphaned ryanodine receptors in the failing heart. Proceedings of the National Academy of Sciences of the United States of America 103, 4305–4310 (2006).

[123] Wei, S. et al. T-tubule remodeling during transition from hypertrophy to heart failure. Circulation Research 107, 520–531 (2010).

[124] Red Cell Membrane Transport in Health and Disease. 373–405 (2003).

[125] Schatzmann, H. J. ATP-dependent Ca++-Extrusion from human red cells. Experientia 22, 364–365 (1966).

[126] Vincenzi, F. F. & Schatzmann, H. J. Some properties of Ca-activiated ATPase in human red cell membranes. Helvetica Physiologica et Pharmacologica Acta 25, CR233–234 (1967).

[127] Schatzmann, H. J. Calcium movements across the membrane of human red cells. J. Physiol. (Lond.) 201, 369–395 (1969).

[128] Lew, V. L. et al. Distribution of plasma membrane Ca^{2+} pump activity in normal human red blood cells. Blood 102, 4206–4213 (2003).

[129] Kubitscheck, U., Pratsch, L., Passow, H. & Peters, R. Calcium-Pump Kinetics Determined in Single Erythrocyte-Ghosts by Microphotolysis and Confocal Imaging. Biophysical journal 69, 30–41 (1995).

[130] Schatzmann, H. The Red Cell Calcium Pump - Annual Review of Physiology, 45(1):303. Annual review of physiology (1983).

[131] Rega, A. F. & Garrahan, P. J. Catalytic Mechanisms of the Ca-Pump ATPase. The Red Cell Membrane 103–121 (1989).

[132] Vincenzi, F. F. Regulation of the Plasma Membrane Ca^{2+}-pump. The Red Cell Membrane 123–142 (1989).

[133] Bennekou, P., Harbak, H. & Simonsen, L. O. Vanadate-induced Ca(2+) and Co(2+) uptake in human red blood cells. Blood Cells, Molecules, and Diseases 47, 214-225 (2011).

[132] Halperin, J. A., Brugnara, C., Van Ha, T. & Tosteson, D. C. Voltage-activated cation permeability in high-potassium but not low-potassium red blood cells. American Journal of Physiology - Cell Physiology 258, 1169–1172 (1990).

[135] Christophersen, P. & Bennekou, P. Evidence for a voltage-gated, non-selective cation channel in the human red cell membrane. Biochimica et Biophysica Acta 1065, 103–106 (1991).

[136] Foller, M. et al. TRPC6 contributes to the Ca(2+) leak of human erythrocytes. Cell Physiol Biochem 21, 183–192 (2008).

[137] Andrews, D. A., Yang, L. & Low, P. S. Phorbol ester stimulates a protein kinase C-mediated agatoxin-TK-sensitive calcium permeability pathway in human red blood cells. Blood 100, 3392–3399 (2002).

[138] Bogdanova, A. et al. NMDA receptors in mammalien erythrocytes. Clinical Biochemistry 42, 1858–1859 (2009).

[139] Kaestner, L., Bollensdorff, C. & Bernhardt, I. Non-selective voltage-activated cation channel in the human red blood cell membrane. Biochimica et Biophysica Acta 1417, 9–15 (1999).

[140] Rodighiero, S., De Simoni, A. & Formenti, A. The voltage-dependent nonselective cation current in human red blood cells studied by means of whole-cell and nystatin-perforated patch-clamp techniques. Biochimica et Biophysica Acta 1660, 164–170 (2004).

[141] Huber, S. M., Gamper, N. & Lang, F. Chloride conductance and volume-regulatory nonselective cation conductance in human red blood cell ghosts. Pflügers Archiv - European Journal of Physiology 441, 551–558 (2001).

[142] Egée, S. et al. A stretch-activated anion channel is up-regulated by the malaria parasite Plasmodium falciparum. J. Physiol. (Lond.) 542, 795–801 (2002).

[143] Bitbol, M., Fellmann, P., Zachowski, A. & Devaux, P. F. Ion regulation of phosphatidylserine and phosphatidylethanolamine outside-inside translocation in human erythrocytes. Biochimica et Biophysica Acta 904, 268–282 (1987).

[144] Leinders, T., van Kleef, R. G. & Vijverberg, H. P. Single Ca^{2+}-activated K^+ channels in human erythrocytes: Ca^{2+} dependence of opening frequency but not of open lifetimes. Biochimica et Biophysica Acta 1112, 67–74 (1992).

[145] Woon, L. A., Holland, J. W., Kable, E. P. & Roufogalis, B. D. Ca^{2+} sensitivity of phospholipid scrambling in human red cell ghosts. Cell Calcium 25, 313–320 (1999).

[146] Kohout, S. C., Corbalán-García, S., Torrecillas, A., Goméz-Fernandéz, J. C. & Falke, J. J. C2 domains of protein kinase C isoforms alpha, beta, and gamma: activation parameters and calcium stoichiometries of the membrane-bound state. Biochemistry 41, 11411–11424 (2002).

[147] Murakami, T., Hatanaka, M. & Murachi, T. The cytosol of human erythrocytes contains a highly Ca^{2+}-sensitive thiol protease (calpain I) and its specific inhibitor protein (calpastatin). Journal of Biochemistry 90, 1809–1816 (1981).

[148] Hellem, A. J., Borchgrevink, C. F. & Ames, S. B. The role of red cells in haemostasis: the relation between haematocrit, bleeding time and platelet adhesiveness. Br J Haematol 7, 42–50 (1961).

[149] Livio, M. et al. Uraemic bleeding: role of anaemia and beneficial effect of red cell transfusions. Lancet 2, 1013–1015 (1982).

[150] Mackman, N. Triggers, targets and treatments for thrombosis. Nature 451, 914–918 (2008).

[151] Noh, J.-Y. et al. Procoagulant and prothrombotic activation of human erythrocytes by phosphatidic acid. AJP: Heart and Circulatory Physiology 299, H347–355 (2010).

[152] Luvira, V., Chamnanchanunt, S., Thanachartwet, V., Phumratanaprapin, W. & Viriyavejakul, A. Cerebral venous sinus thrombosis in severe malaria. Southeast Asian J. Trop. Med. Public Health 40, 893–897 (2009).

[153] Eldor, A. & Rachmilewitz, E. A. The hypercoagulable state in thalassemia. Blood 99, 36–43 (2002).

[154] Taher, A. T., Otrock, Z. K., Uthman, I. & Cappellini, M. D. Thalassemia and hypercoagulability. Blood Rev 22, 283–292 (2008).

[155] Carafoli, E. Calcium signaling: a tale for all seasons. Proceedings of the National Academy of Sciences of the United States of America 99, 1115–1122 (2002).

[156] Splawski, I. et al. Ca_V1.2 Calcium Channel Dysfunction Causes a Multisystem Disorder Including Arrhythmia and Autism. Cell 119, 19–31 (2004).

[157] Brini, M. & Carafoli, E. Calcium signalling: a historical account, recent developments and future perspectives. Cellular and Molecular Life Sciences 57, 354–370 (2000).

[158] Cheng, H.-P., Wei, S., Wei, L.-P. & Verkhratsky, A. Calcium signaling in physiology and pathophysiology. Acta Ppharmacologica Sinica 27, 767–772 (2006).

[159] Knot, H. J. et al. Twenty years of calcium imaging: cell physiology to dye for. Molecular Intervention 5, 112–127 (2005).

[160] Bush, A. I. Kalzium Ist Nicht Alles. Neuron 65, 143–144 (2010).

[161] Powell, T. & Twist, V. W. A rapid technique for the isolation and purification of adult cardiac muscle cells having respiratory control and a tolerance to calcium. Biochemical and biophysical research communications 72, 327–333 (1976).

Non-linear and ultra high-speed imaging for explorations of the murine and human heart

Lars Kaestner & Peter Lipp

Reprint from Optics in Life Science, J. Popp & G. von Bally (eds.): Proc. of SPIE-OSA Biomedical Optics, SPIE (2007) Vol. 6633, pp. 66330K-1 – 66330K-10.

■ Abstract

Cardiac failure is still one of the mayor reasons for death in the Western population but the pathophysiology of the molecular processes in the heart is far from being completely understood. Therefore further basic research is necessary. With recent developments of optical technologies novel tools to investigate cardiac physiology and pathophysiology became available. They comprise non-linear imaging techniques such as second harmonic generation imaging and fast two-photon excitation imaging of cardiac tissue. In addition, high-speed multi-beam two-photon imaging as well as ultra-high speed single beam single photon 2D-confocal imaging offer novel approaches to study cellular and subcellular signalling events in cardiac tissue and/or single cardiac myocytes. Here we introduce and discuss these new technologies and their practical application to study cardiac physiology and pathophysiology.

12.1   Introduction

The physiological functions as well as pathophysiological processes in the heart can be attributed to the operation of single cardiac myocytes or alterations in their regular function. The dimension of a ventricular myocyte is conserved over most mammalian species and is roughly 100 μm x 20 μm x 20 μm. In order to visualize individual cells, their subcellular structure and function, classical clinical imaging methods such as computer tomography (CT), magnet resonance imaging (MRI) and ultrasound are approaching cellular dimensions but still very often lack the resolution power necessary for fast imaging of subcellular signalling events. In the high resolution modes of CT and MRI repetitive measurements are often necessary to achieve an image of acceptable signal-to-noise ratio and consequently time resolution is low[1]. From this it follows that it is impossible to visualise individual, isolated and non-repetitive cardiac events caused e.g., by arrhythmic behaviour are almost impossible to visualise. Nevertheless, classical optical methods lack the ability for deep penetration into tissue. Even with the extended penetrating of near infrared light image formation is restricted to typically 1 mm in depth, albeit this depends on the particular tissue. Therefore, optical imaging of cardiac myocytes requires either (1.) endoscopic imaging, (2.) imaging in the open chest, (3.) organ or tissue withdrawal, (4.) cell isolation or (5.) model cells.

1. Endoscopic imaging is not yet advisable for application in the heart, because the image formation speed of an endoscopic confocal scanner such as OptiScan FIVE1 (Victoria, Australia) is too slow to resolve the heart beat.

2. Imaging in the open chest is not appropriate for diagnosis or research on human patients because of unjustifiable risks. However, from the sole technical point of view this setup for imaging is rather similar to imaging of the perfused Langendorff heart.

3. Imaging of organs or tissues ex vivo is a widely used method to investigate human and animal heart function. Entire animal hearts can be perfused retrogradely (retrograde Langendorff perfusion) after explantation and their beating behaviour is conserved. In such setups hearts can be investigated for more than one day[2]. The advantages of tissue or organ withdrawal are: visualisation of myocytes in their native

environment, visualisation of myocyte – non-myocyte interactions and visualisation of extracellular components (e.g., extracellular matrix). One of the disadvantages with respect to fluorescence imaging is the difficult and time consuming delivery of fluorescent dyes.

4. Cell isolation is probably the most established method to investigate cellular functions since isolated cells can be kept under optimal experimental control and even long term culturing with minimized dedifferentiation of adult myocytes has been achieved (see e.g., [3]). Moreover, isolated cells display almost perfect optical as well as pharmacological accessibility.

5. Model cells include tumour derived cell lines, such as the rat H9c2 cells[4] or the mouse HL-1 cells[5]. Despite their general ease of use these cells unfortunately display gene expression, morphology and physiology that are far away from the native cells and consequently interpretation of experimental results obtained with these cells is problematic.

Here we provide examples for non-linear tissue microscopy on samples derived from human appendages, cellular imaging in a beating mouse heart and ultra high speed imaging of elementary calcium signals, so called calcium sparks, in isolated murine cardiomyocytes.

12.2 Methodology

12.2.1 Human heart tissue preparation

Human tissue was prepared from atrial appendages, which were obtained after informed consent from patients undergoing cardiac surgery with extracorporeal circulation. The use of the tissue was approved by the local ethics committee (approval number 76/05). The appendages were collected in PBS within 15 seconds after excision. The tissue was transferred to the cell culture lab in ice cold Ca^{2+}-free solution (CFS) containing (in mM): NaCl 134, Glucose 11, KCl 4, $MgSO_4$ 1.2, Na_2HPO_4 1.2, HEPES 10 (pH adjusted to 7.35 with NaOH). Documentation of the atrial appendages was performed using a digital camera (DN100,

Nikon, Japan) attached to a stereo-microscope (SMZ800, Nikon, Japan). For second harmonic generation (SHG) imaging samples were placed on a petri dish and were immerged in CFS solution.

12.2.2 Preparation of the mouse heart

The animals (4 to 6 weeks old, 20-25 g) were handled and sacrificed in accordance with the »Guide for the Care and Use of Laboratory Animals« published by the US National Institutes of Health (NIH Publication No. 85-23, revised 1996). Animals were anaesthetised by an intraperitoneal (i.p.) injection of pentobarbital sodium, 400 mg/kg body weight (Narcoren; Merial, Germany). Directly afterwards, we injected 250 µl of citrate (40 mM) i.p. Approximately five minutes later, the animal was killed by decapitation. After removal from the animal the heart was connected to a peristaltic Micro Perfusion Pump (Bioptechs Inc., USA) and placed in a specialised measuring chamber (cp. section 12.3.2 and ◘ Fig. 12.2A) for retrograde perfusion.

12.2.3 Isolation of murine adult cardiomyocytes

Mouse atrial myocytes were isolated from C57BL6/N mice. The animals were sacrificed as described above (section 12.2.2). The heart was flushed with 3 ml of ice-cold CFS while still in the animal. After that, the arrested heart was removed, attached to a Langendorff apparatus and retrogradely perfused with O_2-saturated CFS containing 200 µM EGTA for 10 min. The perfusate was then changed to O_2-saturated CFS with Liberase Blendzyme 4 (Roche, Germany) at a final concentration of 167 µg/ml (duration: 10-12 min).

The auricles were removed and placed in O_2-saturated CFS containing 167 µg/ml Liberase Blendzyme 4 (at 37°C in a water bath for 15 min). The resulting supernatant was discarded and each of the auricles were re-suspended in 1 ml of O_2 saturated CFS supplemented with 0.09% of DNAse and incubated for 5 min. Now mouse atrial myocytes were released from the soft tissue by gentle trituration after which the extracellular Ca^{2+} concentration was carefully increased by repetitively adding 150 μl aliquots of high-Ca^{2+} solution (HCS); 5 times every 5 min. HCS consisted of CFS supplemented with 0.09% of DNAse and 200 μM of Ca^{2+}.

Cells were allowed to settle for 10 min, then the cell suspension was plated on extracellular matrix (ECM, 1.11 mg/ml; Harbor Bioproducts, USA) coated coverslips. The myocytes were allowed to settle down for approximately 1 h in medium M199 with Earle's modified salts, glutamine (Biowest; Nuaille, France), 100 μg/ml Penicillin/Streptomycin and 50 μg/ml Kanamycin (PAA Laboratories, Austria).

The isolation of rat ventricular myocytes was performed as described recently[3].

12.2.4 Loading of fluorescent Ca^{2+} indicator into cardiac myocytes

Fluo-4 loading was performed as previously described[6], in short: An aliquot of 1 mM stock solution of Fluo-4 AM (Molecular Probes, USA) in dimethylsulfoxide containing 20% pluronic acid (Molecular Probes, USA) was dissolved in tyrode solution (containing in mM: NaCl 135, Glucose 10, KCl 5.4, $MgCl_2$ 2, $CaCl_2$ 1.5, HEPES 10 (pH adjusted to 7.35 with NaOH)) giving a final concentration of 1 μM Fluo-4 AM. Cells on coated cover slips were incubated for 30 min at room temperature (20-22°C). Then the cardiomyocytes were washed and additional 15 min were allowed for deestrification.

12.2.5 Imaging techniques

For the various modes of optical imaging four different set-ups were used:

1. a galvo-driven scanhead (Yanus I, TILL Photonics GmbH, Germany) was attached to an upright microscope (BX50WI, Olympus, Japan). The illumination source was a software-controlled Titan:Sapphire laser (Chameleon-XR, Coherent, USA). Imaging was performed under the control of »FluoView« software (Olympus, Japan). For a detailed description of the setup see Quintana *et al.*, 2006[7]. For second harmo-

nic generation (SHG)-imaging the reflectance mode was applied in all experiments. The laser was tuned to 830 nm and the SHG signal was detected after reflection on a 440 nm dichroic longpass mirror and passing a 395 nm – 418 nm bandpass filter. The light passing the dichroic mirror and a 470 nm – 550 nm bandpass filter was considered as autofluorescence.

2. a 64-beam multi-photon scanner (TriM-Scope, LaVision BioTec, Germany) was attached to an inverted microscope (IX71, Olympus, Japan). The illumination source was a software-controlled Titan-Sapphire laser (Mai Tai, Spectra Physics, USA). Detection unit was a VGA-CCD camera (PixelFly, PCO AG, Germany). Imaging was performed with the »Imspector« software (LaVision BioTec, Germany). For a detailed description see Kaestner *et al.*, 2004[8].

3. a kilobeam array scanner ($VT_{infinity}2$, VisiTech Intl. Ltd., UK) was attached to an inverted microscope (TE2000, Nikon, Japan). The illumination source was a directly-coupled AOTF-controlled solid-state laser (Sapphire 488-20, Coherent, USA; 488 nm / 20 mW). As detection unit a back-illuminated electron multiplying CCD camera (iXon DV887, Andor Technology Plc, Ireland) was used running under VoxelScan software (VisiTech Intl. Ltd., UK). Rendering of the 3D-surface was performed under the Imaris software (Bitplane, Switzerland).

4. an acousto-optical deflector (AOD) based scanhead (VT^{eye}, VisiTech international Ltd., UK) was attached to an inverted microscope (TE2000, Nikon, Japan). The illumination source was the 488 nm line of a 35 mW Argon gas laser (model 161C, Spectra Physics, USA). For a detailed description see Kirchhefer *et al.*, 2006[9].

12.2.6 Measurements of fluorescence spectra

For measuring a fluorescence spectrum of a single cell, a fluorescence spectrometer (USB-2000FLG, Ocean Optics, Netherlands) was attached to an inverted microscope (TE2000-U, Nikon, Japan) as previously described[10]. The illumination source was a monochromator (Polychrome IV, TILL Photonics GmbH, Germany). The spectra were spectrally corrected for the dichroic mirror and the emission filter. Transmission spectra of these optical components were measured using an absorption spectrometer (Genesys 6, Thermo Scientific, USA).

12.3 Results

12.3.1 Non-linear imaging of human heart tissue - Second harmonic generation and autofluorescence imaging in living cardiac tissue

For SHG imaging we employed a filter set that allowed simultaneous SHG and autofluorescence detection. Human atrial appendages as those illustrated in ◘ Fig. 12.1A were examined. A representative pair of SHG/autofluorescence images is depicted in ◘ Fig. 12.1B (Ba - SHG; Bb - autofluorescence). The arrow in image Ba points to a small capillary vessel characterised by the increased presence of collagen fibers (SHG-signal, ◘ Fig. 12.1Ba) in the surrounding tissue. Such small vessels are commonly present in human atrial tissue.

A five-fold enlargement (rescanned as shown by the dashed square in Bb) is depicted in ◘ Fig. 12.1Bc and ◘ Fig. 12.1Bd. ◘ Fig. 12.1C depicts the spectrum of the autofluorescence of the same specimen excited at 410 nm.

12.3.2 High speed two-photon multi-beam imaging of a beating mouse heart

For imaging individual cells in a beating heart we used a mouse heart as depicted in ◘ Fig. 12.2A. In the intact heart the cardiac myocytes-containing myocardium is covered by the epicardium comprising epithelial cells and extracellular matrix. In contrast to the pericard, the epicard is adhered to the myocard and can not be removed easily. Thus imaging of myocytes requires an extended light

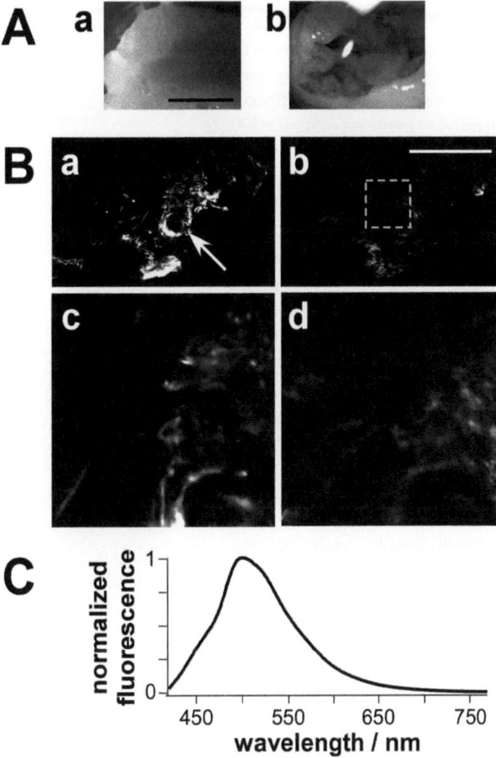

◘ **Fig. 12.1** Non-linear imaging of human auricles. Panel A shows the macroscopic appearance of the atrial appendages; Aa – from outside the heart (epicardial side) and Ab from inside the heart, where the surgical cut was performed. The scale bar represents 5 mm. Panel B depicts laser scanning images taken from inside an atrial appendage as shown in image Ab. Images Ba and Bc display the SHG reflection signal indicating elastin and collagen fibres of the extracellular matrix, while images Bb and Bd represent the same image section viewed at the autofluorescence wavelengths. The spherical structure in Ba marked by the white arrow resembles a small vessel. The dashed square in image Bb shows the section that was scanned at a higher magnification for panels Bc and Bd. The scale bar represents 50 µm for panels Ba and Bb, and 10 µm for panels Bc and Bd. Panel C provides a fluorescence spectrum of the autofluorescence from panels Bb and Bd upon excitation with 410 nm light.

penetration depth. We thus utilised two-photon excitation of the tissue. In order to minimise artefacts due to sample movements we developed a measuring chamber as depicted in ◘ Fig. 12.2A. The heart is impaled on a needle from top to almost the bottom. The aorta needs to be on top in order to allow retrograde perfusion. If the needle

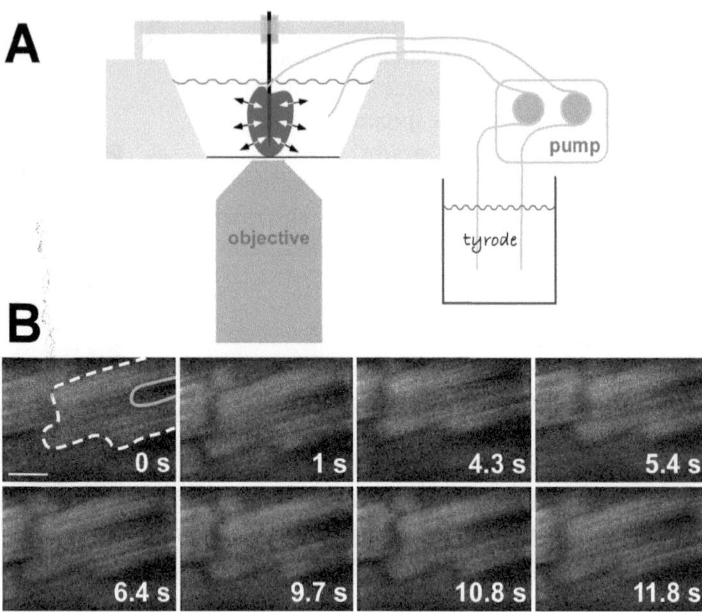

■ **Fig. 12.2** Imaging of a perfused beating mouse heart. Panel A shows the design of the measuring chamber for use with an inverted microscope. In order to minimise motion artefacts the heart was fixed by a needle (black vertical line in panel A), the needle was placed exactly in the optical axis of the microscope objective. Since the heart was »beating around« the needle (illustrated by the arrows) a virtually motion-free image could be recorded. A peristaltic pump was employed for continuous retrograde perfusion. Panel B depicts representative autofluorescence images of an identified myocyte throughout a recording period of 12 seconds (at video-rate). The scale bar represents 10 μm. The shape of the cell and the nucleus is highlighted in the first image by the dashed white line and the solid grey line, respectively.

is spiked in the middle it is very likely to keep the main vessels and important electrical conductance structures, such as the sinus node and the atrio-ventricular (AV)-node, intact. The heart is beating around the needle that is placed in the optical axis of the objective generating a virtually »motion-free« part of the tissue, i.e., minimising movement artefacts. Representative images of an identified cell are depicted in ■ Fig. 12.2B. Although there is still movement of the cells it is possible to identify the very same cell throughout repetitive heart beats. This not only holds true for the x-y-place-ment of the myocyte, but also for the z-position. In all 8 images of ■ Fig. 12.2B the nucleus of the cell is clearly visible. Taking into account the inherent optical sectioning of two-photon excitation imaging and the thickness of cardiac myocyte nuclei of around 5 μm the cell returned to the same x-, y- and z-position within a range of less than 5 μm.

12.3.3 Ultra high-speed imaging of Ca^{2+} signals in isolated cardiomyocytes

Imaging of fast Ca^{2+} signals such as Ca^{2+} sparks requires scanning processes that allow a high fre-quency image generation and acquisition. Con-ventional galvo-mirror driven scanning mecha-nisms such as those in common single beam confocal scan heads are generally not suitable for such fast or ultra-fast imaging. For an »escape route« out of that problem researchers have employed the line-scan mode by sacrificing one optical dimension for speed, e.g., Lipp and Niggli, 1993[11]. Unfortunately, especially for localised Ca^{2+} signals their subcellular distribution and detailed spatio-temporal characteristics are often difficult to analyse from such line-scan images since they only represent fluorescence signals along a single line. We thus followed two alternative approaches beside the single beam, galvo-driven technol-ogy. The first approach utilised a kilo-beam array scanner in which approximately 2,500 laser-beams simultaneously excited the sample and the detec-tion unit was a highly sensitive back-illuminated electron multiplication CCD camera. Scanning in such a device could be as fast as 1 ms per image (1 kHz) but the rate at which images with a suitable optical resolution could practically be acquired was usually limited to approximately 200 images per second. Example traces and representative sections of image acquisition in a single mouse atrial myocyte loaded with the fluorescent Ca^{2+}

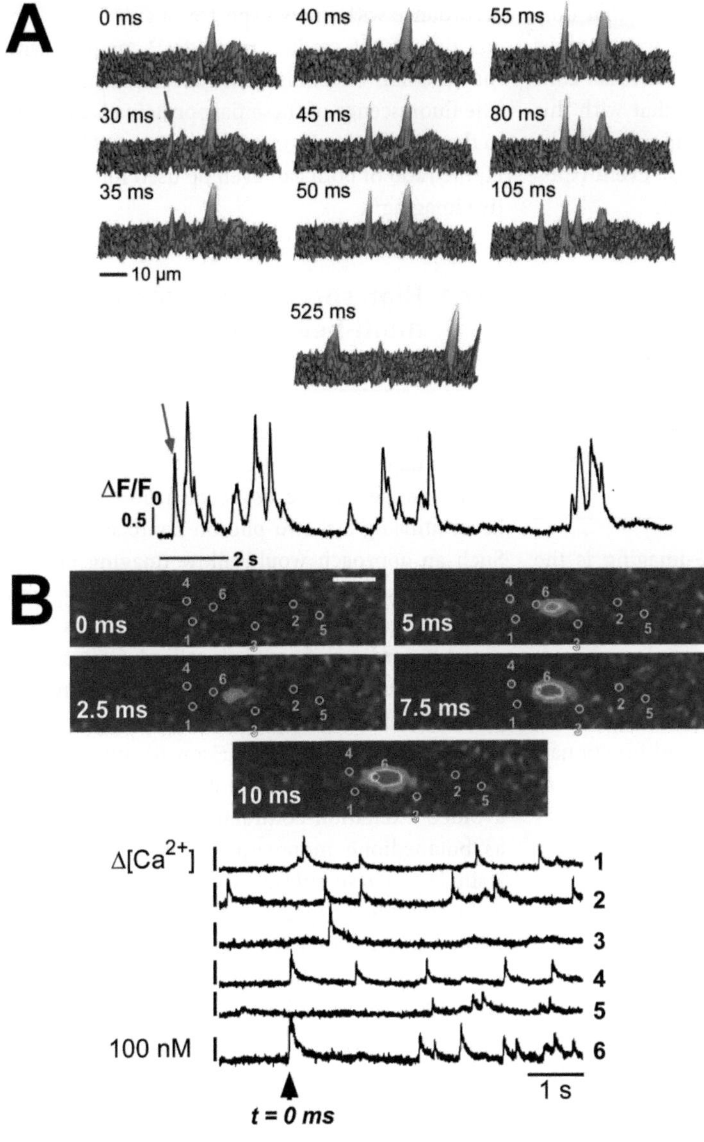

Fig. 12.3 Ca^{2+} spark recordings in mouse atrial (A) and rat ventricular (B) myocytes. (A) Single mouse atrial myocytes were loaded with fluo-4 AM and their unstimulated, spontaneous activity was imaged with a kilo-beam array scanner at an image acquisition rate of 200 Hz. Exemplified single time points are illustrated in the upper part of (A). For this the fluorescence distribution was rendered into a 3-dimensional surface plot which was colour coded in addition. The time course of fluorescence changes at an individual identified Ca^{2+} spark (red arrow) was re-plotted at the bottom of panel A. For (B) single rat ventricular myocytes were loaded with fluo-4 AM and the spontaneous activity was imaged with an AOD-driven single beam scanner at a constant imaging rate of 400 Hz. The upper part of panel B depicts a series of consecutive images recorded from within a single cell. The original grey-scale images were colour-coded with warmer colours representing higher fluorescence values (i.e. higher Ca^{2+} concentrations). Fluorescence from the regions of interests was re-plotted in the bottom part of this panel. Numbers to the right of the traces correspond to the numbering of the regions of interest in each image. The scale bar in panel B represents 5 µm. (For interpretation of the references to color in this figure legend, the reader is referred to the web version of the original article.)

indicator fluo-4 are depicted in **Fig. 12.3A**. The distribution of the fluorescence at each time point had been rendered into 3-dimensional surface plots illustrating multiple active Ca^{2+} spark sites (increases in the Ca^{2+} concentration are depicted as peaks in the surface, additional colour coding was applied). One of these sites has been marked with a red arrow and the time-dependent changes of the fluorescence have been re-plotted in the lower part of **Fig. 12.3A** indicating that during the recording period (approx. 20 s at 200 Hz) virtually no bleaching of the fluorescent dye was observed. The second approach is based on a single beam scanner, where the acquisition rate is increased by using an acousto-optical deflector (AOD) for scanning in the time-critical x-direction. Scanning in the y-direction is still achieved with a galvo mirror. **Fig. 12.3B** shows an example recording obtained from a fluo-4 AM loaded single rat ventricular myocyte at an acquisition rate of 400 Hz. It has to be mentioned that during the recording period bleaching was obvious (bottom part of ◨ Fig. 12.3B). The individual consecutive confocal sections illustrate that with this imaging 2D-scanning imaging technique even the upstroke phase of a local Ca^{2+} release event (Ca^{2+} spark) can be visualised and analysed.

12.4 Discussion

12.4.1 Non-linear imaging of human heart tissue - Second harmonic generation and autofluorescence imaging in living cardiac tissue

The prevalent advantage of SHG-imaging is the rapid and non-invasive examination of the tissue sample without pre-staining that fosters further investigation of the sample such as electrophysiological examination, cell isolation or biochemical analysis. In the past, SHG-imaging has been used to investigate cardiac morphology and functionality ranging from macroscopic structures such as cardiac valves[12] to subcellular muscle structures like myofibrils[13].

Here, we could provide evidence for a useful extension of this approach to samples from human atrial appendages. As shown before[12,14,15] the structures made visible using SHG in this report are most likely collagen and elastin containing extracellular matrix. This notion is supported (i) by the visibility of the spherical structure in the confocal section representing the cut-open volume of a coronary vessel (arrow in ◨ Fig. 12.1Ba), and (ii) by the fact that there is basically no overlap of the SHG-reflection signal with the autofluorescence image obtained simultaneously. The autofluorescence is often attributed to apoptotic activity of the cells[16]. Indeed the apoptosis inducing factor (AIF) is a fluorescent flavoprotein[17]. When we compared the spectral distribution of the autofluorescence described here (◨ Fig. 12.1C) with those published for the AIF[18] there was hardly any overlap. The fluorescence maximum of AIF is approximately at 340 nm while our measured spectrum peaks at 500 nm. The latter peak is in accordance with known spectra of autofluorescent constituents of cardiac myocytes[19], namely with the emission of oxidised metabolic flavoproteins. The fluorescence of these flavoproteins is different to the more bathochromic fluorescence of NAD(P)H; the ratio of both can even be used for quantitative imaging[20].

12.4.2 High speed two-photon multi-beam imaging of a beating mouse heart

Here, we provided strong evidence for our ability to image individual cardiac mycoytes in the intact, native cardiac tissue of a beating mouse heart utilising fast two-photon confocal imaging. Such an approach would allow imaging of cells in their tissue context while maintaining their physiological environment. Nevertheless, the direct recording of small, localised Ca^{2+} signals in the intact heart tissue is still challenging and has not been shown up to now. It should be mentioned here, that in comparison to other studies of cardiac tissue or intact hearts, we especially avoided excitation-contraction uncouplers such as butanedione monoxime (BDM) or cytochalasin D[21]. These substances are known to exert unforeseeable unspecific effects on the cardiac tissue and individual myocytes.

The limited quality of the images presented in ◨ Fig. 12.2B can most likely be attributed to at least three main reasons: (i) limited signal to noise ratio of the fast camera, (ii) optical aberrations due to the use of an oil objective for imaging deep in an aqueous medium, and (iii) emission light scattering since imaging was performed with an integrating CCD camera where scattered light is spatially detected at the scattering object. Such processes are avoided when using single-beam PMT based scanners, since scattered emission light at any given time point only originates from within the excitation volume and is therefore registered to a particular voxel by the frame-, line-, and pixel-clock.

In similarity to the kilo-beam array scanner, the multi-photon multi beam approach allows scanning speeds that are not yet reached by frame rates of sensitive CCD-cameras. Therefore this technology is regarded to have an immense future potential. For imaging of mammalian hearts the technique can be improved when the topics discussed above are addressed.

12.4.3 Ultra high-speed imaging of Ca²⁺ signals in isolated cardiomyocytes

Here we introduced two different techniques to perform two-dimensional confocal recordings of elementary Ca^{2+} signals in cardiomyocytes at frame rates between 200 and 400 Hz. Interestingly, for both approaches the acquisition rate is not limited by the scanning process itself: for the kilo-beam array scanner the detector, namely the pixel read out of the camera is the limiting factor. For the AOD-based scanner the limit in the acquisition speed is no longer the limitation of the confocal microscope (which can achieve frame rates higher than 600 Hz), but the amount of photons emitted by the fluorphore in the sample. For image sizes of 50x200 pixel scanned at a frame rate of 400 Hz the pixel dwell time is as low as 0.25 µs (not considering beam fly back times). In contrast the virtual pixel dwell time in the kilo-beam array scanner under the recording conditions used for the measurements in ◘ Fig. 12.3A is approximately 100 µs, i.e., for half the acquisition speed the pixel dwell time is 400 times increased. Taking into account that the quantum efficiency of a back-thinned CCD chip is almost three times as high as the quantum yield of a PMT, the future potential of the kilo-beam array scanner becomes evident. Nevertheless, AOD-based scanners still remain the more flexible devices so far. This includes the choice of the size of the confocal aperture, image-zoom functions and the compatibility with two-photon excitation as shown recently[22,23].

For ultra-fast high resolution confocal imaging of calcium signals in cardiomyocytes the two approaches are superb technologies where the kilo-beam array scanner is favourable for »long-term« measurements with reduced bleaching[24], while the AOD-based scanner represents the faster and more flexible technology.

12.5 Conclusion

We probed state-of-the-art optical imaging technologies for their usefulness in investigations of the heart and its constituents. It was shown that each technique is applicable to particular questions in cardiac research.

Furthermore these methods can be used in combination to explore the phenomenon of cardiac arrhythmias. Arrhythmic behaviour might be caused by »cellular arrhythmias«, which can be triggered by local Ca^{2+}-signals, namely Ca^{2+}-sparks and Ca^{2+}-waves, which generate delayed after depolarisations and/or propagating action potentials. These elementary Ca^{2+}-signals are best acquired by subcellular calcium-imaging using fluo-4 with frame rates between 200 and 400 Hz in isolated cardiac myocytes, although adaptation of some of the other techniques might allow imaging of these important signals in the intact organ. In contrast, structural remodelling of the heart as it occurs during various cardiac pathologies such as atrial and ventricular arrhythmias (e.g., atrial flutter or fibrillation; ventricular hypertrophy) can be investigated by the non-invasive method as second harmonic generation imaging. This novel imaging technique allows the visualisation of extracellular matrix components such as collagen and elastin without the need of any staining procedures. Interestingly, SHG imaging can be combined with autofluorescence imaging in the unstained state or probably also with fluorescence imaging using fluorescent indicators (assuming a compatible excitation spectrum) simultaneously. With such approaches it will thus be possible to simultaneously gain information about the amount and distribution of the extracellular matrix and at least one more parameter of cellular function.

12.6 References

[1]   C. T. Badea, E. Bucholz, L. W. Hedlund, H. A. Rockman and G. A. Johnson, »Imaging methods for morphological and functional phenotyping of the rodent heart,« Toxicologic pathology 34(1), 111-117 (2006)

[2]   M. Skrzypiec-Spring, B. Grotthus, A. Szelag and R. Schulz, »Isolated heart perfusion according to Langendorff–still viable in the new millennium,« Journal of

pharmacological and toxicological methods 55(2), 113-126 (2007)

[3] C. Viero, U. Kraushaar, S. Ruppenthal, L. Kaestner and P. Lipp, »A primary culture system for sustained expression of exogenous proteins in conserved adult rat vetricular myocytes,« Cell Calcium 43(1), 59-71 (2008)

[4] B. W. Kimes and B. L. Brandt, »Properties of a clonal muscle cell line from rat heart,« Exp. Cell Res. 98(2), 367-381 (1976)

[5] S. M. White, P. E. Constantin and W. C. Claycomb, »Cardiac physiology at the cellular level: use of cultured HL-1 cardiomyocytes for studies of cardiac muscle cell structure and function,« Am J Physiol Heart Circ Physiol 286(3), H823-829 (2004)

[6] L. Kaestner, W. Tabellion, P. Lipp and I. Bernhardt, »Prostaglandin E2 activates channel-mediated calcium entry in human erythrocytes: an indication for a blood clot formation supporting process,« Thromb Haemost 92(6), 1269-1272 (2004)

[7] A. Quintana, E. C. Schwarz, C. Schwindling, P. Lipp, L. Kaestner and M. Hoth, »Sustained activity of CRAC channels requires translocation of mitochondria to the plasma membrane,« J Biol Chem 281(52), 40302-40309 (2006)

[8] L. Kaestner, A. Juzeniene and J. Moan, »Erythrocytes-the 'house elves' of photodynamic therapy,« Photochem Photobiol Sci 3(11-12), 981-989 (2004)

[9] U. Kirchhefer, G. Hanske, L. R. Jones, I. Justus, L. Kaestner, P. Lipp, W. Schmitz and J. Neumann, »Overexpression of junctin causes adaptive changes in cardiac myocyte Ca(2+) signaling,« Cell Calcium 39(2), 131-142 (2006)

[10] L. Kaestner, W. Tabellion, E. Weiss, I. Bernhardt and P. Lipp, »Calcium imaging of individual erythrocytes: problems and approaches,« Cell Calcium 39(2), 13-19 (2006)

[11] P. Lipp and E. Niggli, »Microscopic spiral waves reveal positive feedback in subcellular calcium signaling,« Biophys J 65(6), 2272-2276 (1993)

[12] K. Schenke-Layland, I. Riemann, U. A. Stock and K. Konig, »Imaging of cardiovascular structures using near-infrared femtosecond multiphoton laser scanning microscopy,« Journal of biomedical optics 10(2), 024017 (2005)

[13] F. Vanzi, M. Capitanio, L. Sacconi, C. Stringari, R. Cicchi, M. Canepari, M. Maffei, N. Piroddi, C. Poggesi, V. Nucciotti, M. Linari, G. Piazzesi, C. Tesi, R. Antolini, V. Lombardi, R. Bottinelli and F. S. Pavone, »New techniques in linear and non-linear laser optics in muscle research,« J Muscle Res Cell Motil 27(5-7), 469-479 (2006)

[14] N. D. Kirkpatrick, J. B. Hoying, S. K. Botting, J. A. Weiss and U. Utzinger, »In vitro model for endogenous optical signatures of collagen,« Journal of biomedical optics 11(5), 054021 (2006)

[15] P. J. Campagnola, A. C. Millard, M. Terasaki, P. E. Hoppe, C. J. Malone and W. A. Mohler, »Three-dimensional high-resolution second-harmonic generation imaging of endogenous structural proteins in biological tissues,« Biophys J 82(1 Pt 1), 493-508 (2002)

[16] M. Ranji, S. Kanemoto, M. Matsubara, M. A. Grosso, J. H. Gorman, 3rd, R. C. Gorman, D. L. Jaggard and B. Chance, »Fluorescence spectroscopy and imaging of myocardial apoptosis,« Journal of biomedical optics 11(6), 064036 (2006)

[17] E. Daugas, D. Nochy, L. Ravagnan, M. Loeffler, S. A. Susin, N. Zamzami and G. Kroemer, »Apoptosis-inducing factor (AIF): a ubiquitous mitochondrial oxidoreductase involved in apoptosis,« FEBS Lett 476(3), 118-123 (2000)

[18] M. J. Mate, M. Ortiz-Lombardia, B. Boitel, A. Haouz, D. Tello, S. A. Susin, J. Penninger, G. Kroemer and P. M. Alzari, »The crystal structure of the mouse apoptosis-inducing factor AIF,« Nature structural biology 9(6), 442-446 (2002)

[19] S. Huang, A. A. Heikal and W. W. Webb, »Two-photon fluorescence spectroscopy and microscopy of NAD(P)H and flavoprotein,« Biophysical journal 82(5), 2811-2825 (2002)

[20] J. V. Rocheleau, W. S. Head and D. W. Piston, »Quantitative NAD(P)H/flavoprotein autofluorescence imaging reveals metabolic mechanisms of pancreatic islet pyruvate response,« J Biol Chem 279(30), 31780-31787 (2004)

[21] L. C. Baker, R. Wolk, B. R. Choi, S. Watkins, P. Plan, A. Shah and G. Salama, »Effects of mechanical uncouplers, diacetyl monoxime, and cytochalasin-D on the electrophysiology of perfused mouse hearts,« Am J Physiol Heart Circ Physiol 287(4), H1771-1779 (2004)

[22] R. D. Roorda and G. Miesenbock, »Beam-steering of multi-chromatic light using acousto-optical deflectors and dispersion-compensatory optics,« in Patent Application Publication, United Staates (2002).

[23] A. Bouzid and J. Lechleiter, »Laser scanning fluorescence microscopy with compensation for spatial dispersion of fast laser pulses,« in Patent Application Publication, United States (2002).

[24] P. Lipp and L. Kaestner, »Image based high content screening – A view from basic science,« in High-Throughput Screening in Drug Discovery J. Hüser, Ed., pp. 129-149, Wiley VCH, Weinheim (2006).

A system for optical high resolution screening of electrical excitable cells

Oliver Müller, Qinghai Tian, Roman Zantl, Valentin Kahl, Peter Lipp, Lars Kaestner

Reprint from Cell Calcium (2010) **47**, 224-233.

■ Abstract

The application of primary excitable cells for high content screening (HCS) requires a multitude of novel developments including cell culture and multi-well plates. Here we introduce a novel system combining optimised culture conditions of primary adult cardiomyocytes with the particular needs of excitable cells for arbitrary field stimulation of individual wells. The major advancements of our design were tested in calcium imaging experiments and comprise (i) each well of the plate can be subjected to individual pulse protocols, (ii) the software driving electrical stimulation can run as a stand-alone application but also as a plug-in in HCS software packages, (iii) the optical properties of the plastic substrate (foil) resemble those of glass coverslips fostering high resolution immersion-based microscopy, (iv) the bottom of the foil is coated with an oleophobic layer that prevents immersion oil from sticking, (v) the top of the foil is coated with an elastic film. The latter enables cardiomyocytes to display loaded contractions by mimicking the physiologically occurring local elastic network (e.g., extracellular matrix) and results in significantly increased contractions (with identical calcium transients) when compared to non-elastic substrates. Thus, our novel design and culture conditions represent an essential further step towards the application of primary cultured adult cardiomyocytes for HCS applications.

13.1 Introduction

Optical measurements of cellular functions have been established on the laboratory level for many years (e.g., [1-3]). They were always driven by the technological developments towards real-time high resolution imaging devices throughout the recent two decades[4]. These low-throughput techniques have proven great value and have fostered our understanding of a great variety of physiological but also pathophysiological processes in living cells[5,6]. One of the areas of cell physiology that has benefited greatly from optical advancements was our understanding of cellular and sub-cellular calcium signalling[7].

Such investigations include imaging molecular events such as calcium blips[8] and quarks[9] and cellular calcium transients[10,11] but also *in vivo* calcium imaging by means of genetically encoded biosensors (GEBs) expressed in transgenic mouse lines[12,13]. Recently, these developments were complemented by major advancements in the genetic manipulation of primary cells with viral gene transfer[14]. The establishment of a large variety of animal models for human pathologies[15] can be seen along the same line. In addition to *in vivo* analysis that will always represent ultra-low throughput experimental series, cellular and/or sub-cellular studies of the physiology and pathophysiology of individual cells complement the *in vivo* data. These high resolution characterisations of cellular responses are often highly repetitive and laborious tasks.

It would thus be desirable to transfer such single-cell experiments to a screening environment allowing for higher throughput and higher reproducibility of the measurements. However, so far optical high content screening (oHCS) has largely been restricted to cultured cell lines[16]. Such specimens are relatively easy to handle and genetic manipulation is straightforward. They have represented and will most likely be the prime tool in ultra-high throughput screening (uHTS) since they allow for extremely reproducible experimental results, a requirement for screening large chemical libraries.

In contrast, the utilisation of primary cells in oHCS does not only appear desirable, instead it seems almost mandatory in the further development of pharmacological targets and safety screens. This appears essential because the optimal proximity to the *in vivo* situation is a prime prerequisite and goal. In particular screening based measurements of excitable cells are very demanding and have thus not made it into automised screening applications. The major reasons are: (i) limited reproducibility in the yield and quality of cell isolation and the following primary culture; (ii) lack of controllable trigger devices to excite cells in multi-well plates; (iii) low photon detection efficiency due to the use of low numerical aperture (NA) air objectives. Especially high NA oil objectives were excluded due to thick plastic well bottoms and insufficient oil handling on multi-well plates, but they are necessary for high resolution sub-cellular imaging. Taken together, up to now these challenges have been largely unsolved and have thus limited the use of primary cells, especially primary electrical excitable cells such as cardiac myocytes or neurones, for the application in oHCS.

Recently, we introduced an improved cell isolation and primary culture method that resolved the issue mentioned in (i) for cardiac myocytes[17,18]. Here we present a complementary system that overcomes the restrictions mentioned above. Going even further it adds more benefits especially for screening of primary isolated cardiomyocytes, through mimicking a native-like extracellular environment. We will demonstrate the use of such a system for global calcium and contraction measurements but also for high resolution calcium imaging such as total internal reflection fluorescence (TIRF) microscopy.

13.2   Materials and methods

13.2.1 Cell isolation and culture

Isolation and culture of adult rat cardiomyocytes was performed as described previously[18]. Adult male Wistar rats (6–12 weeks old, 200–400 g) were handled and sacrificed in accordance with the Guide for the Care and Use of Laboratory Animals published by the US National Institutes of Health (NIH Publication No. 85-23, revised 1996). Animals received an intraperitoneal injection of a mixture of ketaminehydrochloride (Ursotamin, Serumwerk, Bernburg, Germany) and xylazinhydrochloride (Rompun, Bayer Health Care, Leverkusen, Germany) at a final dose of 137 mg/kg body weight and 6.6 mg/kg body weight, respectively. When anaesthetised the rat was killed by decapitation. The heart was flushed with 10 ml of ice-cold Ca^{2+}-free solution (CFS) containing (in mM): NaCl 134, glucose 11, KCl 4, $MgSO_4$ 1.2, Na_2HPO_4 1.2, HEPES (Merck, Darmstadt, Germany) 10 (pH adjusted to 7.35 with NaOH). After that, the heart was removed, attached to a Langendorff apparatus and perfused retrogradely with O_2 saturated CFS containing 200 µM EGTA at a rate of 4 ml/min for 5 min. The perfusate was then changed to O_2 saturated CFS containing Liberase Blendzym IV (Roche Diagnostics Corp., Indianapolis, USA) at a final concentration of 335 µg/ml for 25 min.

The ventricles were removed, minced and placed in O_2 saturated CFS containing 335 µg/ml Liberase Blendzym IV (at 37 °C in a water bath for 2 min). After sedimentation, the resulting supernatant was discarded and the pellet was mixed and resuspended in 20–25 ml of O_2 saturated CFS and incubated as above. The supernatant was discarded again and the pellet was mixed and resuspended in 20–25 ml of O_2 saturated low-Ca^{2+} solution containing 50% of CFS and 50% of Ca^{2+} containing solution (CCS) and incubated as above. CCS is composed of CFS supplemented with 0.09% of DNAse and 200 µM of Ca^{2+}. Furthermore, the supernatant was discarded, the pellet was resuspended in 20–25 ml of O_2 saturated CCS and incubated as above. Now, rat ventricular myocytes were released from the soft tissue by gentle trituration. The cell suspension was plated into the measuring cavities, the internal bottom surface of which was coated with extracellular matrix proteins (ECM gel from Engelbreth-Holm-Swarm mouse sarcoma, Sigma–Aldrich, St. Louis, USA), they were allowed to settle down for approximately 1 h in medium M199 with Earle's modified salts, glutamine (Biowest, Nuaillé, France), 100 µg/ml Penicillin/Streptomycin and 50 µg/ml Kanamycin (PAA Laboratories, Linz, Austria). The medium was supplemented with 870 nM insulin, 65 nM transferrin and 29 nM Na-selenite (Sigma–Aldrich, St. Louis, USA) (ITS supplemented medium). Myocytes were cultured in an incubator at 37 °C with a 5% CO_2 atmosphere. One hour after plating, the medium was changed for fresh medium supplemented with ITS. This procedure was repeated at the first and third day *in vitro* (DIV).

For TIRF microscopy and related measurements DIV1 cells were loaded with 1 µM Fluo-4 AM and for photometric measurements cardiomyocytes were loaded with 5 µM Indo-1 AM (both dyes, Molecular Probes, Eugene, USA). In all instances loading time was 30 min and 10 min were allowed for deesterification. All experiments were carried out at room temperature (23 °C).

13.2.2 Imaging, photometry and cell length measurements

Experiments involving TIRF microscopy were conducted as previously described[19], in short: cells were placed on an inverted microscope (IX70, Olympus, Tokyo, Japan) equipped with a 100× TIRF objective (Plan Apo 1.45 NA, Olympus, Tokyo, Japan). Excitation of Fluo-4 was achieved using a 20 mW 488 nm laser (Cyan Scientific, Spectra Physics, Mountain View, USA) and a monochromator (VisiChrome, VisiTron Systems GmbH, Puchheim, Germany) for TIRF and epifluorescence imaging, respectively. In both cases fluorescence was recorded with a back-thinned electron multiplying charge coupled device (CCD) camera (QuantEM:512SC, Photometrics, Tucson, USA). Acquisition was performed with Meta-Morph software (Molecular Devices, Downingtown, USA). Image processing was carried out with ImageJ (Wayne Rasband, National Institute of Mental Health, Bethesda, USA).

Documentation of the cells in phase contrast images was performed on an inverted microscope (TS100) equipped with a 20× air objective and a CCD-camera (DN100; all Nikon, Tokyo, Japan).

Global Ca^{2+} transients were measured using Indo-1. For this, the myocytes were transferred to an inverted microscope (TE2000U, Nikon, Tokyo, Japan) attached to a combined fluorescence/cell length imaging system using a 40× oil-immersion objective (S Fluor 1.4 NA, Nikon, Tokyo, Japan). The system comprised two avalanche photo diodes (APD) and a monochromator (Polychrome IV) for fluorescence acquisition (both: TILL Photonics, Gräfelfing, Germany). The cells were excited at 360 nm while simultaneously recording the fluorescence signal at 415 ± 30 nm and 470 nm longpass, respectively (sampling rate 1 kHz). The ratio and further semi-automatic peak detection was determined in Igor Pro software (WaveMetrics, Inc., Lake Oswego, USA) running custom-made macros.

Real-time cell length changes were monitored with a fast camera (sampling rate 240 Hz, Myo-Cam, IonOptix Corp., Milton, USA) from electrically stimulated cells maintained on coverslips or elastic surfaces by using an edge-detection approach by the Ion Wizard software (IonOptix Corp., Milton, USA). The system directly stores cell length changes that were further analysed in Igor Pro software with custom-made macros.

The measuring cavities referred to above are chambers of a 24-well plate. Since the design of this multi-well plate was a major aim of this paper, it is described in great detail in Section 13.3.3.

13.2.3 Measurements of plate movement

In order to investigate the interaction between the culture substrate, immersion oil and objective we utilised a fully motorised and software controlled microscope (uiMic, TILL Photonics GmbH, Gräfelfing, Germany) equipped with oil-immersion objectives (UPLSAPO 20×oil 0.85 and SPLANAPO 100×oil 1.4, Olympus, Tokyo, Japan) which both displayed flat top surfaces (area around 35 mm^2). We tested two different oils: Type N immersion liquid (Leica Microsystems CMS GmbH, Wetzlar, Germany) with a viscosity of 600 cSt at 23 °C and Nikon immersion oil NF (Nikon, Tokyo, Japan) with a viscosity of 800 cSt at 23 °C. For our experiments we used bottles immediately after initial opening. The volume of the oil drops applied was 60 µl for both types of oil.

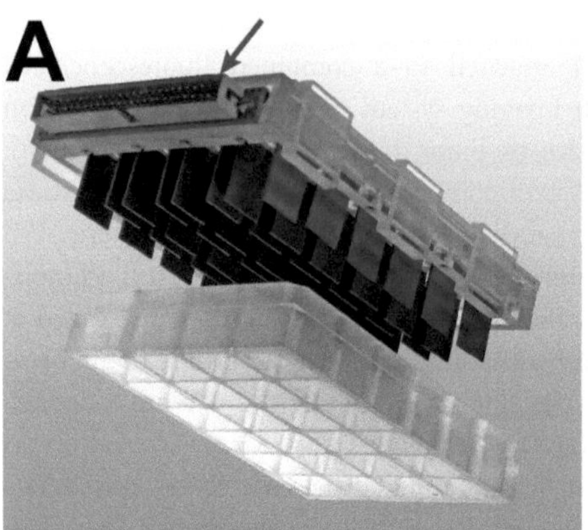
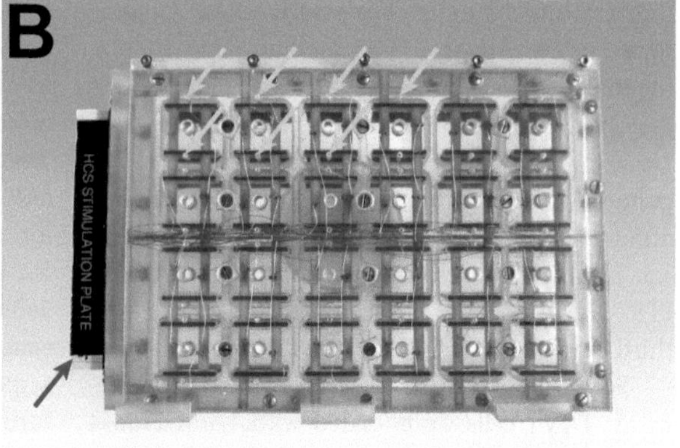

☐ Fig. 13.1 General overview of the front end of the system described in this paper. (A) Depicts how the electrode containing lid is fitting on the 24-well plate. The top view in (B) shows the wiring of the carbon electrodes with platinum wires, the wholes for pipetting substances/drugs to the cells in the middle of each well and the small wholes (indicated by yellow arrows) that allow cannulae to perform a background perfusion. The red arrows point to the electrical interface, a 50 pin ribbon cable connector. (For interpretation of the references to colour in this figure legend, the reader is referred to the web version of the original article.)

13.2.4 Electrical measurements and analysis software

Voltage traces were measured and recorded using a 200 MHz oscilloscope (TDS 2024B, Tektronix Inc., Beaverton, USA).

Statistical analysis was performed with Graph-Pad Prism 5 (GraphPad Software Inc., La Jolla, USA).

13.3 Results and discussion

13.3.1 General concept

In contrast to other multi-well plates for screening experiments we had to take into account some of the exceptional properties of cardiomyocytes. In particular we ought to incorporate the ability to apply electrical stimuli. Moreover, due to the brick-like geometry of the cells we had to ensure that myocytes were indeed isolated and not clustered or superimposing each other. The latter prerequisite forced us to maximise the area of each well by still adhering to the microtiter plate form factor. This ensured compatibility with existing automation hardware.

We thus constructed a 24-well plate/lid combination that maximised the surface area per well, included an optimised electrode design and enabled the arbitrary and independent pulsing of each individual compartment of the plate. An overview of the resulting design is given in ☐ Fig. 13.1. Encompassing square-shaped compartments resulted in an optimal ratio of the area available for imaging and dead area around and behind the electrodes.

13.3.2 The lid

From ☐ Fig. 13.1 it can be deducted that the lid was designed from plastic (polycarbonate) that combined two important advantages: (i) it was transparent for visible light and thus enabled white light transmission imaging and (ii) durability against temperatures that occur during heat-sterilisation. The latter is necessary for reusing the lid.

The concept to use electrodes for electrical field stimulation of adult cardiac myocytes of different species had been appreciated for many years, e.g., [20-24]. Such early reports mostly used platinum as material for the electrodes. However, platinum

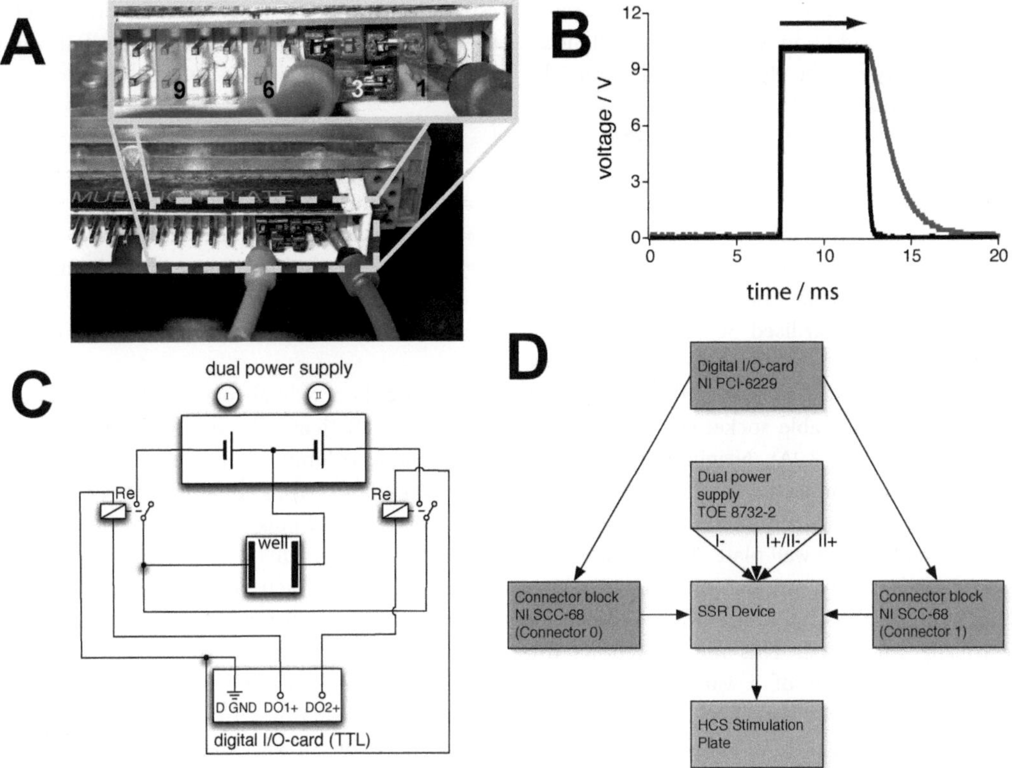

◘ Fig. 13.2 Hardware of the electrical field stimulation procedure. If the number of supplying channels is limited, in panel (A) is demonstrated how electronic jumpers can be used to synchronise any number of wells by switching the »jumper« directly into the ribbon cable socket. The insert shows that 3 jumpers are used to connect a column of 4 wells. The greenish transparent block marks the electrical mass, while the orange blocks display examples for single well contacts and are labelled with corresponding numbers. (B) is a plot of a voltage trace if the pulse is generated by a DC voltage supply and switched by solid-state relays. The black arrow indicates the length of the trigger signal and the black graph depicts the switching when a Panasonic AQY225R1S relay was used (details see text). In comparison the grey curve indicates the pulse shape if a relay was taken with properties of approximately one order of magnitude worse in switching time and leak current. Panel (C) is a scheme of the circuit diagram used for each well of the plate. Such a design ensures electrically alternating pulses. The diagram in (D) depicts which hardware components are involved in total and how they are connected. (For interpretation of the references to color in this figure legend, the reader is referred to the web version of the original article.)

electrodes were soon replaced by more cost-effective carbon electrodes[25]. In addition, carbon electrodes ensure an improved biocompatibility since these electrodes were inert with respect to the cells and the culture medium even in long-term cultures (1 week in duration). Such a concept proofed successful and is commercialised by IonOptix Corp. (Milton, USA). This company offers equipment to utilise 4- to 8-well plates of selected providers for an electrical field stimulation in culture. However, the standard plastic plates lack most of the proper-

ties described below, such as e.g., the possibility for high resolution imaging.

It is important to note here that in our preliminary experiments during the fine tuning of the well/lid combination we found that the minimum distance between the electrodes and the well wall was 1 mm. This design minimised capillary effects between the wall and the electrode. As shown in ◘ Fig. 13.1B for each compartment of the 24-well plate there is a corresponding circular hole on the lid allowing the application of substances during the experiment without the necessity of removing the lid. In addition there were small holes positioned between the electrodes and the neighbouring wall that allow a constant flow of solution at the bottom of each compartment (◘ Fig. 13.1B, marked by yellow arrows).

While all wells share a common electrical mass on one of their carbon electrodes the other electrode is connected individually. All 25 resulting electrical contacts can be accessed via the common and standardised 50 pin ribbon cable socket (marked with a red arrow in ◘ Fig. 13.1A and B). Each contact is provided by a pair of pins of the ribbon cable socket (for details refer to legend of ◘ Fig. 13.2A). Neighbouring pins in a row could be connected by electrical »jumpers« and such electrically connect two adjacent compartments of the well plate. This »mode« can be extended to any number of compartments. ◘ Fig. 13.2A displays an image were 4 wells are electrically connected. This method allows the electrical stimulation of a varying number of compartments in the case just a single stimulation channel (e.g., by a MyoPacer, IonOptix Corp., Milton, USA) is available.

13.3.3 Hardware for electrical stimulation

Although the number of stimulation channels can be minimised with the »jumper« design outlined above, we constructed hardware that allowed programming of individual pulse protocols for each compartment of the 24-well plate. There were two important issues that needed to be considered during the design process: (i) long-term electrical stimulation of cells was best achieved when applying alternating pulses and (ii) the demand on the power supply can be minimised by ensuring that at any given time only a maximum number of four compartments receive current for electrical stimulation. The former requirement diminished the accumulation of electrolytic by-products in the proximity of the electrodes. The latter property established a cost-effective design by using standard, low-demand power supplies that could be purchased from standard stores. In order to estimate how much current has to be provided by the power supplies, we considered the following equation:

$$I = \frac{U \cdot A}{\rho \cdot l}$$

U = upper limit of voltage (25 V)
A = effective area of electrodes (150 mm^2)
ρ = specific electrical restistance of medium (125 Ωcm)
l = distance between the electrodes (12 mm)

In our case the maximal current per well was 250 mA. For an efficient pulse protocol it is desirable to pulse one column of the well plate (equals 4 wells) simultaneously, leading to a power demand of 1 A at a desired pulse duration of 5 ms. Based on these requirements the power source TOE 8732-2 (Toellner Electronic Instrumente GmbH, Herdecke, Germany) was chosen, that could deliver a maximal current of 1 A (even at the maximum voltage of 32 V). It provided a dual voltage output (positive and negative voltages) that could be regulated by an analogue control voltage of 0–10 V (translating into 0–32 V output) enabling even the external control of the output voltage (see below).

For generation of the desired pulses from the supplied voltages we envisaged a design that involved the application of fast switching solid-state relays (AQY225R1S, Panasonic Corp., Kadoma City, Japan). They were controlled by TTL-signals from a digital input/output (I/O)-card (NI PCI-6229, National Instruments, Austin, USA). The relay offered switching times of maximal 0.75 ms (on) and 0.2 ms (off) enabling an almost rectangular voltage profile as depicted in ◘ Fig. 13.2B. In addition, these electrical components also displayed an almost negligible leak current of 10^{-8} A. This was of particular importance since higher leak currents would have resulted in a constant accumulation of electrolytic by-products in the proximity of the electrodes. Such a combination of relays and digital output channels was designed for each well. The electrical circuit diagram for one well is depicted in ◘ Fig. 13.2C. ◘ Fig. 13.2D displays an overview of the entire electrical circuit.

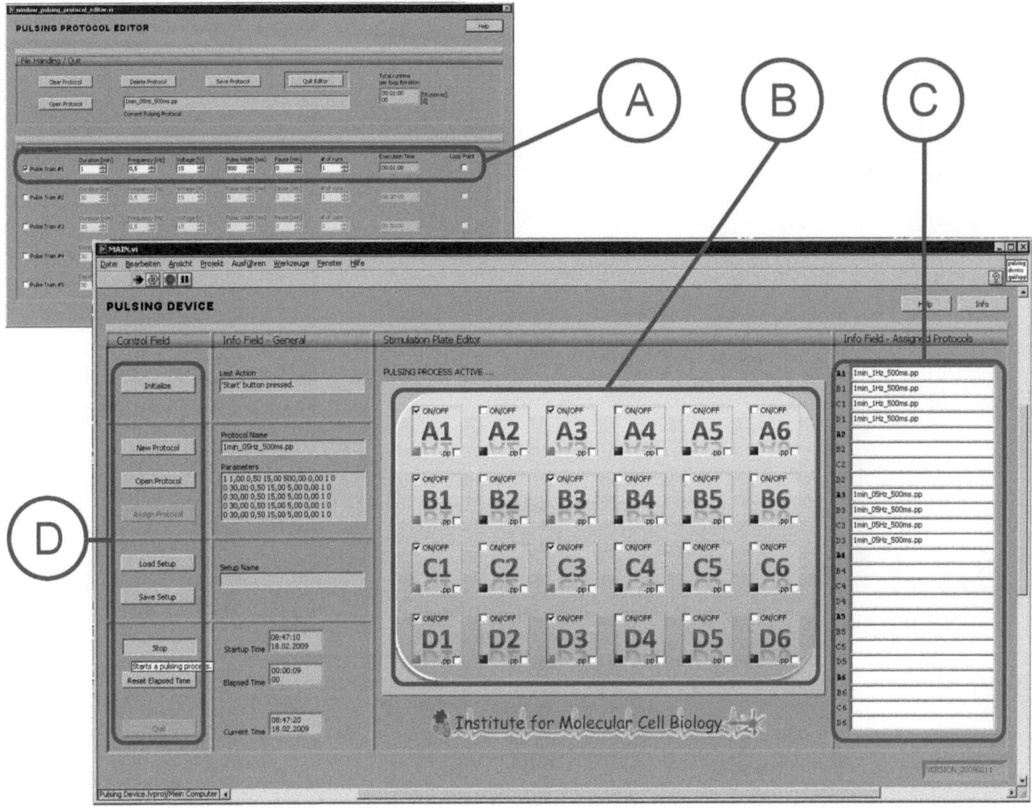

Fig. 13.3 Graphical user interfaces of the pulse protocol editor (top) and the main application window (bottom). (A) The parameters that can be adjusted to define a pulse train. (B) The control panel of the 24-well plate allows for selecting individual wells for incorporation in the pulsing process as well as for protocol change. Highlighted green indicators mark active wells. (C) Overview of assigned protocols. (D) Control panel for accessing instrumentation hardware in real-time.

13.3.4 Software for electrical stimulation

As described above, one of the major aims of our design was to achieve the ability for arbitrary pulse protocols to be assigned to individual wells of the plate. For this we have set up software in LabVIEW (National Instruments, Austin, USA) that allowed (i) programming of arbitrary pulse protocols and (ii) arbitrary connection of such pre-defined protocols to individual or groups of compartments (◾ Fig. 13.3). The protocol editor (◾ Fig. 13.3A) enabled the free programming of complex and repetitive pulse protocols of arbitrary length that itself can contain loops of pulse trains. These protocols will be pre-defined by the user and can be linked to individual wells of the 24-well plate later (◾ Fig. 13.3B). It should be noted here that our software contains templates for various formats (e.g. single compartments, 4-well plates or 24-well plates) that will be graphically displayed accordingly. It is thus possible to also change the stimulation regime for individual wells from a chronic mode to an experimental mode while all other compartments remain in their »chronic« stimulation mode.

In order to enable integration of this pulsing software into larger screening software packages we have wrapped it into a dynamic-link library (DLL), as shown in ◾ Fig. 13.4. This software interface is implemented using the American National Standards Institute C (ANSI C) programming environment LabWindows/C for virtual instrumentation

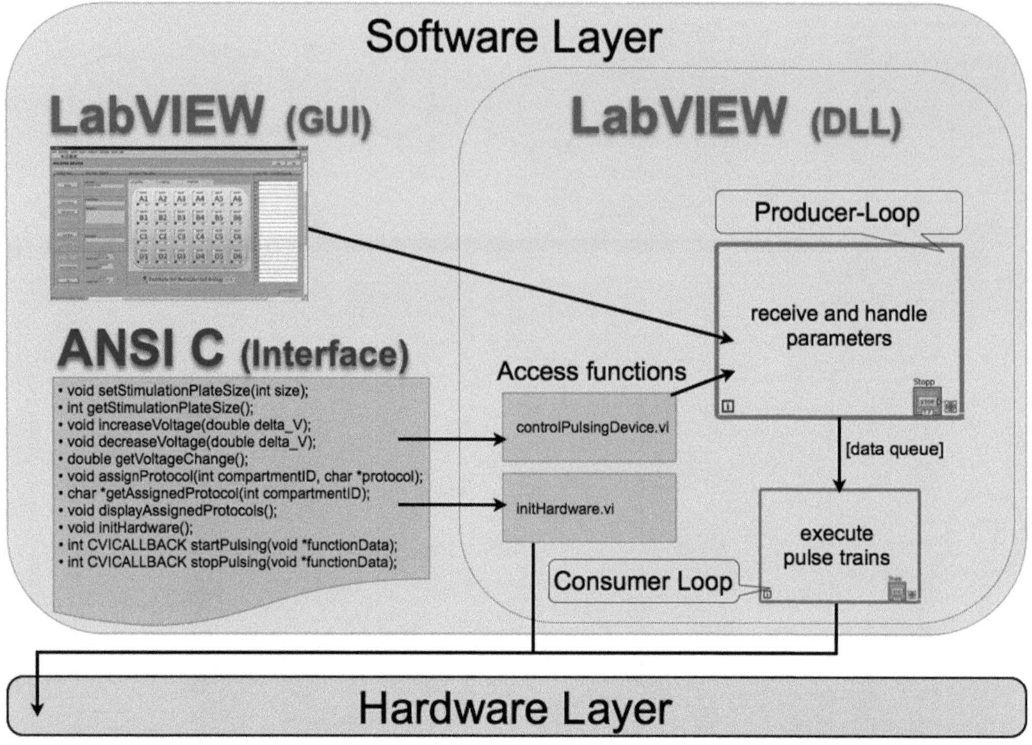

◘ Fig. 13.4 Software architecture and data flow of the electrical stimulation software. The software can be controlled either from the GUI or from the ANSI C interface. The latter allows for accessing the dynamic-link library (DLL) from external programs supporting ANSI C interfaces. Software- and hardware layer are de-coupled using the producer-consumer design pattern in order to maintain real-time accessibility.

(CVI), which is part of the LabVIEW programming environment. Since LabWindows/CVI uses the same libraries as LabVIEW, it bridges the gap between LabVIEW and ANSI C. The C software interface thus provides the same functionality as the stand-alone software. Every external software package that supports an ANSI C interface can thus control the electrical stimulation software by making use of the functions defined in the software interface. In this case, the software interface entirely replaces the graphical user interface (GUI).

13.3.5 The 24-well plate

The well plate is designed as a disposable item (dimensions of the body given in ◘ Fig. 13.5A). Since high resolution imaging in a screening envi-

ronment requires objectives with high NA, the bottom of the well plate entailed a design that permitted the use of immersion media. This, in turn required material of the bottom of the multi-well plate that resembled optical properties similar or identical to glass coverslips.

For this purpose different kinds of plastic foils were tested, including polymethylmethacrylate (PMMA) foil, cyclo olefin polymer (COP) foil and cyclo olefin copolymer (COC) foil (all foils available from ibidi GmbH, Martinsried, Germany). The three foils met the basal optical requirements (specification of number 1.5 glass coverslips, i.e. thickness between 160 μm and 190 μm and refractive index of 1.51)[26]. Especially the COP and COC foils displayed a very low autofluorescence over the spectral range analysed (300–900 nm). Furthermore, the foils exhibited a high chemical resistance

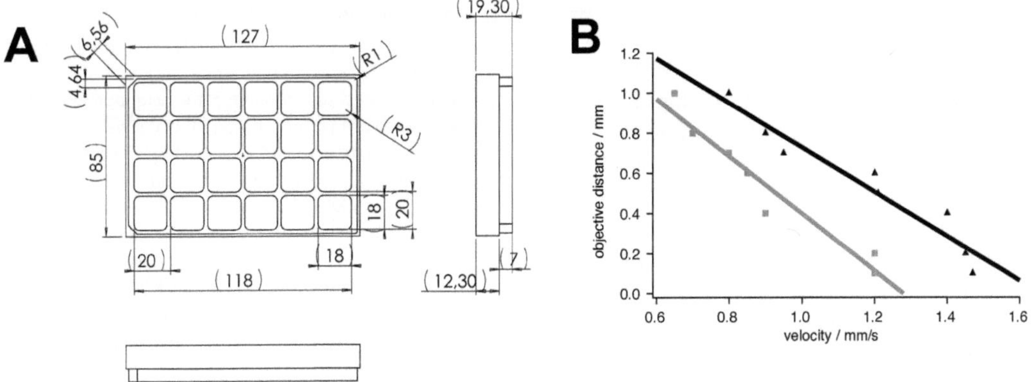

◘ Fig. 13.5 Design of the 24-well plate. (A) The blueprint of the chassis of the multi-well plate. The quadratic footprint of the wells is to ensure the compatibility with the carbon electrodes of the lid (cp. Fig. 13.1). On the bottom of the chassis a number 1.5 foil is gluelessly bonded. In contrast to conventional plastic bottoms this foil resamples optical properties of glass coverslips (details see text). Additionally the foil can be covered with an oleophobic coating to allow the use of oil-immersion objectives in optical screens. (B) The relationship between the velocity of the well plate and the distance between objective and well plate. The measurement points indicate maximal distance between objective and well plate for a given velocity where it was just possible to move the entire length of the well plate (11 cm) without a break of the oil drop on the objective. The grey squares and black triangles represent measurements of low viscosity (600 cSt) and high viscosity (800 cSt) immersion oil, respectively. The fitted lines are linear regressions. Measurements were performed at room temperature (23 °C).

against, e.g. popular organic solvents such as ethanol and dimethyl sulfoxide. The foil was welded to the bottom of the well plate. This glue-less technology ensured a high biocompatibility. The flatness of the bottom was better than 10 μm in a single well and better than 50 μm over the range of the entire plate.

In order to use objectives with the highest NA possible, we picked oil as the immersion medium of choice. Furthermore, it was the only medium available for TIRF objectives. In addition to enabling the use of a large variety of oil immersion objectives, oil displayed a diminished evaporation rate when compared to water (0.3 vs. 0.01 with butyl acetate set to 1). This was favourable for long-term measurements.

The smearing of oil at the bottom of the well plate was regarded as a problem. However, it could be avoided by an oleophobic coating of the foil. The coating led to a reduced adhesion of the immersion oil to the foil. Thus allowing the oil to preferentially stick to the objective. To quantify this, we tested this parameter at different objective-plate distances, varying velocities and two different oils. The results of these measurements

are displayed in ◘ Fig. 13.5B and ◘ Tab. 13.1. As depicted in ◘ Fig. 13.5B, when driving the well plate along its entire length, the maximal velocity that could be used without breaking the oil droplet was very slow when compared to the maximal velocity technically offered by microscope stages (75–80 mm/s). Since this experimental procedure was not reflecting experimental procedures very well (experiments are rather characterised by »jumping« from well to well), we re-evaluated our approach. We performed shorter movements of 8 mm long stretches, which resembled the closest distance of points within the field stimulated area of two adjacent wells. ◘ Tab. 13.1 summarises the results of such an experimental series in which we tested the maximal number of 8 mm »jumps« before the oil drop broke. This number was related to the waiting time between the jumps for both immersion oils tested. We moved the well plate with standard working velocity of the microscope of 13 mm/s. This was almost 10 times faster then the fastest velocities achieved in the test from above. In a screening assay arrangement imaging periods in between moving times (from well to well) will last longer then 10–20 s, thus the resting times necessary for the high-viscosity immersion oil (10–20 s) nicely met such requirements. When using the standard moving velocity of the microscope and the oil with the higher viscosity, imaging times and necessary resting times were in the same range. Thus the process would allow an almost uninterrupted screening. Nevertheless, if the ratio between plate movement and imaging is shifted more towards the moving period, i.e. the imaging period becomes shorter, the properties of the oil and the oleophobic coating becomes limiting for the screening speed. We identified this as a property that will need additional attention in the future. Since we are currently limited to commercially available low fluorescence immersion oils we will foster the properties of the foil. Up to now we have tested oleophobic coatings with a surface energy of around 18–22 mN/m. Possibilities are foreseeable to reduce this value to below 15 mN/m that would enable even faster plate movements in between imaging periods.

13.3.6 TIRF calcium imaging

As described above, the foil/coating combination was chosen to enable the use of high NA objectives for high resolution imaging. Here, TIRF microscopy was a particular demanding imaging modality that we intended to test with the substrate, since the generation of the evanescent wave requires homogeneous optical properties of the substrate and a particular flat surface.

We employed TIRF microscopy to study calcium signals in rat ventricular myocytes loaded with Fluo-4 (see also [27,28]). ◼ Fig. 13.6 summarises the results of such experiments. Our initial surprising finding was that each individual myocyte developed a different pattern of surface contact to the substrate, independent of the type of the substrate (◼ Fig. 13.6A(a,b)—glass coverslips, A(c,d)—foil).

In ◼ Fig. 13.6B we compared TIRF microscopy with epi-fluorescence microscopy. ◼ Fig. 13.6B(a) illustrates the principle arrangements for epi-fluorescence and TIRF imaging while B(b) depicts exemplified fluorescence images (B(b left)—TIRF images, B(b right)—epi-fluorescence image).

◼ Tab. 13.1 The oleophobic coated well plate was moved in 8 mm steps (corresponds to the way from the area just between the carbon electrodes of one well to nearest such position of the neighboring well) at a speed of 13 mm/s. The length of the well plate allows for a maximum of 12 steps. The left column indicates the (resting/imaging) time in between the steps. The middle and the right columns give the minimal number of steps for two oils with differing viscosity that can be performed without the break of the oil drop on top of the objective. Distance between objective and plate was the working distance of 170 μm. Tests were performed at room temperature (23 °C).

resting time in seconds between 8 mm steps	minimal number of possible steps with …	
	low viscosity (600 cSt) immersion oil	high viscosity (800 cSt) immersion oil
0.5	2	3
1	2	3
2.5	4	4
5	5	7
10	5	12
15	8	12
20	11	12
25	12	12

When cardiac myocytes are electrically field stimulated near-plasma membrane Ca^{2+} transients can be measured (◼ Fig. 13.6B(c); grey arrowhead denotes the time of field stimulation). These data clearly supported our notion that the optical properties of the foil used with our 24-well plate indeed met even the highest demands of TIRF imaging and will thus enable its application high resolution imaging during high content screening applications.

13.3.7 Elastic coating

In cardiomyocytes global Ca^{2+} signals evoke transient contractions of individual cells, the so-called twitch. In the *in vivo* situation, cardiac muscle cells are embedded into an elastic network comprising

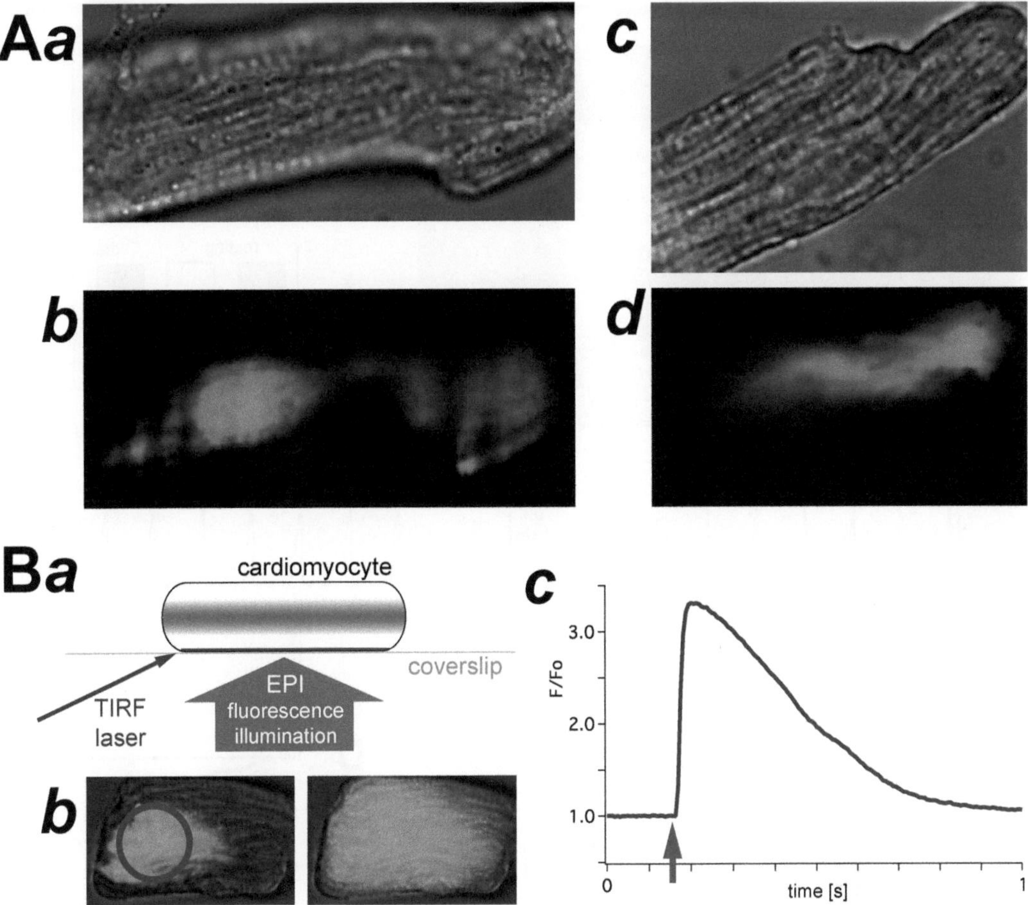

◘ Fig. 13.6 TIRF microscopy of Fuo-4 loaded cardiac myocytes. (A) Provides example images comparing TIRF microscopy on a glass coverslip (Ab) and on olefin polymer foil (Ad). (Aa) and (Ac) are the corresponding white light images. Panel (B) compares the calcium imaging modes of TIRF microscopy and video imaging on a single ventricular myocyte. The transition from TIRF to epi-fluorescence was performed by a refocus and a switch of the excitation source as indicated in (Ba). An overlay of white light images and fluorescence is depicted in (Bb) for TIRF microscopy (left image) and epi-fluorescence (right image). The normalised intensity of the region of interest drawn in (Bb) recorded at an acquisition rate of 160 Hz is plotted in (Bc). The grey arrow depicts the field stimulation pulse of 5 ms.

neighbouring cells and the protein network of the extracellular matrix[29]. The mechanical properties of the environment of individual myocytes are important determinants of signalling events but they also play a vital role for their mechanical performance. During the twitch, part of the energy is »stored« in the elastic properties of the extracellular matrix[30]. During relaxation processes this energy is partially released as a so-called restoring forces. In this way, relaxation of the individual cell is also dependent on its direct environment. During structural remodelling this network undergoes pathological changes that themselves impinge on contractility[31]. Usually, whether plated on glass coverslips or plastic, cardiac myocytes face an inelastic substrate (elastic modulus of about 70 GPa and 1 GPa, respectively), that will neither provide *in vivo* signalling nor an *in vivo*-like mechanical support for the cells. For mimicking such an environment at least to a certain degree, we employed two measures: (i) the foil was coated with an elastic layer of 40 µm having an elasticity modulus of 28 kPa and

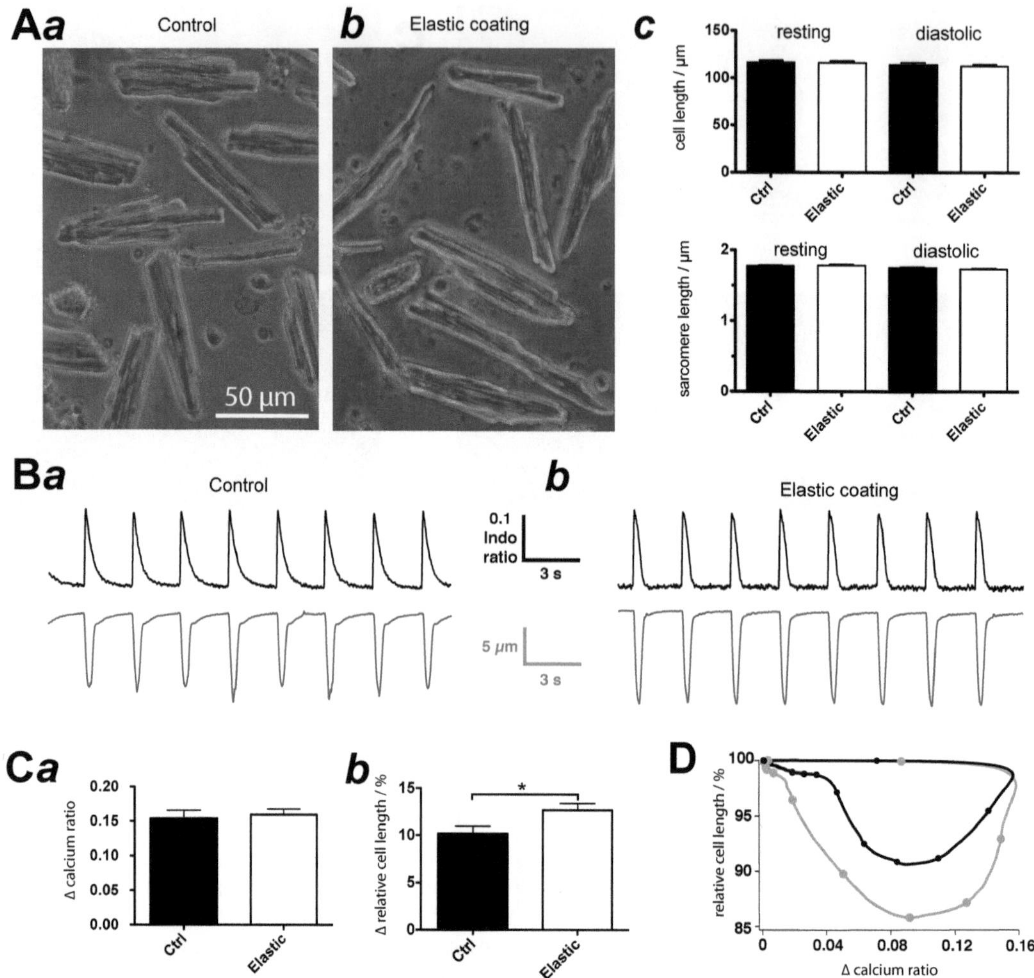

Fig. 13.7 Simultaneous measurements of calcium transients and cell length changes in isolated ventricular myocytes. Images of cardiomyocytes plated on different substrates are shown in (Aa) and (Ab) for glass coverslips and elastic coating on the ibidi foil respectively. The bar graphs in (Ac) reveal the absence of differences concerning the cell length as well as the sarcomere length between the two substrates. There is no significant difference of neither cell length nor sarcomere length with regard to the resting state or the diastolic steady state (pulsing at 0.5 Hz). Each bar represents the mean (±SEM) of 30–40 cells from 4 animals. Part (B) depicts sample traces of both entities for representative cells plated on glass coverslips (control) in (Ba) and on elastic coating (Bb). Cells were electrically field stimulated with 5 ms pulses at a frequency of 0.4 Hz. Statistical analysis of the calcium transients (Ca) and the cell length changes (Cb); comparison between glass substrate (black bars, n = 27 cells) and elastic coating (white bars, n = 43 cells). While there is virtually no change in the calcium signals there is a significant change (p < 0.05) in the relative cell length changes. Part (D) depicts the direct calcium/cell length relationship for representative cells. The curves are means of 10 transients each. To provide temporal orientation points were plotted as circles at a temporal distance of 100 ms for both curves. For interpretation of the differences it is worthwhile to check whether the absolute cell length is different in the two cell populations. In this case the differences could nicely be explained by the Frank–Starling law[33]. Since end diastolic sarcomere lengths are thought to be longer in tissue than in isolated cells and here we mimic the loaded conditions in tissue, we wanted to check the spatial start point of contraction for the two coatings. However, for both conditions, the absolute cell length as well as the sarcomere length before (resting) and in between stimulations (diastolic) depicted no significant difference (Fig. 13.7A(c)).

(ii) this elastic layer was itself was coated with a mixture of extracellular matrix proteins.

The beneficial effect of the ECM coating was already shown in previous reports[17,32]. In order to evaluate the elastic coating we performed simultaneous contraction and calcium recordings in adult rat ventricular myocytes and compared cells plated on rigid substrates with those seeded onto the foil coated with the elastic layer.

For this acutely isolated adult rat ventricular myocytes were electrically stimulated using two field electrodes. ◘ Fig. 13.7A depicts typical images of cells on non-elastic (a) and elastic coating (b). For further measurements cardiomyocytes were loaded with Indo-1 prior to the experiments. Example traces for calcium recordings (black curves) and cell length transients (grey curves) are shown for uncoated substrate (control) in ◘ Fig. 13.7B(a) and for elastic coating in B(b). The statistical analysis of the cells is provided in ◘ Fig. 13.7C and revealed no difference in the amplitude of electrically evoked calcium transient between elastic and non-elastic coating. Surprisingly, despite the unaffected calcium signals, contraction transients were significantly increased in cells seeded on the elastic coating (◘ Fig. 13.7C(b)). Although we discovered a statistical significance, we noticed that the effect could only be recorded nicely in a sub-population of adherent cells. When analysing the TIRF images shown in ◘ Fig. 13.6, it became apparent that many myocytes solely establish a rather focal contact area with their substrates and only a particular sub-population of myocytes displayed more than one contact area. Nevertheless, multiple contact areas were necessary for the elastic coating to exert its effect since myocytes attached to a single contact area do not have anything to contract »against«. We thus assumed that the average result that we recorded was a clear underestimation of the true beneficial properties of the elastic coating. Unfortunately it is not possible to perform TIRF microscopy on elastically coated foils due to the refractive index of 1.41 of the 40 µm coating.

For visualisation of differences in the cell length/calcium relationship we superimposed two contraction/calcium traces we recorded in typical experiments. In ◘ Fig. 13.7D, the black tracing depicts the cell length/calcium relationship in a myocyte seeded on inelastic substrates while the grey trace shows a typical relationship depicted from a myocyte on elastic coating. Although, both calcium transients were of the same amplitude the cell on elastic coating displayed almost 50% increased changes of the cell length in comparison to the cell on the inelastic substrate. From these studies we concluded that the application of elastic coatings as substrates for seeding cardiac myocytes was highly beneficial.

13.4 Conclusion

Here we introduced the hardware, the software and the proof-of-principle of a novel device for cell culture and optical measurements for electrically excitable cells. This system paves the way for primary cells such as adult cardiac myocytes or neurons into optical screening applications. At the same time our approach also meets single experiment laboratory standards in terms of high resolution optical techniques. The entire design provides a scalability for increased throughput considering the inclusion in an automated imaging environment. The sum of these properties bears the potential to foster oHCS of primary cells and such could lead to quickened physiological and pathophysiological understanding, enhanced identifications of new pharmacological targets and lead substances as well as improvements in pharmacological safety screens.

13.5 References

[1] T.J. Starr, Fluorescence microscopy and autoradiography of colchicine-induced micronucleated cells. Nature, 200 (1963), pp. 608–609.

[2] T.A. Ryan, P.J. Millard and W.W. Webb, Imaging $[Ca^{2+}]_i$ dynamics during signal transduction. Cell Calcium, 11 (1990), pp. 145–155.

[3] L. Kaestner and P. Lipp, Non-linear and ultra high-speed imaging for explorations of the murine and human heart, J. Popp, von Bally F. G., Editors Optics in Life Science, vol. 6633 SPIE, Munich (2007) 66330K-1–66330K-10.

[4] L. Kaestner and P. Lipp, Towards imaging the dynamics of protein signalling, L. Spencer, F. Shorte, Frischknecht, Editors , Imaging Cellular and Molecular Biological Functions, Springer, Berlin, Heidelberg (2007), pp. 289–312.

[5]   N. Scheller, P. Resa-Infante, S. de la Luna, R.P. Galao, M. Albrecht, L. Kaestner, P. Lipp, T. Lengauer, A. Meyerhans and J. Diez, Identification of PatL1, a human homolog to yeast P body component Pat1. Biochim. Biophys. Acta, 1773 (2007), pp. 1786–1792.

[6]   P. Weissgerber, B. Held, W. Bloch, L. Kaestner, K.R. Chien, B.K. Fleischmann, P. Lipp, V. Flockerzi and M. Freichel, Reduced cardiac L-type Ca^{2+} current in Ca(V)beta2–/– embryos impairs cardiac development and contraction with secondary defects in vascular maturation. Circ. Res., 99 (2006), pp. 749–757.

[7]   M.D. Bootman, M.J. Berridge and P. Lipp, Cooking with calcium: the recipes for composing global signals from elementary events. Cell, 91 (1997), pp. 367–373.

[8]   M. Bootman, E. Niggli, M. Berridge and P. Lipp, Imaging the hierarchical Ca^{2+} signalling system in HeLa cells. J. Physiol., 499 Pt 2 (1997), pp. 307–314.

[9]   P. Lipp and E. Niggli, Submicroscopic calcium signals as fundamental events of excitation–contraction coupling in guinea-pig cardiac myocytes. J. Physiol., 492 Pt 1 (1996), pp. 31–38.

[10] C.C. Ashley and E.B. Ridgway, Simultaneous recording of membrane potential, calcium transient and tension in single muscle fibers. Nature, 219 (1968), pp. 1168–1169.

[11] L. Kaestner, W. Tabellion, E. Weiss, I. Bernhardt and P. Lipp, Calcium imaging of individual erythrocytes: problems and approaches. Cell Calcium, 39 (2006), pp. 13–19.

[12] Y.N. Tallini, M. Ohkura, B.R. Choi, G. Ji, K. Imoto, R. Doran, J. Lee, P. Plan, J. Wilson, H.B. Xin, A. Sanbe, J. Gulick, J. Mathai, J. Robbins, G. Salama, J. Nakai and M.I. Kotlikoff, Imaging cellular signals in the heart in vivo: cardiac expression of the high-signal Ca^{2+} indicator GCaMP2. Proc. Natl. Acad. Sci. U.S.A., 103 (2006), pp. 4753–4758.

[13] O. Garaschuk, O. Griesbeck and A. Konnerth, Troponin C-based biosensors: a new family of genetically encoded indicators for in vivo calcium imaging in the nervous system. Cell Calcium, 42 (2007), pp. 351–361.

[14] H.L. Heine, H.S. Leong, F.M. Rossi, B.M. McManus and T.J. Podor, Strategies of conditional gene expression in myocardium: an overview. Methods Mol. Med., 112 (2005), pp. 109–154.

[15] A. Ludwig, S. Herrmann, E. Hoesl and J. Stieber, Mouse models for studying pacemaker channel function and sinus node arrhythmia. Prog. Biophys. Mol. Biol., 98 (2008), pp. 179–185.

[16] P. Lipp and L. Kaestner, Image based high content screening—a view from basic science, J. Hüser, Editor, High-Throughput Screening in Drug Discovery, Wiley VCH, Weinheim (2006), pp. 129–149.

[17] C. Viero, U. Kraushaar, S. Ruppenthal, L. Kaestner and P. Lipp, A primary culture system for sustained expression of a calcium sensor in preserved adult rat ventricular myocytes. Cell Calcium, 43 (2008), pp. 59–71.

[18] L. Kaestner, A. Scholz, K. Hammer, A. Vecerdea, S. Ruppenthal and P. Lipp, Isolation and genetic manipulation of adult cardiac myocytes for confocal imaging. J. Vis. Exp., 31 (2009) http://www.jove.com/index/Details. stp?ID=1433.

[19] U. Becherer, M. Pasche, S. Nofal, D. Hof, U. Matti and J. Rettig, Quantifying exocytosis by combination of membrane capacitance measurements and total internal reflection fluorescence microscopy in chromaffin cells. PLoS ONE, 2 (2007), p. e505.

[20] S. Borzak, S. Murphy and J.D. Marsh, Mechanisms of rate staircase in rat ventricular cells. Am. J. Physiol., 260 (1991), pp. H884–H892.

[21] C.T. Ivester, R.L. Kent, H. Tagawa, H. Tsutsui, T. Imamura, Cooper Gt and P.J. McDermott, Electrically stimulated contraction accelerates protein synthesis rates in adult feline cardiocytes. Am. J. Physiol., 265 (1993), pp. H666–H674.

[22] F.G. Spinale, R. Mukherjee, B.M. Fulbright, J. Hu, F.A. Crawford and M.R. Zile, Contractile properties of isolated porcine ventricular myocytes. Cardiovasc. Res., 27 (1993), pp. 304–311.

[23] O. Ellingsen, A.J. Davidoff, S.K. Prasad, H.J. Berger, J.P. Springhorn, J.D. Marsh, R.A. Kelly and T.W. Smith, Adult rat ventricular myocytes cultured in defined medium: phenotype and electromechanical function. Am. J. Physiol., 265 (1993), pp. H747–H754.

[24] H. Tsutsui, Y. Urabe, D.L. Mann, H. Tagawa, B.A. Carabello, Cooper Gt and M.R. Zile, Effects of chronic mitral regurgitation on diastolic function in isolated cardiocytes. Circ. Res., 72 (1993), pp. 1110–1123.

[25] H.J. Berger, S.K. Prasad, A.J. Davidoff, D. Pimental, O. Ellingsen, J.D. Marsh, T.W. Smith and R.A. Kelly, Continual electric field stimulation preserves contractile function of adult ventricular myocytes in primary culture. Am. J. Physiol., 266 (1994), pp. H341–H349.

[26] R. Zantl, U. Rädler and E. Horn, Chemotaxis in µ-channels. Imag. Microsc., 8 (2006), pp. 30–32.

[27] L. Cleemann, G. DiMassa and M. Morad, Ca^{2+} sparks within 200 nm of the sarcolemma of rat ventricular cells: evidence from total internal reflection fluorescence microscopy. Adv. Exp. Med. Biol., 430 (1997), pp. 57–65.

[28] Y. Bai, A. Tang, S. Wang and X. Zhu, Total internal reflection fluorescence microscopy study of spiral Ca^{2+} waves in single heart cell. J. Microsc., 229 (2008), pp. 555–560.

[29] J.L. Sepulveda, V. Gkretsi and C. Wu, Assembly and signaling of adhesion complexes. Curr. Top. Dev. Biol., 68 (2005), pp. 183–225.

[30] M.S. Forbes and N. Sperelakis, The membrane systems and cytoskeletal elements of mammalian myocardial cells. Cell Muscle Motil., 3 (1983), pp. 89–155.

[31] V. Pelouch, I.M. Dixon, L. Golfman, R.E. Beamish and N.S. Dhalla, Role of extracellular matrix proteins in heart function. Mol. Cell Biochem., 129 (1993), pp. 101–120.

[32] L. Kaestner, S. Ruppenthal, S. Schwarz, A. Scholz and P. Lipp, Concepts for optical high content screens of excitable primary isolated cells for molecular imaging, SPIE Biomed. Opt., 7370 (2009) 737008-1r–737008-8r.

[33] D.M. Bers, Excitation–Contraction Coupling and Cardiac Contractile Force, Kluwer Academic Publishers, Dordrecht, Boston, London (2001).

Concepts for optical high content screens of excitable primary isolated cells

Lars Kaestner, Sandra Ruppenthal, Sarah Schwarz, Anke Scholz, Peter Lipp

Reprint from Molecular Imaging II, eds. Kai Licha, Charles P. Lin, Proceedings of SPIE (2009) 7370, 737008-1 - 737008-8.

▪ Abstract

Here we describe the cell- and molecular-biological concepts to utilise excitable primary isolated cells, namely cardiomyocytes, for optical high content screens. This starts with an optimised culture of human adult cardiomyocytes, allowing culture with diminished dedifferentiation for one week. To allow fluorescence based molecular imaging genetically encoded biosensors need to be expressed in the cardiomyocytes. For transduction of end-differentiated primary cells such as neurons or cardiomyocytes, a viral gene transfer is necessary. Several viral systems were balanced against each other and an adenoviral system proofed to be efficient. This adenoviral transduction was used to express the calcium sensors YC3.6 and TN-XL in cardiomyocytes. Example measurements of calcium transients were performed by wide-field video imaging. We discuss the potential application of these cellular and molecular tools in basic research, cardiac safety screens and personalised diagnostics.

14.1 Introduction

Isolated adult cardiac myocytes serve as models somewhere in between embryonic and neonatal cells on one side and the working heart on the other side. Although the technique of adult cardiac cell isolation is established for more than 30 years[1], experiments have vastly been restricted to investigations acute after cell isolation. This had two major reasons: (i) cardiac myocytes, when in culture quickly dedifferentiate and morphologically rather look like fibroblasts (cp. ◼ Fig. 14.1B) or even round up completely to form so called myoballs. These morphological changes come along with functional alterations[2] and then represent no longer adult cardiomyocytes. (ii) visualisation in terms of molecular imaging in living cells either for sub-cellular structures or for functional sensors have – for a long time – been restricted to small molecule dyes. These dyes are not suitable for long-term observations or repetitive staining since the specificity in terms of localisation is restricted to certain time slots (typically in the range of one hour) and over time may even be cytotoxic.

Here we present concepts how to overcome both drawbacks by (i) adapting a recently introduced culture method for adult rat cardiomyocytes[3] to human atrial myocytes and (ii) describing a viral transduction method to express genetically encoded markers and biosensors (GEBs) in cardiac myocytes. In contrast to small molecule dyes, fluorescent proteins and GEBs have the advantage of high biocompatibility since they are »produced« by the cells themselves. Such, they allow repetitive optical measurements in the same time range as the improved culture conditions keep the cells conserved.

14.2 Methodology

14.2.1 Isolation procedure of myocytes from human auricles

Atrial appendages were obtained after informed consent from patients undergoing cardiac surgery with extracorporal circulation. The use of the tissue was approved by the local ethics committee (approval number 76/05). The appendages were collected in phosphate buffered solution (PBS) containing in (mM): NaCl (134), KCl (4), glucose (11), $MgSO_4$ (1.2), Na_2HPO_4 (1.2) and N-2-hydroxyethylpiperazine-N'-2-ethanesulfonicacid (HEPES) (10) within 15 seconds after excision. The tissue was transported to the cell culture lab in ice cold PBS, and the isolation procedure was started within 30 min after dissection. The tissue was cut into 1 mm^3 pieces at the latest 30 min after the dissection and then incubated in PBS. The tissue was digested in an enzymatic solution of collagenase type II and elastase (both Worthington, New Jersey, USA) for approximately 80 min with repetitive gentle trituation and visual/microscopic inspection. Afterwards, the collagenase solution was decanted and cells were washed in PBS supplemented with bovine serum albumin (BSA, Roth, Germany) in order to quickly inhibit the remaining collagenase activity. Then the calcium-concentration was elevated by eventually adding 10 ml culture medium M199 to 10 ml cell suspension. This M199 addition was performed step wise: 50-100 µl every 2-4 min for the first 500 µl, then 5 times 200 µl every 3-4 min, 7 times 500 µl every 3-4 min and finally 5 times 1 ml every 3-4 min. This was performed under repetitive microscopic inspection and the intervals for increasing the calcium concentration were adjusted as soon as excessive spontaneous activity of the myocytes occurred. Cells were then seeded in culture flasks or on cover slips. One hour later the medium was exchanged. If not mentioned otherwise, all chemicals and substances were acquired at Sigma-Aldrich (St. Louis, USA).

14.2.2 Cell culture of human atrial myocytes

For acute experiments the human atrial myocytes were seeded onto coated cover slips, for long-term culture we seeded the cells onto coated culture flasks. Coating of the cover slips and the flasks was achieved by one of the following coats: poly-L-lysine, collagen, fibronectin, gelatine, foetal bovine serum (FBS, Biowest, France) and a mixture of extracellular matrix proteins (ECM, Harber Bio-Products, Norwood, USA). Each coating was applied according to the recommendation of the supplier. Long-term culturing of the cells was performed in culture medium (M199) supplemented with either 5% FBS or a mixture of insulin, transferrin and selenite (ITS). All culturing was carried out in standard culturing conditions: 37°C, 5% CO_2 and 95% humidity.

14.2.3 Adenovirus construction

Generation of recombinant Adenoviruses was accomplished by using Transpose-Ad™ adenoviral vector system (MP Biomedicals, USA) as described in detail elsewhere[4]. pCR259 Adenovirus transfer vectors encoding YC3.6 and TN-XL were transformed in HighQ-1 Transpose-Ad™ 294 competent cells, a bacterial cell line carrying the Transpose-Ad™ 294 plasmid and a plasmid encoding a trans-acting Tn7 transposase. After a Tn7-based transposition, recombinant adenoviral genome was purified from bacteria and transfected into the QBI-HEK 293 cell line using Lipofectamine 2000 (Invitrogen, USA). In this cell line, the recombinant Adenoviruses were generated and propagated (◘ Fig. 14.2).

The pcDNA3-YC3.6 vector as well as the corresponding TN-XL vector were kindly provided by Dr. Atsushi Miyawaki (RIKEN, Wako, Saitama, Japan) and Dr. Oliver Griesbeck (Max-Planck Institute for Neurobiology, Munich, Germany), respectively.

14.2.4 Imaging

For data acquisition the cover slips were mounted in a self constructed experimental chamber in a Tyrode solution containing (in mM): NaCl (135), KCl (5.4), glucose (10), $MgCl_2$ (1), $CaCl_2$ (1.5) and HEPES (10). The pH-value of all solutions was adjusted to 7.35 using NaOH. Cells were placed on the stage of an inverted microscope (NIKON, TE2000U) that was attached to video-imaging hardware (TILL Photonics, Munich, Germany). The system comprised a fast video camera (Imago, TILL Photonics) and a fast monochromator for excitation (Polychrome IV, TILL Photonics). For Förster Resonance Energy Transfer (FRET) mea-

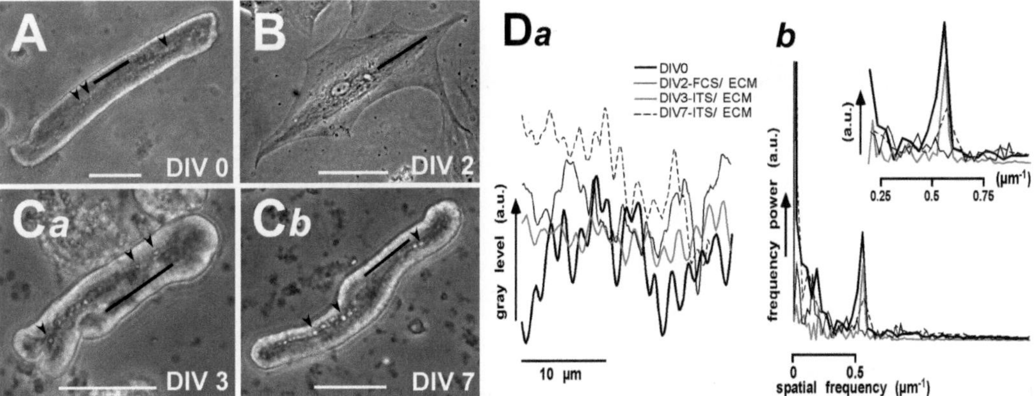

□ Fig. 14.1 Primary culture of isolated myocytes from human atrial appendages. Panel A shows a freshly isolated myocyte and B a myocyte two days in culture (DIV – days in vitro) on an untreated cell culture flask in medium supplemented with fetal bovine serum (FBS). In contrast, Ca&b depict examples of myocytes in long term culture on a surface coated with extra cellular matrix (ECM) proteins and in serum-free medium supplemented with insulin, transferrin and selenite (ITS). The arrow heads point to vesicular structures in the cardiomyocytes that are present in freshly isolated cell (panel A) as well as in all stages of the culture (panel Ca&b). The scale bar is 30 µm in all images. Panel Da shows a grey-value intensity profile along the black lines in panels A,B,Ca&b. The power spectrum of the profile shown in Da is drawn in the graph of panel Db. The peak of interest has been magnified for the inset.

surements we used a simplified method to record the CFP and YFP-related fluorescence upon nominal CFP excitation. For simultaneous detection of the two channels we placed a dual view image splitting system (DV2, Optical Insights, Santa Fe, USA), with a 515 nm dichroic mirror to separate the two channels. Electrical stimulation was performed by pulse generation using a MyoPacer (IonOptix, Milton, USA). All experiments were carried out at room temperature (20-22°C).

14.3 Results

14.3.1 Culture of human cardiac myocytes

A recently introduced method for long-term culture of rat adult cardiac myocytes[3] was adapted for human atrial myocytes as described in section 14.2.1. An important prerequisite for long-term culture of cardiac myocytes is an optimised attachment of the cells to their substrate (either glass or plastic). We tested various coatings in different concentrations, such as poly-L-lysine, collagen, fibronectin, gelantin, FBS and a mixture of extracellular matrix proteins (ECM). All of these coatings are well known to improve cell adhesion to various surfaces[5]. In our hands surface coating with ECM at a concentration of 5.6 µg/cm^2 resulted in an optimised attachment of the cells to both, glass and plastic surfaces. For rodent cardiac myocytes it is well known that within a couple of days culturing of the cells in serum supplemented medium results in rapid de-differentiation into a flat, spread-out phenotype[6]. When using such serum containing medium (5% FCS), we could also observe a similarly behaviour as depicted in □ Fig. 14.1B (compare □ Fig. 14.1A and B, n=78 from 3 hearts).

When substituting a mixture of insulin (870), transferrin (65) and selenite (29; all concentrations in nM; ITS) for the serum supplement such a de-differentiation could be largely reduced or even suppressed for time periods of up to 7 days *in vitro* (□ Fig. 14.1Ca and Cb). As can be read from the typical light microscopic images of atrial human myocytes, the morphology was largely preserved, the cells remained elongated and displayed visible cross-striation. In order to

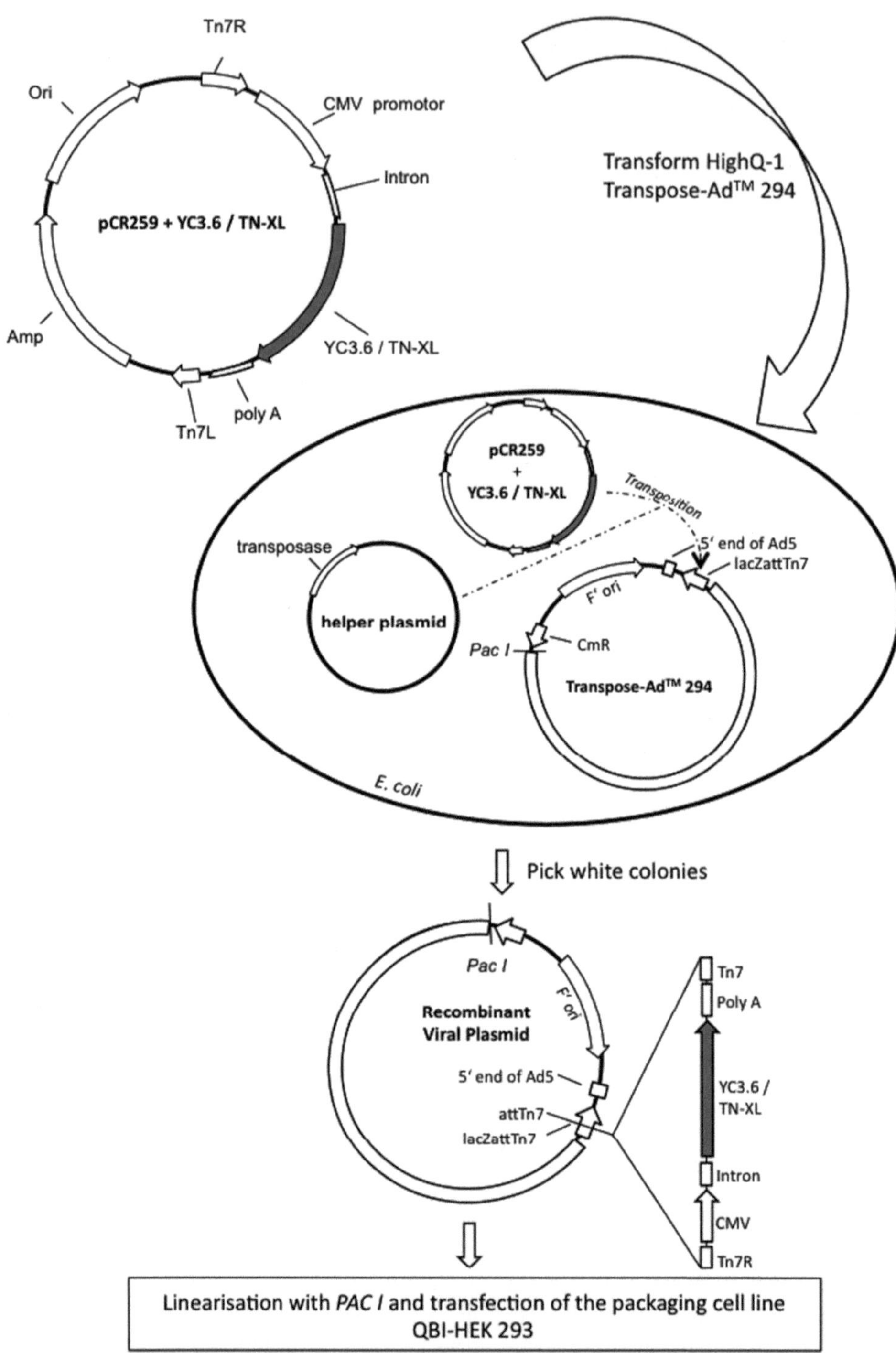

◻ Fig. 14.2 Diagram of the generation of recombinant Adenoviruses using the Transpose-AdTM adenoviral vector system. The gene of interest, the GEBs YC3.6 and TN-XL, are cloned into the pCR259 Adenovirus transfer vector, which is used to transform HighQ-1 Transpose-AdTM 294 competent cells. In these cells the transposition of the GEBs into the recombinant viral genome occurs. The isolated recombinant viral plasmid has to be linearised and transfected into the packaging cell line QBI-HEK 293.

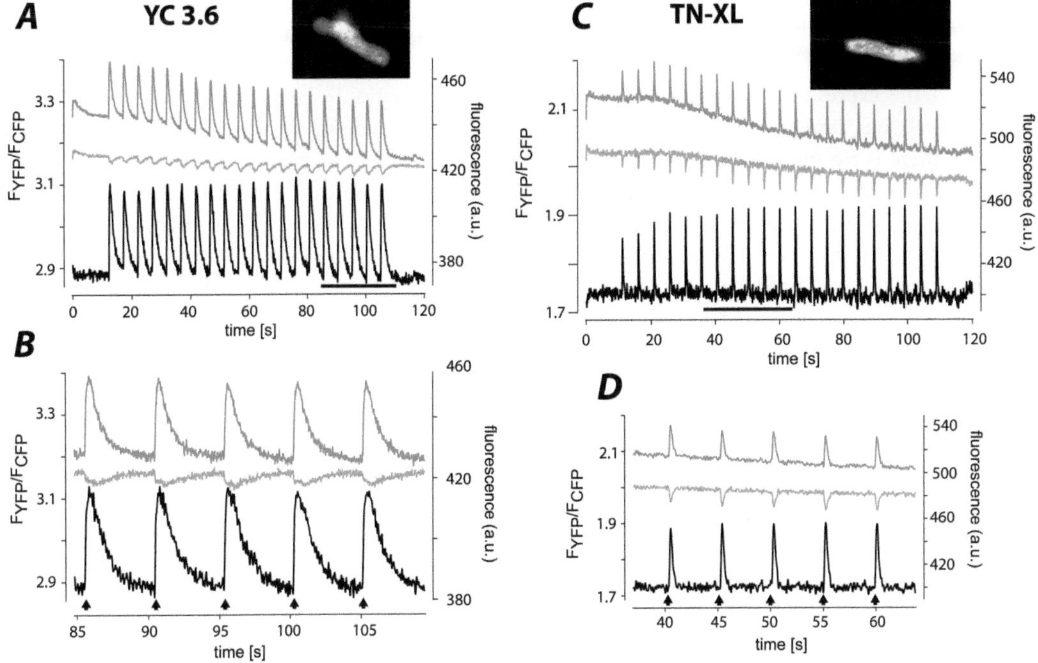

◘ Fig. 14.3 Calcium measurements of Adenoviral transfected GEBs in isolated cardiomyocytes for the sensors YC3.6 (A and B) and TN-XL (C and D). The images in panels A and C depict the measured cell. Graphs A and C plot an overview of traces measured for 2 min, while panels B and D show the calcium transients that are underlined in A and C. The upper grey curves depict the YFP-channel signal, the lower grey curves the CFP signal, the black traces the fluorescence ratio YFP/CFP and the back arrows the time points of a 5 ms electrical field stimulation.

quantify this preservation we analysed the cross striation: Intensity profiles along the longitudinal axis ◘ Fig. 14.1A-C were Fourier transformed and the resulting magnitude was plotted against the spatial frequency (◘ Fig. 14.1.1D*b*). The peaks correspond to the sarcomere length (1.89 μm). The frequency corresponding to 1.89 μm showed the highest peak in freshly isolated cells and was well preserved in the ITS-supplemented culture on days *in vitro* (DIV)3 and DIV7, but totally absent in the cell that were cultured in FBS medium. These data were typical for all myocytes tested (n=4 for ITS and n=5 for FBS, from 3 appendages). Here it was a consistent finding that in comparison to our experience with mouse and rat myocytes, the human atrial cells always contained a varying number of vesicles, independent of their duration in culture (see ◘ Fig. 14.1A and *Ca&b*; marked by arrowheads).

14.3.2 Viral transduction of genetically encoded biosensors

Transfection of end-differentiated cells, such as neurons or cardiomyocytes, when using classical transfection methods like chemical mediated transfection or electroporation is not efficient or not applicable. Therefore viral gene transfer is the method of choice for such cell types. To perform viral transduction we evaluated differing viruses: Semliki Forest virus[7], lentivirus[8] and adenovirus[9].

Semliki Forest virus transduction lead in neurons to a fast expression of the transduced gene within 6 hours but killed all cells within 24 hours. This allows studies of freshly isolated cells but does not allow studies of chronic application of potential drugs. However, in cardiac myocytes we could not – even with high titre of the Semliki Forest virus – achieve a transfection rate exceeding 5%.

In contrast Lentivirus gene transfer required about a weak for protein expression. This then is to long a time for the culture system described in 14.3.1 - the expression may take place within the one week, but similar as for the Semliki Forest virus, there is no time range left for chronical investigations of pharmaceuticals. Therefore, Lentiviruses may well serve for gene transfer in animals, but for gene transfer in primary cultured cells like neurons or cardiac myocytes they are not suitable.

Finally, adenoviral gene transfer was tested. In adult cardiomyocytes the proteins show a sufficient expression within 24 hours for most vectors tested (e.g., pericams, YC3.6, TN-XL and a number of fusion proteins). Furthermore, the adenoviral-mediated expression is fairly stable over at least one week and we could not find any sign of affection on the cells by the virus and the consecutive gene expression (GEBs and fusion proteins exclusively). The design of an Adenovirus as it is used to produce YC3.6 and TN-XL (see below) is depicted in ◘ Fig. 14.2.

YC3.6[10] and TN-XL[11] are both Förster Resonance Energy Transfer (FRET)-based constructs that allow for a ratiometric and therefore rather quantitative measurement of calcium. Examples of calcium measurements with both constructs are plotted in ◘ Fig. 14.3. These are typical examples of 12 measurements (out of 3 hearts) for each sensor.

14.4 Discussion

14.4.1 Culture of human cardiac myocytes

While in animal models genetic manipulation can be performed on the organism level by generating knock-out or knock-in animals[12,13], work in terms of genetic manipulation with human hearts is limited to single cells. Thus it appears important to establish procedures that allow the genetic manipulation of these cells. Such a manipulation, i.e., expression of sensors (as described in 14.3.2), but also expression of exogeneous proteins or expression of short hairpin RNAs (shRNA)[14], requires construction of viral gene transfer techniques as described above. Even the most rapid protein expression achieved with adenoviral expression

systems require the isolated human myocytes to reside in culture for one day for protein expression, but three, preferably more, days for protein down regulation[15]. Here, we provide evidence that the combination of serum-free ITS supplemented medium and ECM-coating of the substrate fosters the preservation of the morphology of the single myocytes in culture. Both approaches minimise the dedifferentiation that usually occurs when culturing adult cardiac myocytes[2,6].

Originally, the addition of ITS was proposed by the Alliance for Cellular Signalling (ACS) for adult mouse cardiac myocytes[16,17]. However, the ACS suggested to additionally supplementing such a culturing medium with 2,3butanedione monoxime (BDM), a chemical phosphatase. BDM will largely suppress contraction by inhibiting proper phosphorylation of the contractile proteins. Although BDM might help in the preservation of gross cell morphology (i.e. leading to a higher proportion of elongated cells) we regard such a supplement as potential problematic due to its unspecificity and known »side effects« on other potentially very important phosphorylation reactions. We thus omitted that supplement from our culture medium.

The ECM-mixture we used was derived from Engelbreth-Holm-Swarm tumor cells and contains amongst other components laminin, collagen, entactin and heparin sulphate proteoglycan. In preliminary studies with human and rodent cardiac cells we have found that the ECM mixture is much more beneficial for the survival and lack of de-differentiation of myocytes than any of the individual components. We thus conclude that the ECM mixture mimics the physiological surrounding *in vivo* much better than any single component of the connective tissue. This is essential because the signalling of the extracellular matrix to intracellular compartments is a complex network assembling multiple protein interactions[18].

14.4.2 Viral transduction of genetically encoded biosensors

The adenoviral transduction machinery is – among the viral system considered – the most appropriate one. It will allow for high content screens on primary isolated cells over several days and thus allow testing chronical application of test substances[19] or alternatively would also allow a protein down regulation by RNAi approaches.

Both constructs used in the example measurements depicted in ◘ Fig. 14.3 contain eCFP as a FRET donor and have very similar FRET acceptors[20]: Venus in the YC3.6 and citrine in TN-XL. The major differences are based on the calcium sensing unit, calmodulin for YC3.6 and troponin C for TN-XL. ◘ Fig. 14.3 reveals long transients (about 5 s) for YC3.6 and short transients (< 1 s) for TN-XL on an identical cellular background. A major reason is the kinetic behaviour of the sensing unit. For TN-XL the dominating calcium binding off rate is as short as 142 ms[11]. Although there is a slower component of 867 ms with a tenth of the amplitude compared to the short exponential it is hidden in the baseline. This holds true since the calcium binding constant of TN-XL at 2.5 μM is a magnitude higher than the one for YC3.6. Although the 2.5 μM of the TN-XL corresponds to the native buffer capacitance in the cytosol of cardiomyocytes[21], ◘ Fig. 14.3C&D shows just the peak of the calcium transient. However, the YC3.6 compares with 250 nM to the binding properties of the popular small molecule calcium sensors, such as Indo-1 (K_d of 230 nM), Fura-2 (K_d of 145 nM) or Fluo-4 (K_d of 345 nM)[22]. So the advantage of the TN-XL is it's fast kinetics (especially the off-rate), whereas the YC3.6 has a significant higher sensitivity for calcium.

14.5 Conclusion

Here we provide the cellular and molecular tools for a high content screening approach of adult cardiac myocytes. The examples for the measurements of intracellular calcium, provided here, could be extended to further parameters such as cAMP activity[23] or membrane potential[24]. For such high content screens we propose three major applications: (i) basic research to identify signalling cascades and pathways for a better understanding of sub-cellular molecular mechanisms in the heart. This includes investigations on transgenic animal models and may lead to the identifica-

tion of novel pharmacological targets. (ii) cardiac safety screens on a cellular level may provide an equivalent to presently used QT-screens in animal experiments[25]. (iii) a screen on human cardiac myocytes of a particular patient that underwent cardiac surgery with extracorporal circulation. If such a patient needs further medication (e.g., anti-arrythmica – almost half of all cardiac surgery patients develop post operative atrial fibrillation[26]) the appendage could be used for cell isolation and an optical screen of the cells would allow identifying *ex vivo* on living cardiomyocytes the most appropriate drug. Thus the screening approach might be used as a personalised diagnostic tool.

14.6 References

[1] Powell, T. and Twist, V. W., »A rapid technique for the isolation and purification of adult cardiac muscle cells having respiratory control and a tolerance to calcium,« Biochem Biophys Res Commun 72, 327-333 (1976).

[2] Bugaisky, L. B., and Zak, R., »Differentiation of adult rat cardiac myocytes in cell culture,« Circ Res 64, 493-500 (1989).

[3] Viero, C., Kraushaar, U., Ruppenthal, S., Kaestner L., and Lipp, P., »A primary culture system for sustained expression of a calcium sensor in preserved adult rat ventricular myocytes,« Cell Calcium 43, 59-71 (2008).

[4] Kaestner, L., Scholz, A., Hammer, K, Vecerdea, A., Ruppenthal, S, and Lipp, P, »Isolation and genetic manipulation of adult cardiac myocytes for confocal imaging,« J Vis Exp, 31, 1433 (2009).

[5] Lundgren, E., Terracio, L., Mardh, S., and Borg, T. K., »Extracellular matrix components influence the survival of adult cardiac myocytes in vitro,« Exp Cell Res 158, 371-381 (1985).

[6] Claycomb, W. C., Burns, A. H., and Shepherd, R. E., »Culture of the terminally differentiated ventricular cardiac muscle cell. Characterization of exogenous substrate oxidation and the adenylate cyclase system,« FEBS Lett 169, 261-266 (1984).

[7] Lundstrom, K., »Semliki Forest virus vectors for gene therapy,« Expert opinion on biological therapy 3, 771-777 (2003).

[8] Delenda, C., »Lentiviral vectors: optimization of packaging, transduction and gene expression,« The journal of gene medicine 6 Suppl 1, S125-138 (2004).

[9] Russell, W. C., »Update on adenovirus and its vectors,« The Journal of general virology 81, 2573-2604 (2000).

[10] Nagai, T., Yamada, S., Tominaga, T., Ichikawa, M., and Miyawaki, A., »Expanded dynamic range of fluorescent indicators for Ca(2+) by circularly permuted yellow fluorescent proteins,« Proc Natl Acad Sci U S A 101, 10554-10559 (2004).

[11] Mank, M., Reiff, D. F., Heim, N., Friedrich, M. W., Borst, A., and Griesbeck, O., »A FRET-based calcium biosensor with fast signal kinetics and high fluorescence change,« Biophys J 90, 1790-1796 (2006).

[12] Weissgerber, P., Held, B., Bloch, W., Kaestner, L., Chien, K. R., Fleischmann, B. K., Lipp, P., Flockerzi, V., and Freichel, M., »Reduced cardiac L-type Ca2+ current in Ca(V)beta2-/- embryos impairs cardiac development and contraction with secondary defects in vascular maturation,« Circ Res 99, 749-757 (2006).

[13] Kirchhefer, U., Hanske, G., Jones, L. R., Justus, I., Kaestner, L., Lipp, P., Schmitz, W., and Neumann, J., »Overexpression of junctin causes adaptive changes in cardiac myocyte Ca(2+) signaling,« Cell Calcium 39, 131-142 (2006).

[14] Rossi, J. J., »Expression strategies for short hairpin RNA interference triggers,« Human gene therapy 19, 313-317 (2008).

[15] Kasahara, H., and Aoki, H., »Gene silencing using adenoviral RNAi vector in vascular smooth muscle cells and cardiomyocytes,« Methods in molecular medicine 112, 155-172 (2005).

[16] Sambrano, G. R., Fraser, I., Han, H., Ni, Y., O'Connell, T., Yan, Z., and Stull, J. T., »Navigating the signalling network in mouse cardiac myocytes,« Nature 420, 712-714 (2002).

[17] O'Connell, T. D., Ni, Y. G., Lin, K.-M., Han, H., and Yan, Z., »Isolation and Culture of Adult Mouse Cardiac Myocytes for Signaling Studies,« AfCS Research Reports 1, 1-9 (2003).

[18] Sepulveda, J. L., Gkretsi, V., and Wu, C., »Assembly and signaling of adhesion complexes,« Curr Top Dev Biol 68, 183-225 (2005).

[19] Lipp, P., and Kaestner, L., »Image based high content screening – A view from basic science,« in High-Throughput Screening in Drug Discovery, J. Hüser, ed. ,Wiley VCH, Weinheim, pp. 129-149 (2006).

[20] Palmer, A. E., and Tsien, R. Y., »Measuring calcium signaling using genetically targetable fluorescent indicators,« Nature protocols 1, 1057-1065 (2006).

[21] Bers, D. M., [Excitation-Contraction Coupling and Cardiac Contractile Force], Kluwer Academic Publishers, Dordrecht, Boston & London, (2001).

[22] Haugland, R. P., [Handbook of Fluorescent Probes and Research Products], Molecular Probes, Eugene, (2002).

[23] Salonikidis, P. S., Zeug, A., Kobe, F., Ponimaskin, E., and Richter, D. W., »Quantitative measurement of cAMP concentration using an exchange protein directly activated by a cAMP-based FRET-sensor,« Biophys J 95, 5412-5423 (2008).

[24] Tsutsui, H., Karasawa, S., Okamura, Y., and Miyawaki, A., »Improving membrane voltage measurements using FRET with new fluorescent proteins,« Nature methods 5, 683-685 (2008).

[25] Arrigoni, C. and Crivori, P., »Assessment of QT liabilities in drug development,« Cell biology and toxicology 23, 1-13 (2007).

[26] Mayson, S. E., Greenspon, A. J., Adams, S., Decaro, M. V., Sheth, M., Weitz, H. H., and Whellan, D. J., »The changing face of postoperative atrial fibrillation prevention: a review of current medical therapy,« Cardiology in review 15, 231-241 (2007).

A primary culture system for sustained expression of a calcium sensor in preserved adult rat ventricular myocytes

Cedric Viero, Udo Kraushaar, Sandra Ruppenthal, Lars Kaestner, Peter Lipp

Reprint from Cell Calcium (2008) **43**, 59-71.

■ **Abstract**

For studying heart pathologies on the cellular level, cultured adult cardiac myocytes represent an important approach. We aimed to explore a novel adult rat ventricular myocyte culture system with minimised dedifferentiation allowing extended experimental manipulation of the cells such as expression of exogenous proteins. Various culture conditions were investigated including medium supplement, substrate coating and electrical pacing for one week. Adult myocytes were probed for (i) viability, (ii) morphology, (iii) frequency dependence of contractions, (iv) Ca^{2+} transients, and (v) their tolerance towards adenovirus-mediated expression of the Ca^{2+} sensor »inverse pericam«. Conventionally, in either serum supplemented or serum-free medium, myocytes dedifferentiated into flat cells within 3 days or cell physiology and morphology were impaired, respectively. In contrast, myocytes cultured in medium supplemented with an insulin–transferrin–selenite mixture on substrates coated with extracellular matrix proteins showed an increased cell attachment and a conserved cross-striation. Moreover, these myocytes displayed optimised preservation of their contractile behaviour and Ca^{2+} signalling even under conditions of continuous electrical pacing. Sustained expression of inverse pericam did not alter myocyte function and allowed long lasting high speed Ca^{2+} imaging of electrically driven adult myocytes. Our single-cell model thus provides a new advance for high-content screening of these highly specialised cells.

15.1 Introduction

The acute isolation of adult cardiac myocytes has been established decades ago[1] to investigate the cells' physiological behaviour. In contrast, studies requiring extended culture periods, e.g., for protein expression or knock-down, have always been limited to a couple of days in culture due to extensive morphological and physiological alterations of the adult myocytes occurring shortly after isolation[2]. This restriction could not be compensated for adequately by the creation of cardiac cell lines since they do not represent cardiac myocyte physiology well enough[3]. Currently, neonatal myocytes serve as a limited model for the adult cell, but it has to be noted, that in comparison to adult myocytes neonatal cells display a different phenotype and genotype. Nevertheless, long-term culturing of these cells even in larger quantities is routine.

In conventional culture, isolated adult rat cardiomyocytes rapidly change from a »brick-like« structure towards a more stellated, neonatal-like shape. Moreover, their size increases considerably[4]. In serum-free culture medium, adult cardiac myocytes from guinea-pigs, rats, rabbits and mice are usually quiescent and retain their viability and unique rod-shaped morphology for at least a couple of days[5-7]. These cells maintain highly organised membrane and myofibrillar structures that support contractions induced by electrical stimulation. Thus, they appear suitable to short-term (1–3 days) virus-mediated expression of exogenous proteins[8]. For future studies requiring long-term expression of exogenous proteins or vector-based RNA interference (RNAi) to knock-down protein expression it appears essential to employ longer culture periods without a loss of morphology and physiology of the freshly isolated cells. Moreover, experimental manoeuvres inducing »slow-onset« cellular responses will also entail long-term culturing of the myocytes. Additionally, molecular biology techniques such as Western blotting demand large amounts of proteins from homogeneous cell populations. Thus, culturing set-ups are needed that offer the possibility to electrically stimulate large homogeneous populations of cells simultaneously. A decade ago, an adult rat ventricular myocytes culture system was developed with conditions that allow short-term (3 days) culture together with the ability to impose arbitrary electrical pulse protocols[9].

The goal of the present study was to use that basic approach and refine it to a long-term culture system (1 week) with diminished cellular dedifferentiation. We tested the suitability of our system in multiple ways including morphology, survival rate, contractile behaviour, Ca^{2+} signalling and success for adenoviral mediated expression of an exogenous protein (inverse pericam, a fluorescence calcium indicator based on calmodulin[10]).

15.2 Methods

15.2.1 Isolation and primary culture of adult rat ventricular myocytes

We adopted a protocol for cell isolation based on established procedures in rabbit and mouse[11,12] for the rat heart. Adult male Wistar rats (6–12 weeks old, 200–400 g) were handled and sacrificed in accordance with the »Guide for the Care and Use of Laboratory Animals« published by the US National Institutes of Health (NIH Publication No. 85-23, revised 1996). Animals were anaesthetised by an intraperitoneal injection (i.p.) of pentobarbital sodium, 160 mg/kg body weight (Narcoren; Merial, Germany). Directly afterwards, we injected (i.p.) 0.5–1 ml (according to the body weight) of a citrate (40 mM) solution to prevent formation of blood clots. Ten minutes later, the animal was killed by decapitation. The heart was flushed with 10 ml of ice-cold Ca^{2+}-free solution (CFS) containing (in mM): NaCl 134, Glucose 11, KCl 4, $MgSO_4$ 1.2, Na_2HPO_4 1.2, HEPES 10 (Merck, Germany) (pH adjusted to 7.35 with NaOH). After that, the heart was removed, attached to a Langendorff apparatus and perfused retrogradly with O_2 saturated CFS containing 200 μM EGTA at a rate of 4 ml/min for 5 min. The perfusate was then changed to O_2 saturated CFS containing collagenase type I (Worthington, New Jersey, USA) at a final concentration of 1 mg/ml for 25 min.

The ventricles were removed, minced and placed in O_2 saturated CFS containing 1 mg/ml collagenase (at 37 °C in a water bath for 2 min). After sedimentation, the resulting supernatant was discarded and the pellet was mixed and resuspended in 20–25 ml of O_2 saturated CFS and incubated as above. The supernatant was discarded again and the pellet was mixed and resuspended in 20–25 ml of O_2 saturated low-Ca^{2+} solution (LCS) containing 50% of CFS and 50% of high-Ca^{2+} solution (HCS) and incubated as above. HCS is composed of CFS supplemented with 0.09% of DNAse and 200 μM of Ca^{2+}. Furthermore, the supernatant was discarded, the pellet was resuspended in 20–25 ml of O_2 saturated HCS and incubated as above. Then rat ventricular myocytes were released from the soft tissue by gentle trituration. The cell suspension was plated into »peel-off« culture flasks (Techno Plastic Products AG, Switzerland), the internal bottom surface of which were coated with poly-L-lysine (500 μg/ml; Sigma, USA) or with a mixture of extracellular matrix proteins (ECM, 1.11 mg/ml; Harbor Bio-Products, Norwood, MA, USA). The myocytes were allowed to settle down for approximately 1 h in medium M199 with Earle's modified salts, glutamine (Biowest; Nuaille, France), 100 μg/ml penicillin/streptomycin and 50 μg/ml kanamycin (PAA Laboratories, Austria). In addition to the control condition (pure medium), the medium was supplemented with either 5% fetal calf serum (FCS supplemented medium) or 870 nM insulin, 65 nM transferrin and 29 nM Na–selenite (Sigma, USA) (ITS supplemented medium). Myocytes were cultured in an incubator at 37 °C with a 5% CO_2 atmosphere. After plating the medium was changed at 1 h, day *in vitro* (DIV) 1, 3 and 6 with warm fresh medium, in order to remove the loosely attached cells.

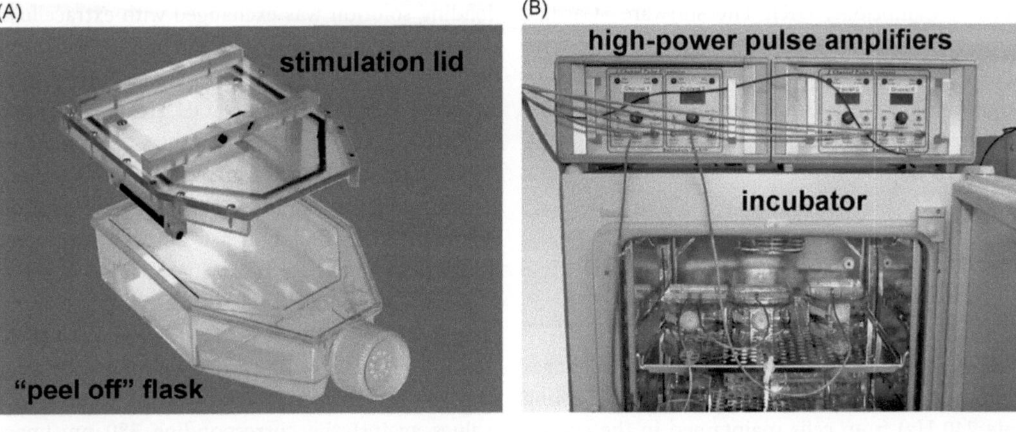

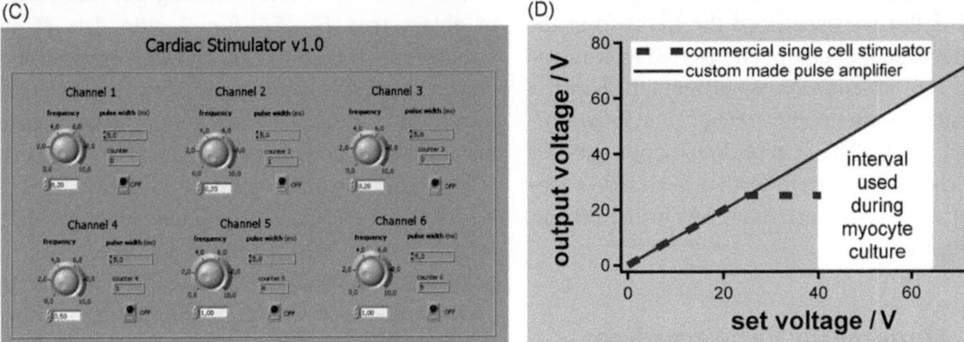

⬛ Fig. 15.1 Electrical stimulation and prolonged culture of adult cardiomyocytes. Panels (A–C) illustrate components of the optimised culturing system. (A) Shows the custom-made stimulation lid (top) and »peel-off« flask (bottom), (B) two high-power amplifiers on top of a standard cell incubator connected to several »peel-off« flasks for electrical stimulation of myocyte populations. Panel (C) depicts the graphical user interface of the LabVIEW based software »Cardiac stimulator« controlling the electrical pacing of the myocytes in the flasks. Panel (D) displays the relationship of set voltage (x-axis) and output voltage (y-axis) of a commercial single cell stimulator (dashed line) and the custom made pulse amplifier (solid line) connected to an individual culture flask filled with medium as shown in panels (A) and (B). The typically required voltages used during myocyte culture are highlighted.

For the experiments involving the viral gene transfer the cells were plated on ECM-coated cover slips, placed in 12-well plates and kept in M199 medium supplemented with ITS. Adenovirus-mediated gene transfer was initiated 1 h after cell plating to allow a fast protein expression. The myocytes were transfected with a multiplicity of infection (MOI) of 5–20 plaque-forming units/cells. The regime for exchanging the culture medium was as described above.

The continuous electrical stimulation was performed at 37 °C. All other experiments were carried out at room temperature (20–22 °C).

15.2.2 Electrical stimulation

For electrical stimulation of entire cell populations we designed and built Plexiglas lids, resistant to heat sterilisation, as shown in ⬛ Fig. 15.1A with the following features: (i) two parallel carbon electrodes for electrical field stimulation; (ii) standardised connectors for external electrical pulses; and (iii) silicone sealing for taking the culture flask out of the incubator while maintaining sterile internal conditions. The set-up for electrical stimulation comprised a custom-made high-current pulse amplifier (cp. ⬛ Fig. 15.1B; Babraham

Tech[nix], Cambridge, UK). The software »Cardiac Stimulator« was running under LabView software allowing continuous pacing of culture flasks at an adjustable frequency, cp. ◘ Fig. 15.1C. We used 0.2 Hz throughout all culture conditions involving pacing of cardiac myocytes.

15.2.3 Measurements of cell length change

Electrical stimulation induced cell-length changes were recorded with a fast video camera (sampling rate 240 Hz) from cells maintained in the culture flask. For this we transferred the flasks from the incubator onto the stage of an inverted microscope (Eclipse TS100, NIKON, Japan) equipped with a cell-length measurement system (IonOptix Corporation, USA). The system directly put out cell-length changes that were further analysed in Igor Pro software (Wavemetrics, USA) with custom-made macros.

15.2.4 Fluorimetric Ca²⁺ recordings

Global Ca^{2+} transient were measured with either fura-2 or the Ca^{2+} sensitive fluorescent protein inverse pericam[10]. To perform such recordings, cardiac myocytes were seeded on coated glass cover slips that were placed into culture flasks (for fura-2) or into wells of a 12-well plate (for inverse pericam) before seeding. For fluorescence recordings the cover slips were mounted in a custom made chamber on the stage of an inverted microscope (TE2000U, NIKON; Japan) attached to video-imaging hardware. Imaging was carried out through a 20× oil-immersion objective (Planfluor 0.75 NA, NIKON, Japan). The system comprised a video camera (for fura-2: Imago, TILL Photonics, Germany; for inverse pericam: iXon DV887, Andor Inc., Ireland) and a monochromator for excitation (Polychrome IV, TILL Photonics, Germany).

For the fura-2 recordings cover slips were loaded with dye (fura-2-AM, 0.4–0.75 µM, from a stock of 1 mM in DMSO/20% pluronic) for 30 min at room temperature. Prior to recording, the loading solution was exchanged with extracellular solution (ES) composed of (in mM): NaCl 135, KCl 5.4, $MgCl_2$ 1, glucose 10, $CaCl_2$ 2, HEPES 10 adjusted to pH 7.35 with NaOH. Imaging was performed by exciting the cells at the Ca^{2+}-dependent wavelength (380 nm) and recording the fluorescence signal (>440 nm; image exposure duration: 15 ms). The excitation at 380 nm was interrupted every 50th image by recording a single image at the Ca^{2+}-independent, isosbestic excitation wavelength of 355 nm (◘ Fig. 15.4A). For calculating ratiometric data we linearly interpolated between the 355 nm-images and ratioed the fluorescence values against the corresponding 380 nm-image to obtain true F_{355}/F_{380}-fura-2 ratio data at an acquisition frequency of 66 Hz. This ratioing and further semi-automatic peak detection was performed in Igor Pro software running custom made macros after averaging the fluorescence of regions of interest in the imaging software.

Inverse pericam is a chimeric protein comprising a circularly permuted green fluorescent protein and calmodulin[10]. Imaging of the inverse pericam fluorescence was performed by exciting the fluorophore at 490 nm and recording the fluorescence through a 510 nm long-pass filter (image exposure duration 15–20 ms, resulting in an imaging frequency of 50–66 Hz). Single fluorescence images were obtained by exporting entire movies as multi-page TIFF files and processing them in ImageJ (W. Rasband, NIH, USA). For self ratio traces we calculated the $F_0/\Delta F$ ratio since the emitted fluorescence of the inverse pericam decreased with increasing Ca^{2+} concentrations, thus the term »inverse«.

15.2.5 Adenovirus construction

Generation of recombinant adenoviruses was accomplished using the Transpose-Ad™ Adenoviral Vector System (MP Biomedicals, USA) according to the manufacturer's instructions. A pCR259 adenovirus transfer vector encoding for the calcium-sensitive fluorescence protein inverse pericam was transformed in HighQ-1 Transpose-Ad™ 294 competent cells, a bacterial cell line carrying the Transpose-Ad™ 294 plasmid and a plasmid encoding a trans-acting Tn7 transposase. After a Tn7-based transposition, recombinant adenoviral genome was purified from bacteria and transfected into the QBI-HEK 293 cell line using Lipofectamine 2000 (Invitrogen, Germany). In this cell line, the recombinant adenoviruses were generated and propagated.

The pcDNA3-inverse pericam vector was kindly provided by Dr. Atsushi Miyawaki (Institute of Physical and Chemical Research (RIKEN), Wako, Saitama, Japan).

15.2.6 Data analysis

Results were analysed using a Mann–Whitney rank sum test (SigmaStat software, USA). Effects were regarded as significant when $p < 0.05$ (marked with an asterisk). The results are expressed as mean values ± S.E.M.

15.3 Results

15.3.1 Electrical field stimulation of cardiac myocytes

In our peel off flask/lid system (■ Fig. 15.1A and Section 15.2.2) a field voltage of 40–65 V at pulse durations of 5 ms (rectangular pulses) was necessary to trigger a visible contraction in at least 75% of the isolated myocytes. In order to generate these pulses, commercially available pulse generators such as the MyoPacer (IonOptix Corp., USA) were not sufficient, because (i) their voltage output is limited to 40 V and (ii) the electric current necessary for the peel off flask/lid system was higher than the limit of the total output power of the MyoPacer (compare ■ Fig. 15.1D). This restricted the highest achievable output voltage to 25 V (measured with two independent MyoPacers). It has to be mentioned here that such amplifiers had been designed solely for single cell experiments and our findings might simply indicate design specific limitation. We thus obtained custom made high power and fast switching pulse amplifiers (cp. ■ Fig. 15.1B and Section 15.2.2), which delivered enough power to simultaneously drive four of our peel of flasks (per output channel) at a maximal voltage of more than 80 V (voltage stability confirmed; voltage change during the pulse <5%). For this the hardware generated an electrical current of approximately 2.1 A (calculation based on a specific electrical resistance of culture medium of approximately 125 Ω cm, an electrode distance of 8 cm and a cylindrical electrode geometry of a length of 7.3 cm and a diameter of 6 mm).

The comparison between a single cell stimulator and our custom made pulse amplifier is depicted in ■ Fig. 15.1D. For this we connected a single culture flask filled with medium to the amplifier and measured the actual output voltage for a range of set voltages.

During the myocyte culture the voltage was re-adjusted on a daily basis. With this we ensured to always drive at least 75% of the muscle cells which was inspected visually through a microscope. We observed that the voltage necessary for that increased from about 40 V at DIV0 to approximately 65 V at DIV6. In our studies we applied the pulses (square-shaped; 5 ms in duration) at a constant frequency of 0.2 Hz for the entire culture period.

15.3.2 Long-term culture of cardiomyocytes: morphology and survival rates

Isolation of adult rat ventricular myocytes yielded more than 80% living cells of which more than 70% displayed a rod-shaped morphology (data not shown). Initially, we evaluated various culture conditions based on the light microscopic morphology of the myocytes during a week of culture. ■ Fig. 15.2A summarises our findings for five different culture conditions. Each row of panels depicts the culture conditions while the columns represent successive DIVs.

The first two rows show typical results when culturing adult rat ventricular myocytes in a supplement-free medium (■ Fig. 15.2A). Over the time course of 7 days, less than 50% of the cells were able to largely retain their elongated phenotype. We found that after a few days the myocytes developed numerous small vesicles or vacuoles as indi-

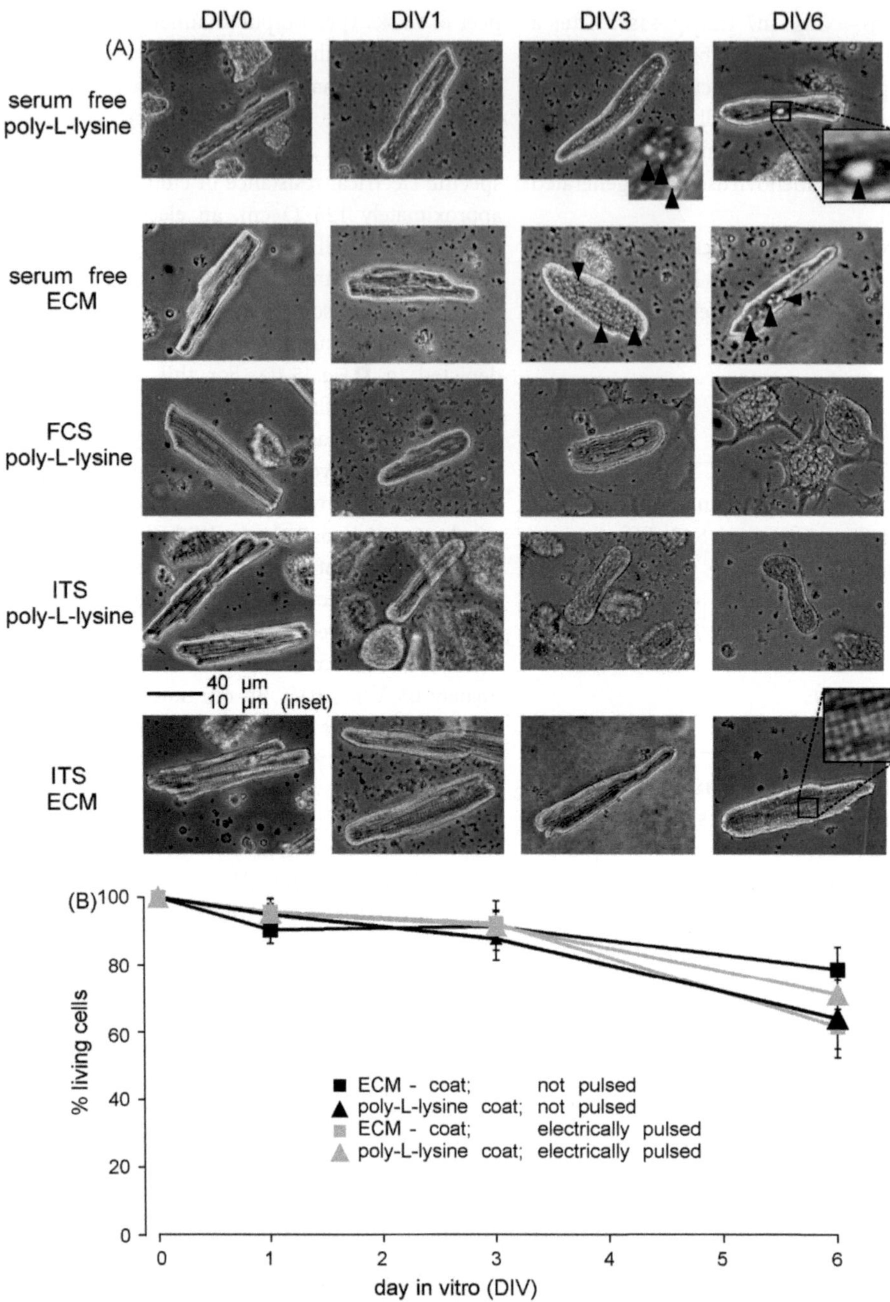

◘ **Fig. 15.2** Morphological properties and survival rates of cardiomyocytes in long-term culture. Panel (A) depicts transmission images of typical adult rat ventricular myocytes under various culturing conditions. The rows represent different culture conditions for the time points shown (columns). For selected combinations part of the cell body has been redrawn in a magnified inset (the same reference length bar indicates 40 µm for the images and 10 µm for the magnified insets). The black arrowheads in the panels of the first two rows highlight the occurrence of numerous vesicles and vacuoles. In panel (B), we have plotted the normalised yield of living cells (as percentage of the total number of cells, i.e. living and dead) versus the time of culturing. The data were taken from 4 to 8 different rat heart preparations for the two most promising culture conditions. Rectangular symbols indicate ITS supplemented culture conditions while triangles refer to media without serum.

cated by the two insets in the first row of images, the development of which were independent of the substrate coating (◘ Fig. 15.2A 1st and 2nd rows).

When cultured in medium supplemented with 5% FCS, the myocytes rapidly »de-differentiated« in their morphology (◘ Fig. 15.2A). This process was so fast that from DIV3 onwards, reliable contraction measurements based on edge detection (c.p. 2.3) were difficult due to massive changes in the cell geometry, i.e. the majority of cells were rounded up (>65%). Furthermore, at DIV6 most cells started to develop lamellipodia-like structures and adopted a flattened »fried egg« shape (◘ Fig. 15.2A, third row, rightmost image). In comparison to serum conditions, for myocytes cultured under serum-free and ITS-supplemented conditions (◘ Fig. 15.2A, two lower rows) such morphological »de-differentiation« was significantly reduced regardless of the substrate coating (poly-L-lysine: ◘ Fig. 15.2A 4th row or ECM-coated substrates: ◘ Fig. 15.2A bottom row). Even after 6 days in culture >32% elongated myocytes were present with the poly-L-lysine coating, without any lamellipodia-like structures. The rate of elongated cells on the flask surfaces coated with ECM was even exceeding those rates (>42%).

From these results we concluded that the two most favourable culture conditions so far were either without any medium supplements on poly-L-lysine coating or with ITS-supplement on ECM-coated substrates. We thus investigated those two conditions further to identify the superior one with respect to cell survival and conservation of the morphology.

◘ Fig. 15.2B compares the survival rates of the myocytes under the two most promising culture conditions, i.e. ITS/ECM and no serum/poly-L-lysine for pulsed and non-pulsed cells. From these data it became apparent that the overall survival rates of ITS/ECM cultured cells were not significantly different compared to the non supplemented culture conditions, a finding observed for pulsed and non-pulsed conditions. This obviously indicated that electrical pacing did not exert a detrimental effect on cell survival.

Interestingly a higher total number of cells was found on the ECM-coated surfaces after the isolation, plating and initial washing steps (data

not shown) in comparison to the poly-L-lysine substrate coating. This might indicate a stronger interaction between the cells and the coating when seeded onto ECM-coatings. Moreover, we found that reliable cell length measurements with poly-L-lysine were difficult during the first 4 h after seeding, because a large proportion of the plated cells displayed highly increased spontaneous activity that ceased over the time course of a few hours after plating (data not shown).

When we visually inspected the myocytes under both culturing conditions at DIV6 we found that in comparison to the ITS/ECM condition the cells in serum free medium displayed (i) numerous vesicles and/or vacuoles (see DIV6, first row in ◘ Fig. 15.22A) and (ii) a loss of apparent cross-striation (compare DIV6 first and last row in ◘ Fig. 15.2A).

15.3.3 Long-term culture of cardiomyocytes: analysis of cross-striation

In order to quantify the presence of cross-striation as an indication for the conservation of highly organised contractile filaments and structures we calculated spatial power spectra from elongated adult rat cardiac myocytes under serum free and ITS/ECM conditions (◘ Fig. 15.3). For this, we generated intensity profiles (◘ Fig. 15.3Aa and b, black lines) along the longitudinal axis of the myocytes at DIV0 and 6 and constructed power spectra (◘ Fig. 15.3B). For cells at DIV0 we consistently found a peak at the spatial frequency of 0.56 ± 0.015 μm^{-1} (n = 6 cells, translating to a regular structure with a repetition every 1.78 μm) regardless of the particular culture condition. This value for the spatial frequency was very close to the one expected for sarcomeric structures (i.e. 1.8 μm sarcomeric length[13]). From this we concluded that the regular banding visually identified in the myocytes at DIV0 was indeed caused by the typical cross-striation generated by the regular arrangement of the contractile filaments and t-tubules. A similar analysis was performed with cells at DIV6 either in ITS/ECM or in serum-free/poly-L-lysine conditions. We found that the cells in the former con-

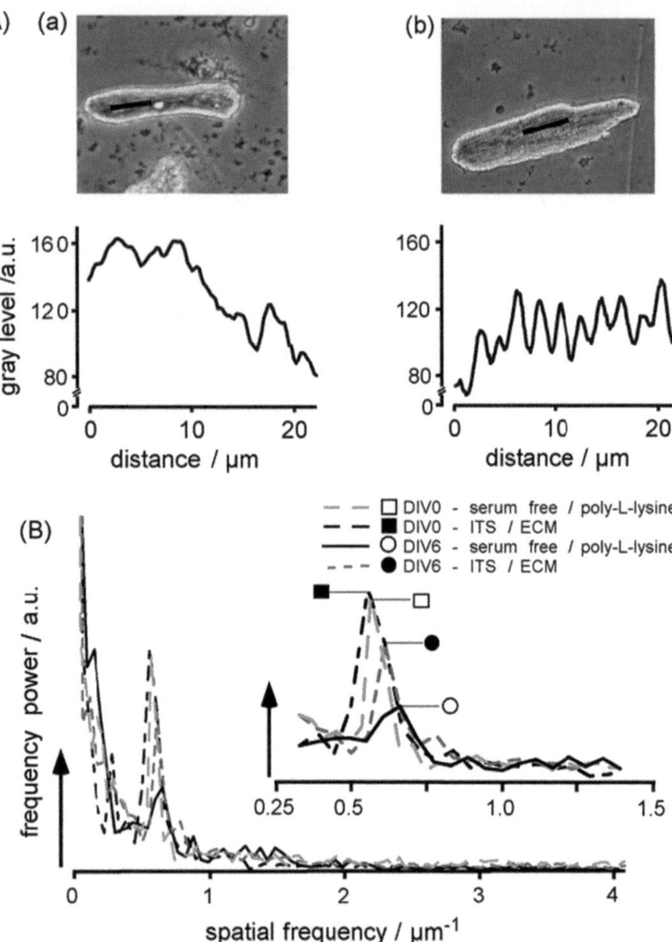

◧ Fig. 15.3 Cross-striation of cardiomyocytes in long-term culture. Panels (A–B) compare rat ventricular myocytes at DIV0 and DIV6 under various culturing conditions. In order to quantify cross striation we recorded intensity profiles (panel A) of myocyte images taken with a regular phase-contrast transmission microscope along their longitudinal axis. The left image depicts a myocyte at DIV6 cultured in serum free medium on poly-L-lysine (Aa) whilst the right image represents a myocyte at DIV6 cultured in ITS supplemented medium on ECM (Ab). For (B) we performed a powerspectral analysis of such line profiles taken from myocytes at DIV 0 and DIV 6. The inset illustrates a magnified view onto the power peak around spatial frequencies of 0.56 μm⁻¹. Symbols highlight the particular peak. Details for the construction of the power-spectra can be found in the Section 15.2. These results were typical for all cells analysed (n = 9 at DIV0 and n = 6 for each DIV6 condition; cells were taken from three rat hearts).

dition displayed a frequency peak that appeared slightly shifted towards higher frequencies (0.61 ± 0.027 μm⁻¹ for DIV6 versus 0.56 ± 0.018 μm⁻¹ for DIV0, n = 6 for each DIV, translating into 1.64 μm for DIV6 versus 1.78 μm for DIV0). The amplitude in that peak was also reduced to 68.9% ± 10% (n = 6). Even for the cells cultured in serum free/poly-L-lysine conditions at DIV6 we could identify a spectral frequency peak in the very same region, although as described above cross-striation was often absent when analysed by visual inspection only. Nevertheless, these peaks were shifted towards higher spatial frequencies even further (0.66 ± 0.03 μm⁻¹, n = 5 cells, translating into 1.5 μm). In addition the amplitude in that peaks was significantly reduced to 24.5 ± 15% (n = 5) when

compared to the DIV0 condition (◧ Fig. 15.3B, inset; compare open circle with other symbols).

After this initial analysis of the culture conditions we set out to comprehensively investigate the physiology of the cultured cells. For this we analysed the frequency dependence of their contractility and Ca²⁺ transients during electrical pacing from DIV0 to DIV6 under various culture conditions.

15.3.4 Shortening-frequency relationship

◧ Fig. 15.4A exemplifies the stimulation protocol used for the cell length and for the Ca²⁺ measurements described below. ◧ Fig. 15.4B depicts the

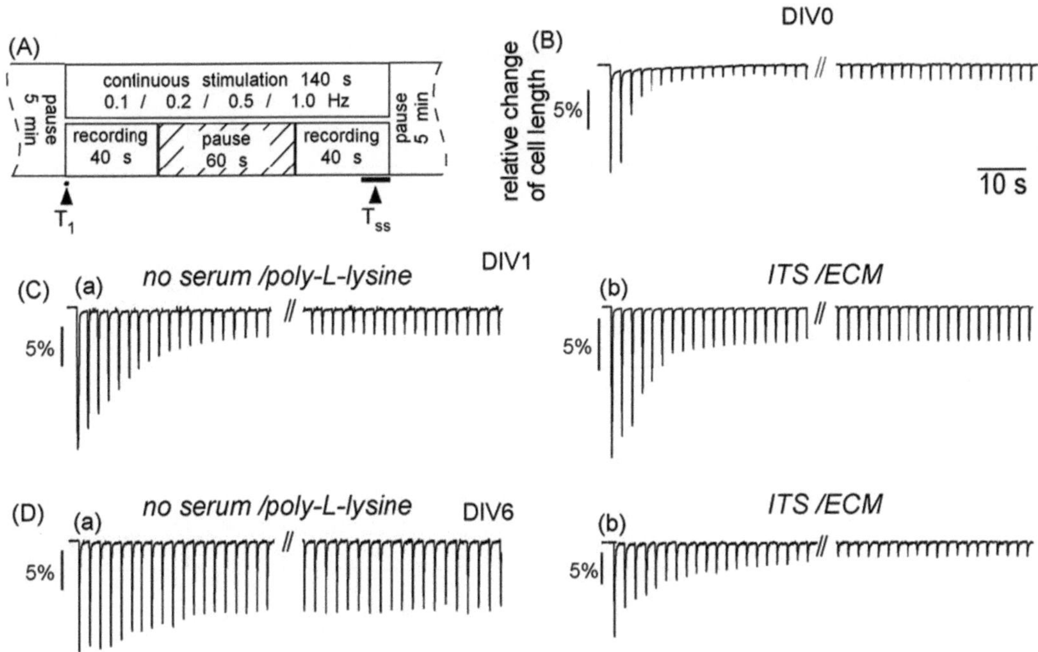

◘ Fig. 15.4 Post-rest behaviour of contraction in cultured adult rat ventricular myocytes. The basic stimulation protocol for characterising the post-rest behaviour of the cultured myocytes is depicted in panel (A). Panel (B) illustrates the typical time course of cell length changes during such trains of stimulations at DIV0. In panels (C) and (D) cell length changes were plotted for cells at DIV1 (C) and DIV6 (D) that were cultured in serum free/poly-L-lysine (a) or ITS/ECM (b) conditions.

time course of cell length changes (0.5 Hz, pulse duration 5 ms) at DIV0. While the first contraction was strong, a typical progressive decay in the contraction amplitude could be observed, a phenomenon termed post-rest potentiation[13]. In ◘ Fig. 15.4C and D traces are exemplified for two different culture conditions (left: no medium supplement on poly-L-lysine; right: ITS supplemented medium on ECM coating) at DIV1 (◘ Fig. 15.4C) and DIV6 (◘ Fig. 15.4D). While at DIV1 both cells displayed post-rest potentiation, the myocyte cultured without supplement showed a greatly diminished potentiation at DIV6 while the cell with ITS/ECM still revealed post-rest potentiation.

It is noteworthy that the absolute maximal cell length changes decreased over time from DIV3 onwards in all conditions tested (data not shown). Most likely this reduction of the absolute amplitudes of cell shortening is attributed to an increase of the interaction between cells and substrates (see details in Section 15.4). Because of this we analysed changes of the post-rest behaviour of contraction (normalised to the pre-stimulation contractions) rather than absolute twitch amplitudes.

In order to quantify the degree of post-rest potentiation we calculated the relative change of contractility by ratioing the twitch amplitude under steady-state conditions (mean value of the last five peaks; T_{ss}) by the initial contraction amplitude (T_1) as depicted in ◘ Fig. 15.4A. ◘ Fig. 15.5 summarises the frequency dependence of that ratio and its relation to the culture conditions. We found that under serum free/poly-L-lysine, FCS/poly-L-lysine and ITS/poly-L-lysine conditions, the negative frequency dependence of the T_{ss}/T_1 ratio was lost between DIV3 and DIV6. In contrast, myocytes cultured in ITS-supplemented medium on ECM coated substrates largely retained the negative frequency dependence (◘ Fig. 15.5Ad, Bd, Cd, Dd for DIV6 data). We observed a particular dramatic

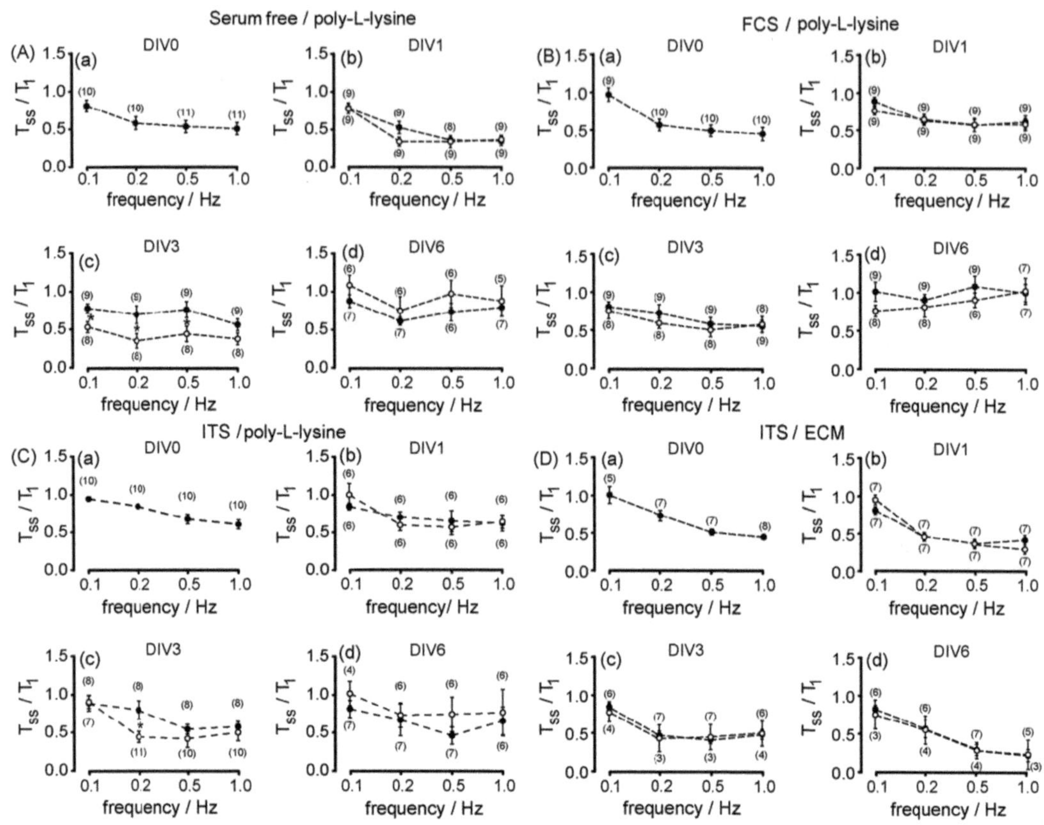

■ **Fig. 15.5** Frequency dependence of the post-rest behaviour of contraction in cultured adult rat ventricular myocytes. For panels (A–D) populations of cells (taken from between 2 and 6 rat hearts) have been analysed and their post-rest behaviour has been quantified as T_{ss}/T_1 ratios (see Fig. 15.2A and text for details). This ratio has been plotted against the stimulation frequencies (0.1–1.0 Hz; random order, 5 ms duration). Open and closed symbols represent data from myocytes that had been continuous pulsed (0.2 Hz, 5 ms duration) or not-pulsed, respectively. The number adjacent to each data point gives the number of myocytes observed. Pairs of data points with a significant difference have been marked with an asterisk.

change for cells cultured in FCS-supplemented medium. The negative frequency dependence turned into a positive relationship termed post-rest decay (■ Fig. 15.5Bd, closed symbols).

When we compared data from cells not stimulated during the culture period with those derived from pulsed cell populations (■ Fig. 15.5 open symbols) we found no major changes in the contractile behaviour of the myocytes regardless of their other culture conditions. Nevertheless, we observed a significant change at DIV3 with a stimulation frequency of 0.2 Hz for cells cultured in ITS supplemented medium with a poly-L-lysine coating (■ Fig. 15.5Cc, marked with an asterisk). The T_{ss}/T_1 ratio displayed a 45% decrease in pulsed myocytes ($n = 11$; non-pulsed cells, $n = 8$). Furthermore, significant differences between paced and non-paced cells were apparent for myocytes cultured under serum free/poly-L-lysine conditions for DIV3 (■ Fig. 15.5Ac). Nevertheless, the rather flat frequency dependence was still preserved during pacing.

From these data we concluded that the contractile behaviour at DIV0 was best preserved in a culture medium supplemented with ITS when cells were grown on ECM coated substrates regardless of whether they were continuously paced or not. In the following we conducted

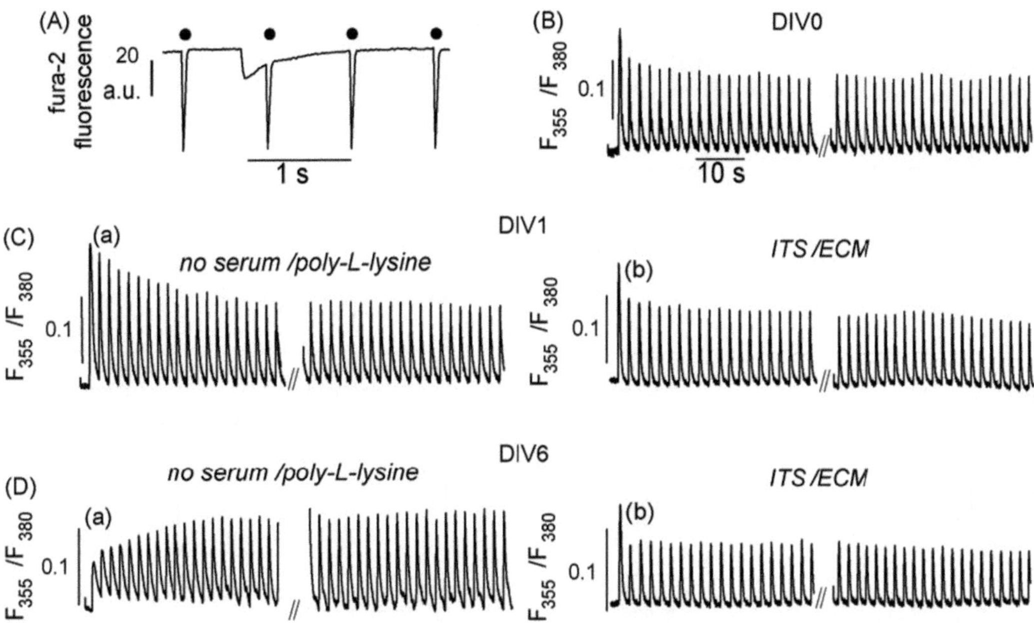

◘ Fig. 15.6 Post-rest behaviour of Ca^{2+} transients in cultured adult rat ventricular myocytes. Panel (A) illustrates the mode of fura-2 recording (downward deflections marked by a filled circle correspond to the 355 nm excitation images; for further details see text). The typical time course of fura-2 ratio transients during such trains of stimulations at DIV0 is depicted in panel (B). For the stimulation regime see Fig. 15.2A. In panels (C) and (D) fura-2 ratio transients were plotted for cells at DIV1 (C) and DIV6 (D) that were cultured in serum free/poly-L-lysine (a) or ITS/ECM (b) conditions.

a similar series of experiments analysing global Ca^{2+} transients with the Ca^{2+} sensitive fluorescent probe fura-2.

15.3.5 Calcium–frequency relationship

◘ Fig. 15.6 and ◘ Fig. 15.7 summarise experiments performed under similar experimental conditions as for ◘ Fig. 15.4 and ◘ Fig. 15.5 using cells from the same preparations in order to be able to correlate the data with each other. ◘ Fig. 15.6A illustrates the method of fura-2 imaging that we used (for a detailed description see Section 15.2). Similar to the twitch data presented in ◘ Fig. 15.4A, the global Ca^{2+} transients also displayed post-rest potentiation when measured in freshly isolated rat ventricular myocytes (◘ Fig. 15.6B for DIV0). We compared time-dependent changes of the Ca^{2+} transient amplitude under conditions of serum free medium and poly-L-lysine coating with the

behaviour of cells in ITS supplemented medium and on ECM coating (◘ Fig. 15.6C and D, a and b, respectively). As a result we also found a loss of post-rest potentiation only in the former condition while under ITS/ECM conditions post-rest potentiation was largely conserved. At DIV6 in the absence of serum the initial post-rest potentiation even turned into a strong post-rest decay (◘ Fig. 15.6Da).

Since in almost 50% of all cells tested at DIV0 the Ca^{2+} transients fused together (relaxation was not complete between the transients leading to a gradual diastolic build-up of the Ca^{2+} concentration) when stimulation frequencies exceeded 0.5–0.7 Hz we omitted the 1 Hz data in the further analysis. All other conditions were similar to those described in ◘ Fig. 15.5. In contrast to the relationships of the twitch amplitude the height of the Ca^{2+} transients only displayed a modest post-rest potentiation with a basically flat frequency dependence at DIV0 (◘ Fig. 15.7Aa). This flat amplitude–

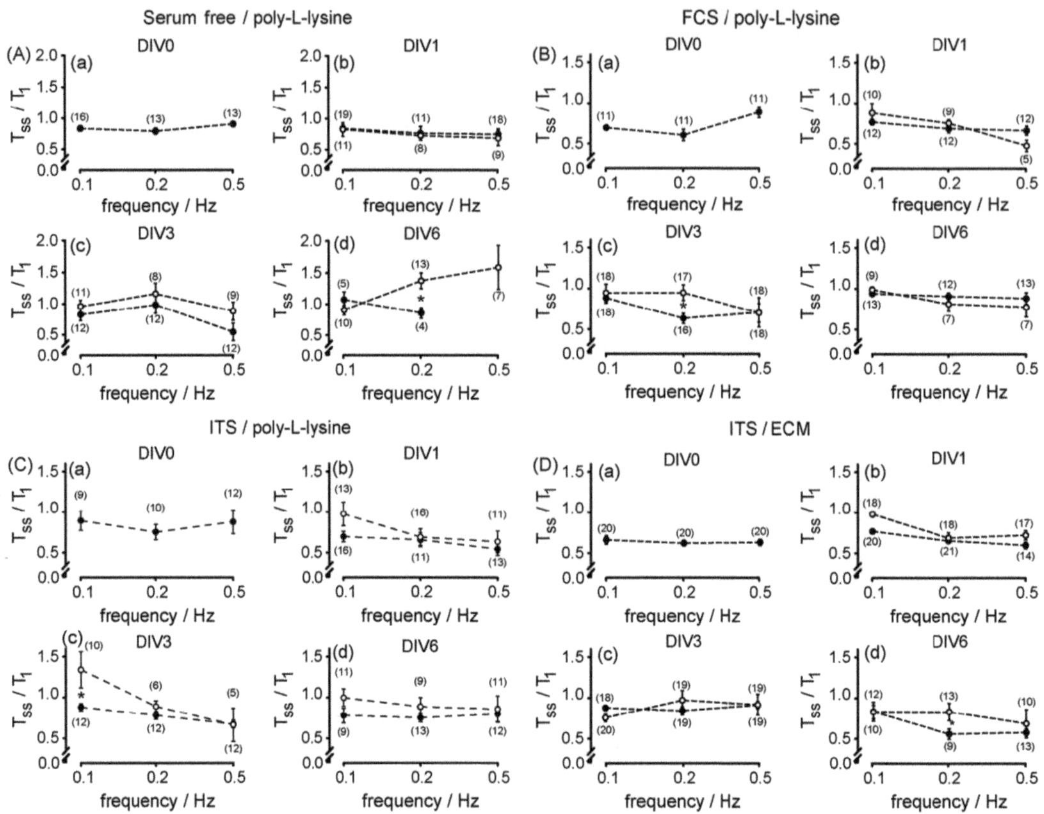

Fig. 15.7 Frequency dependence of the post-rest behaviour of Ca^{2+} transients in cultured adult rat ventricular myocytes. For panels (A–D) a population of cells (taken from 2 to 6 rat hearts) have been analysed and the amplitude ratios T_{ss}/T_1 have been plotted against the stimulation frequencies (0.1–0.5 Hz, random order, 5 ms duration) for various culture conditions as depicted in each panel. Open and closed symbols display mean values for the continuously pulsed and non-pulsed myocytes respectively (0.2 Hz, 5 ms pulse length). The numbers adjacent to each value give the number of myocytes analysed. Pairs of data points with a significant difference have been marked with an asterisk.

frequency relationship was basically preserved for all DIVs and for all culture conditions with the exception of a sole set of conditions. Here, the myocytes at DIV6, that were paced continuously in the absence of any medium supplement on poly-L-lysine coating, displayed a dramatic shift from modest post-rest potentiation at 0.1 Hz to a significant post-rest decay at 0.5 Hz (open symbols in Fig. 15.7Ad). Similarly to the results we obtained for the twitch measurements (Fig. 15.5) continuous pacing did not make any difference to the frequency relationships at any DIV nor under any culture condition apart from the ITS/poly-L-lysine combination at DIV3 (Fig. 15.7Cc).

Thus, myocytes cultured in ITS-supplemented medium and growing on ECM coated substrates most closely retained their morphology, contractility and Ca^{2+} handling when compared to their properties at DIV0.

We thus conducted the final series of experiments to investigate whether cardiomyocytes cultured under ITS/ECM conditions were a good system to perform long-term expression of exogenous proteins. From our fura-2 data we knew that under our culture conditions, Ca^{2+} handling was largely conserved through the 1-week period of culturing. We thus tested expression of a genetically coded Ca^{2+} indicator by adenoviral gene

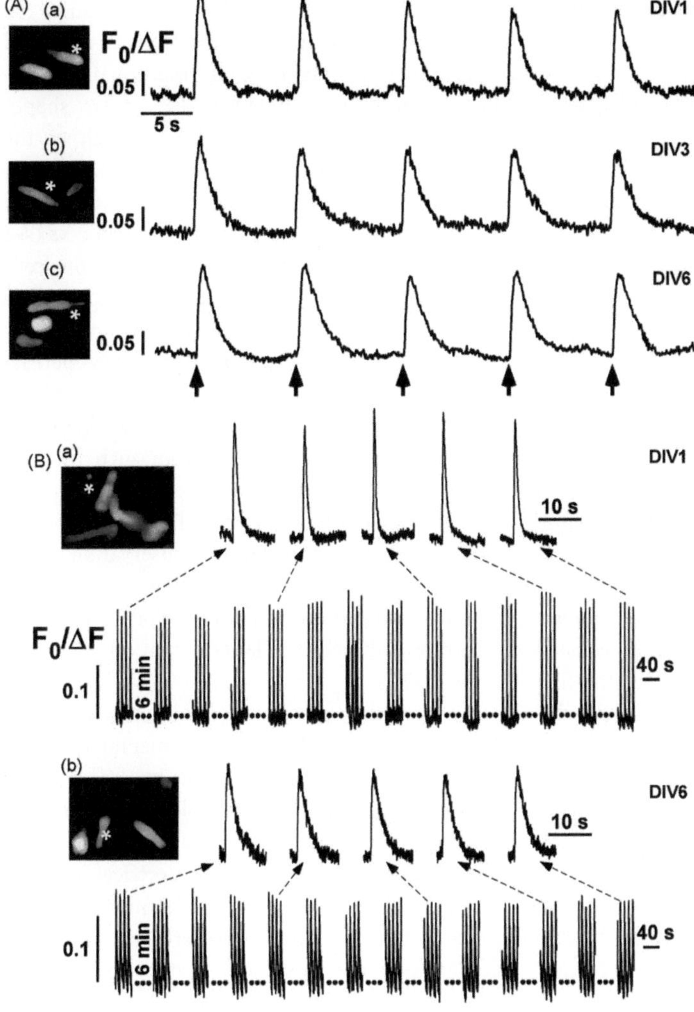

Fig. 15.8 Fluorescence transients of inverse pericam in cultured rat ventricular myocytes after adenoviral gene transfer. Panel (A) depicts typical self-ratio traces of the inverse pericam fluorescence recorded from cultured myocytes at the time points given. Panel (Ac) illustrates the stimulation regime used (black arrows indicate electrical impulses, 5 ms duration). Cell images are added for each example and the cell used for calculation of the ratio traces has been marked with a white asterisk. Panel (B) shows ratio traces of adult ventricular myocytes 1 day (Ba) and 6 days (Bb) after virus infection. Dashed arrows depict recording periods for which exemplified individual ratio transients have been replotted. In the fluorescence images an asterisk marks the myocyte from which the ratio was calculated. Here, we calculated $F_0/\Delta F$ (F_0—fluorescence at the beginning of the recording at rest) to obtain positive changes of the ratio, since the inverse pericam displays a decrease of the fluorescence with increasing Ca^{2+} concentrations. Traces shown here were typical for all cells analysed under these experimental conditions (n > 60 from three rat hearts).

transfer (inverse pericam as originally described by Nagai *et al.*[10]).

15.3.6 Inverse pericam expression and Ca^{2+} measurements

The freshly isolated ventricular myocytes were infected with the virus 1 h after plating and first fluorescence could already be recorded within 24 h. Typically we started experiments 18 h after transfection. Already at that early stage, more than 95% of all viable myocytes displayed sufficient levels of fluorescence for recording Ca^{2+}-de-

pendent fluorescence changes (data not shown). **Fig. 15.8A** exemplifies such changes as recorded in response to electrical stimulations. In all cells at DIV1 we could record a maximal relative fluorescence change of 25 ± 3.1% (*n* = 65). Since excitation of inverse pericams does not require the application of UV light we tested the possibility to perform long-term recordings as depicted in **Fig. 15.8B**. We designed an electrical stimulation regime employing constant pulsing at 0.1 Hz and optical recording for 40 s periods separated by 6 min without light excitation. For the recordings in **Fig. 15.8B** we continued this stimulation regime for a total of 90 min without a detectable loss in pericam self-ratio amplitude or signal quality. During the same time period the absolute inverse pericam fluorescence was decreased by less then 15%, most probably due to bleaching. In some experiments we recorded for more than 2 h with a similar experimental regime without a noticeable decrease in signal quality (data not shown).

15.4   Discussion

The aim of this study was to develop and explore a culture system for adult rat cardiac myocytes and procedures that allow extended experimental manipulation of these cells under conditions of diminished dedifferentiation. For that we setup an experimental system for a long-term culture of adult cardiac myocytes that largely suppressed dedifferentiation, best maintained the morphological and physiological properties of freshly isolated cells and allowed (to our knowledge) for the first time long-term expression of a genetically encoded Ca^{2+} indicator.

One of the biggest problems with long-term cultures of adult cardiac myocytes is the rapid development of morphological and physiological dedifferentiation. The structural changes occurring during extended culturing times of rat ventricular myocytes were studied extensively (e.g. [14,15]). In parallel there were numerous attempts to modulate culture conditions towards minimising dedifferentiation. Such approaches included omitting or substituting the medium supplement FCS (e.g., [11]) and electrical pacing of the adult cells (e.g., [9]). All of those approaches have provided progress towards conditions allowing extended culture periods and reduced dedifferentiation of the adult cells.

We omitted serum from our medium and substituted that by an ITS mixture. In addition, we coated the substrates (plastic and glass) with ECM. Both steps revealed major improvements in both, the number of viable cells (i.e. the loss of viable cells was greatly reduced between DIV0 and DIV6) and the preservation of cell morphology, subcellular microarchitecture (contractile filaments) and physiology (contractility and Ca^{2+} handling) of the cells over the time course of culturing.

As described before, also in our hands the presence of serum in the culture medium appeared to promote dedifferentiation resulting in a flattened morphology. Interestingly, similar shapes are »normal« for cardiac cell lines such as the H9C2 cell line investigated earlier (e.g., [16]). We found that simply omitting the serum supplement is slowing down the dedifferentiation process (see also [11] for rabbit cardiomyocytes) and more cells survive the culture period with an elongated cell body. Unfortunately, adult rat cells cultured under these conditions displayed a progressively increasing number of subcellular vacuoles and/or vesicles, a gradual loss of cross-striation and alterations of their post-rest behaviour. From such findings we concluded that this culture condition was in fact not optimal. In contrast, ITS-supplemented medium greatly improved the conservation of the cellular properties found at DIV0 whether the cells were seeded on poly-L-lysine or ECM coated substrates, whereby the latter condition gave even better results.

In the vast majority of conditions, there was neither a beneficial nor a detrimental effect of electrical pacing unlike the report by Berger *et al.*[9]. This held true for both physiological parameters that we analysed, the contractile performance and Ca^{2+} transients in response to electrical stimulation as well as the stimulation frequency dependence of both parameters. Nevertheless, this does not exclude the possibility that certain culture conditions combined with particular stimulation protocols might increase or decrease survival of the cells or their morphological/physiological state. Interestingly other studies described hypertrophic responses of myocytes when pacing frequencies were raised to higher stimulation frequencies in adult myocyte cultures (3 Hz[17]). This is puzzling since these stimulation rates are rather physiological for rodent hearts (3–10 Hz) at 37 °C.

We made the observation that the absolute maximal cell length change was progressively decreasing over the culturing time. This reduced twitch amplitude can most likely be attributed to a progressively increasing mechanical interaction between the cells and the substrate. Indeed, during regular washing of our culture flasks relative loss of cells was reduced at later DIVs, a finding that might hint to a tighter substrate interaction of the myocytes (data not shown). Such an increased mechanical coupling will simply decrease absolute twitch amplitudes despite a constant contraction force. A mechanism that might be responsible for this was recently found and described the production of extracellular matrix proteins and modulation of the matrix by myocytes themselves[18]. Therefore, we investigated our cells using their post-rest behaviour at various stimulation frequencies: (i) post-rest contractile behaviour of isolated myocytes is attributed to their ability to adjust Ca^{2+} handling to changes (frequency, rest) of the excitation–contraction coupling[13], (ii) the rational behind this protocol was to work with relative shortening changes and to avoid individual differences and changes of cell-substrate interactions as described above, and (iii) to compare an entire range of frequency dependences rather than selected electrical stimulation regimes.

Genetic manipulation of adult cardiac myocytes is often limited to short term procedures and to generation of genetically modified donor animals. While the latter is a very elegant way of introducing proteins to or knocking proteins out from cardiac myocytes, the generation of genetically modified animals is expensive and tissue specific, inducible genetic manipulation is not an easy routine work. Thus, genetic manipulation of isolated cardiac myocytes might be a feasible intermediate step towards genetically modified animals. Traditionally, genetic manipulation of cardiac myocytes has been performed on neonatal ventricular cells from the rat (e.g., [19]). Neonatal cells can be transfected with traditional means (i.e., commercially available transfectants[20]) but in adult cells the yield of such an approach is usually <10%. Better expression or knock-out rates can be achieved with viral gene transfer systems, e.g., adenovirus. Unfortunately, results obtained from neonatal systems are often difficult to transfer to the adult situation, due to significant developmental differences in the expression pattern and signalling. Therefore, it is desirable to perform such experiments on adult cardiac myocytes. Indeed, there are some examples where genetic manipulation, e.g., with an adenoviral system has been applied successfully (e.g., [21]). Mostly, such approaches have been limited to short-term expression (1–2 DIV) and have thus avoided unwanted dedifferentiation of the cells. Nevertheless, longer expression would be highly desirable since many effects of genetic manipulations and/or stimulation regimes will require longer culture periods.

To our knowledge we described for the first time the expression of a genetically encoded Ca^{2+} sensor (inverse pericam) in adult rat ventricular myocytes for extended periods of time. In our hands, the myocytes cultured under ITS/ECM conditions were not negatively affected by the gene transfer through an adenovirus. Expression levels were already high enough for fluorescence recording less than 24 h after infection and throughout our culture period of 7 days expression levels were steadily increasing. This will certainly require additional investigations and titration of the best MOI for rapid onset of fluorescence and stable expression levels. In less than 24 h expression levels allowed extended recording regimes of Ca^{2+} transients without detrimental effects in the signal to noise ratio. In some experiments we were able to record for more than 2 h and the recording was not limited by bleaching or a decrease in the cell quality. Moreover, the frequency of data acquisition was not reduced in favour of a prolonged recording time; we collected images at a rate of 50–60 frames/s that allowed, e.g., the recoding of Ca^{2+} waves (data not shown). Thus, extended expression of genetically encoded Ca^{2+} sensors is a promising approach for long-term, continuous observation of Ca^{2+} handling in isolated cardiac myocytes, especially when working on species for which an appropriate transgene does not exist. A potentially interesting transgenic mouse expressing a Ca^{2+} indicator was introduced recently[22].

In the present report we have introduced a potentially important approach for culturing rat ventricular myocytes combined with expression of exogenous proteins for extended periods of time. Such an approach will allow further explorations of *in vitro* models for studying long-term signal transduction in cardiac myocytes enabling quasi continuous supervision of physiological and morphological parameters such as contractility and Ca^{2+} handling. We envisage our report to be a foundation for the further development of high-content screening systems[23] employing adult cardiac myocytes because the application of adenoviral gene transfer allows the application of genetically encoded reporters such as Ca^{2+} sensors but also sensors of other signal transduction processes (e.g., phosphorylation) in cardiac myocytes.

15.5   References

[1]   T. Powell and V.W. Twist, A rapid technique for the isolation and purification of adult cardiac muscle cells having respiratory control and a tolerance to calcium. Biochem. Biophys. Res. Commun., 72 (1976), pp. 327–333.

[2]   L.B. Bugaisky and R. Zak, Differentiation of adult rat cardiac myocytes in cell culture. Circ. Res., 64 (1989), pp. 493–500.

[3]   K. Kageyama, Y. Ihara, S. Goto, Y. Urata, G. Toda, K. Yano and T. Kondo, Overexpression of calreticulin modulates protein kinase B/Akt signaling to promote apoptosis during cardiac differentiation of cardiomyoblast H9c2 cells. J. Biol. Chem., 277 (2002), pp. 19255–19264.

[4]   B.J. Poindexter, J.R. Smith, L.M. Buja and R.J. Bick, Calcium signaling mechanisms in dedifferentiated cardiac myocytes: comparison with neonatal and adult cardiomyocytes. Cell Calcium, 30 (2001), pp. 373–382.

[5]   J.S. Mitcheson, J.C. Hancox and A.J. Levi, Cultured adult cardiac myocytes: future applications, culture methods, morphological and electrophysiological properties. Cardiovasc. Res., 39 (1998), pp. 280–300.

[6]   G.R. Sambrano, I. Fraser, H. Han, Y. Ni, T. O'Connell, Z. Yan and J.T. Stull, Navigating the signalling network in mouse cardiac myocytes. Nature, 420 (2002), pp. 712–714.

[7]   A. Volz, H.M. Piper, B. Siegmund and P. Schwartz, Longevity of adult ventricular rat heart muscle cells in serum-free primary culture. J. Mol. Cell Cardiol., 23 (1991), pp. 161–173.

[8]   Y.Y. Zhou, S.Q. Wang, W.Z. Zhu, A. Chruscinski, B.K. Kobilka, B. Ziman, S. Wang, E.G. Lakatta, H. Cheng and R.P. Xiao, Culture and adenoviral infection of adult mouse cardiac myocytes: methods for cellular genetic physiology. Am. J. Physiol. Heart Circ. Physiol., 279 (2000), pp. H429–H436.

[9]   H.J. Berger, S.K. Prasad, A.J. Davidoff, D. Pimental, O. Ellingsen, J.D. Marsh, T.W. Smith and R.A. Kelly, Continual electric field stimulation preserves contractile function of adult ventricular myocytes in primary culture. Am. J. Physiol., 266 (1994), pp. H341–H349.

[10] T. Nagai, A. Sawano, E.S. Park and A. Miyawaki, Circularly permuted green fluorescent proteins engineered to sense Ca^{2+}. Proc. Natl. Acad. Sci. U.S.A., 98 (2001), pp. 3197–3202.

[11] J.S. Mitcheson, J.C. Hancox and A.J. Levi, Action potentials, ion channel currents and transverse tubule density in adult rabbit ventricular myocytes maintained for 6 days in cell culture. Pflugers Arch., 431 (1996), pp. 814–827.

[12] R. Hilal-Dandan, J.R. Kanter and L.L. Brunton, Characterization of G-protein signaling in ventricular myocytes from the adult mouse heart: differences from the rat. J. Mol. Cell Cardiol., 32 (2000), pp. 1211–1221.

[13] D.M. Bers, Excitation-Contraction Coupling and Cardiac Contractile Force, Kluwer Academic Publishers, Dordrecht, Boston, London (2001).

[14] H.M. Eppenberger, M. Eppenberger-Eberhardt and C. Hertig, Cytoskeletal rearrangements in adult rat cardiomyocytes in culture. Ann. N. Y. Acad. Sci., 752 (1995), pp. 128–130.

[15] A.C. Nag, M.L. Lee and F.H. Sarkar, Remodelling of adult cardiac muscle cells in culture: dynamic process of disorganization and reorganization of myofibrils. J. Muscle Res. Cell Motil., 17 (1996), pp. 313–334.

[16] B.W. Kimes and B.L. Brandt, Properties of a clonal muscle cell line from rat heart. Exp. Cell Res., 98 (1976), pp. 367–381.

[17] D. Kaye, D. Pimental, S. Prasad, T. Maki, H.J. Berger, P.L. McNeil, T.W. Smith and R.A. Kelly, Role of transiently altered sarcolemmal membrane permeability and basic fibroblast growth factor release in the hypertrophic response of adult rat ventricular myocytes to increased mechanical activity in vitro. J. Clin. Invest., 97 (1996), pp. 281–291.

[18] V. Gupta and K.J. Grande-Allen, Effects of static and cyclic loading in regulating extracellular matrix synthesis by cardiovascular cells. Cardiovasc. Res., 72 (2006), pp. 375–383.

[19] S. Bauer, S.K. Maier, L. Neyses and A.H. Maass, Optimization of gene transfer into neonatal rat cardiomyocytes and unmasking of cytomegalovirus promoter silencing. DNA Cell Biol., 24 (2005), pp. 381–387.

[20] W.C. Heiser, Gene Delivery to Mammalian Cells, Humana Press, Totowa (2003).

[21] A. Rinne, C. Littwitz, M.C. Kienitz, A. Gmerek, L.I. Bosche, L. Pott and K. Bender, Gene silencing in adult rat cardiac myocytes in vitro by adenovirus-mediated RNA interference. J. Muscle Res. Cell Motil., 27 (2006), pp. 413–421.

[22] Y.N. Tallini, M. Ohkura, B.R. Choi, G. Ji, K. Imoto, R. Doran, J. Lee, P. Plan, J. Wilson, H.B. Xin, A. Sanbe, J. Gulick, J. Mathai, J. Robbins, G. Salama, J. Nakai and M.I. Kotlikoff, Imaging cellular signals in the heart in vivo: cardiac expression of the high-signal Ca^{2+} indicator GCaMP2. Proc. Natl. Acad. Sci. U.S.A., 103 (2006), pp. 4753–4758.

[23] P. Lipp and L. Kaestner, Image based high content screening—a view from basic science, J. Hüser, Editor, High-Throughput Screening in Drug Discovery, Wiley VCH, Weinheim (2006), pp. 129–149.

Calcium imaging of individual erythrocytes: Problems and approaches

Lars Kaestner, Wiebke Tabellion, Erwin Weiss, Ingolf Bernhardt, Peter Lipp

Reprint from Cell Calcium (2006) **39**, 13-19.

▪ Abstract

Although in erythrocytes calcium is thought to be important in homeostasis, measurements of this ion concentration are generally seen as rather problematic because of the autofluorescence or absorption properties of the intracellular milieu. Here, we describe experiments to assess the usability of popular calcium indicators such as Fura-2, Indo-1 and Fluo-4. In our experiments, Fluo-4 turned out to be the preferable indicator because (i) its excitation and emission properties were least influenced by haemoglobin and (ii) it was the only dye for which excitation light did not lead to significant autofluorescence of the erythrocytes. From these results, we conclude that the use of indicators such as Fura-2 together with red blood cells has to be revisited critically. We thus utilised Fluo-4 in erythrocytes to demonstrate a robust but heterogeneous calcium increase in these cells upon stimulation by prostaglandin E_2 and lysophosphatidic acid. For the latter stimulus, we recorded emission spectra of individual erythrocytes to confirm largely unaltered Fluo-4 emission. Our results emphasise that in erythrocytes measurements of intracellular calcium are reliably possible with Fluo-4 and that other indicators, especially those requiring UV-excitation, appear less favourable.

16.1   Introduction

Calcium homeostasis plays an important role in the physiology and pathophysiology of erythrocytes[1]. For this, the possibility to perform single cell calcium imaging on erythrocytes has been a desire for more than a decade[2]. However, up to now to our knowledge this had not been achieved so far.

Monitoring the intracellular free calcium concentration in a Fura-2 stained erythrocyte population using a fluorescence spectrometer has been a standard method for more than a decade[3,4]. Nevertheless, already in 1997 Blackwood and colleagues reported problems associated with Fura-2 measurements in human erythrocytes[5]. The problems pointed to an effect of haemoglobin on the spectral properties of Fura-2.

Two independent groups used Fura-2 measurements in an attempt to prove the prostaglandin E_2 (PGE_2)-activation of the non-selective cation channel in human erythrocytes[6] that was previously shown to be calcium permeable by electrophysiological methods[7]. In such patch-clamp measurements, the channel could be activated by PGE_2[8]. However, the Fura-2 based approaches failed to show a calcium entry in human erythrocytes upon PGE_2 stimulation[9,10]. In contrast, we could recently show the PGE_2-activated calcium increases by fluorescence imaging using the calcium indicator Fluo-4[11].

The aim of this paper was to solve the above described discrepancy by evaluating the popular calcium indicators Fura-2, Indo-1 and Fluo-4 for their suitability in erythrocyte calcium measurements. Here, we suggest that single erythrocyte fluorescence measurements are indeed the superior techniques over fluorescence spectrometer measurements as well as radioactive calcium flux measurements[12] especially when taking into account the heterogeneous erythrocyte population.

16.2   Material and methods

16.2.1 Blood, solutions and chemicals

Freshly drawn blood from healthy human donors was used for the experiments. The erythrocytes were washed three times by centrifugation (1500 × g, 8 min) at room temperature in physiological salt solution, pH 7.4. Plasma and buffy coat were removed by aspiration.

The physiological salt solution used throughout all experiments consisted of (in mM): 145 NaCl, 7.5 KCl, 1.5 $CaCl_2$, 10 glucose and 10 HEPES. All experiments were carried out at room temperature.

Haemoglobin was extracted from lysed erythrocytes. Inorganic salts were obtained from Merck (Germany), PGE_2, lysophospatidic acid (LPA), A23187 and poly-L-lysine from Sigma–Aldrich (USA) and the calcium indicators (Fura-2, Indo-1, Fluo-4) from Molecular Probes (USA).

16.2.2 Spectrometers

The absorption spectrum of haemoglobin was measured with a spectrophotometer (Lambda 2, Perkin-Elmer, USA) and the fluorescence and excitation spectra in solutions were recorded using a FluoroMax-2 (Jobin Yvon Inc., USA). The fluorescence spectra of individual erythrocytes were obtained with an USB-2000FLG (Ocean Optics, the Netherlands) attached to the microscope. Fluorescence measurements were restricted to individual erythrocytes with a mechanical mask in the excitation light path. Typical recording times for an individual emission spectrum were about 30 s. The three-dimensional plots of the spectra were constructed by Igor Pro software (Wave Metrics, USA).

16.2.3 Video imaging of individual erythrocytes

The set-up was build around an inverse microscope (TE2000U, Nikon, Japan). The imaging components consisted of a fibre coupled mono-chromator (Polychrome IV), a camera (Imago) and a control software (Tillvision 4, all components: T.I.L.L. Photonics GmbH, Germany). All measurements were done with a 100× S Flour objective. The excitation wavelength for the Fluo-4 measurements was about 480 nm. The luminescence was separated from excitation wavelength by a dicroic/barrier longpass filter set allowing transmission at wavelength above 500 nm. Erythrocytes were loaded with 5 µM Fluo-4 AM for 1 h at 37°C, washed and equilibrated for deesterification for 15 min. For measurements, the cells were placed on poly-L-lysine coated cover slips.

16.3   Results

16.3.1 Interaction between haemoglobin and the fluorometric calcium indicators

◘ Fig. 16.1 provides information about the absorbance spectrum of haemoglobin and highlights excitation maxima (dashed lines) and emission maxima (solid lines) of the calcium indicators Fura-2 (in grey), Indo-1 (thin black lines) and Fluo-4 (thick black lines). The close approximation of the excitation and emission maxima around the peak absorbance of haemoglobin suggested possible spectral interactions between the latter and the calcium indicators. In order to evaluate such an interaction fluorescence spectra of Indo-1 and Fluo-4 as well as the excitation spectrum of Fura-2 were acquired at varying concentrations of haemoglobin and free calcium.

For ◘ Fig. 16.2, we have summarised the relationships between haemoglobin concentration, calcium concentration and fluorescence for Fluo-4, Fura-2 and Indo-1. The left column depicts spectra for different calcium concentrations in the absence of any haemoglobin, while the middle and right columns illustrate the spectra with two increasing haemoglobin concentrations. While at 0.034 g/dl the emission spectrum of Fluo-4 and the excitation spectrum Fura-2 were significantly suppressed, no spectral inhomogeneous effect of the presence of haemoglobin was observed. In contrast to that, the emission spectrum of Indo-1 was altered in a

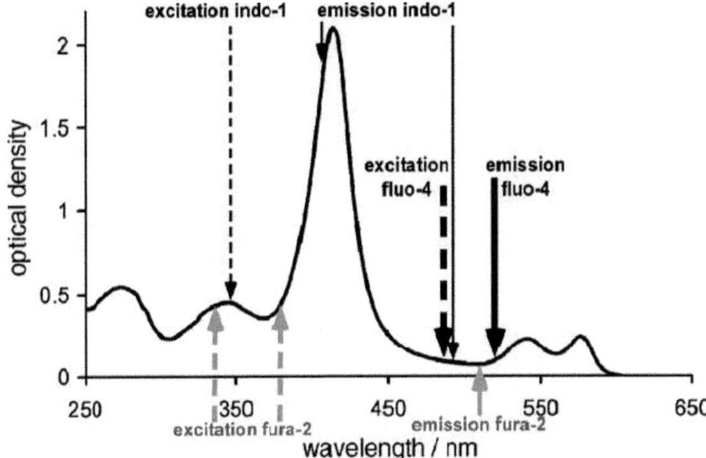

◨ **Fig. 16.1** Absorption spectrum of haemoglobin. The arrows point to the typical excitation wavelength and emission maxima of the calcium fluorophors Fura-2, Indo-1 and Fluo-4.

spectrally heterogeneous way. The Indo-1 emission spectra clearly depicted the effect of the soret-band of haemoglobin around 410 nm (arrow). All these effects were even more pronounced at the higher haemoglobin concentration (right column, 0.34 g/dl). Here, the Indo-1 emission spectrum was almost completely suppressed and the peak emission at around 390 nm was not apparent at all (inset in the Indo-1 row, right column). In addition to the worsening of the haemoglobin effect on Indo-1, at these protein concentrations, the Fura-2 excitation spectra appear largely distorted as can be seen in the inset. Although the Fluo-4 fluorescence was also further suppressed, they were not spectrally shifted or altered in any other manner. From these findings, we conclude that both, Fura-2 and Indo-1 display massively altered spectral properties in the presence of haemoglobin that appear too sever to allow for any meaningful fluorescence recordings of the calcium concentration or even changes, since the interactions between haemoglobin and the indicators appear wavelength and calcium dependent. In contrast, although only representing a single-excitation single-emission indicator, Fluo-4 seems to be best candidate for recordings of calcium changes in individual erythrocytes. However, an *in vivo* calibration for the calcium concentration might be rather difficult. For the functional assays, we thus utilised Fluo-4 despite the fact that in living erythrocytes, the haemoglobin concentration is another 100-fold higher then in our experiments.

16.3.2 UV-irradiation of erythrocytes induces transient strong auto-fluorescence

In addition to the spectral problems with Indo-1 and Fura-2 given above, we have found another major drawback of these indicators that is caused by their requirement of UV-excitation: ◨ Fig. 16.3 shows fluorescence images (A) of a group of native erythrocytes that were subjected to continuous UV-irradiation at an energy typically required for imaging experiments. In all experiments (n = 214 erythrocytes of three donors) we could record a massive increase in the erythrocytes' autofluorescence as depicted in ◨ Fig. 16.3B for the field of view detailed in ◨ Fig. 16.3A, the amplitude of which was variable between individual cells. The increase in autofluorescence for 360 nm irradiation peaked at about 14 min (13.9 ± 0.8 min STD; n = 8) and then declined to about one third of the maximum another 20 min later. Although the individual amplitudes of this effect were variable, the principle time-course was similar between individual cells. The 30 min kinetic in ◨ Fig. 16.3 covers the time range of most calcium sensing experiments in erythrocytes. However, even after 60 min irradiation the fluorescence decline is not back to nought (data not shown).

In an attempt to identify the origin of this autofluorescence transient we illuminated eryth-

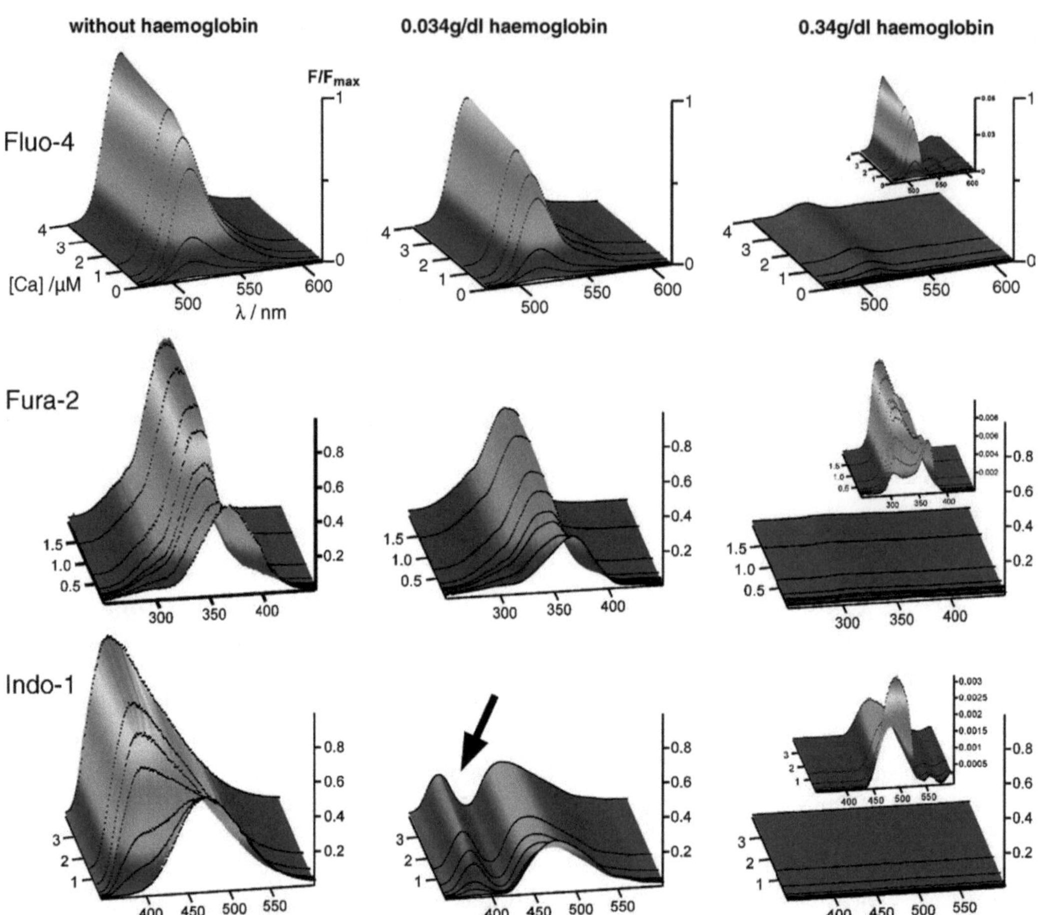

◘ Fig. 16.2 Fluorescence spectra of the calcium fluorophors Indo-1 and Fluo-4 as well as the excitation spectra of Fura-2. The normalised fluorescence intensity is plotted against the wavelength and the calcium concentration. For each dye (rows) the spectra were measured without haemoglobin, with 0.034 g/dl and with 0.34 g/dl haemoglobin (columns). The arrow points to the heavy spectral alteration of the Indo-1 spectrum in the presence of 0.034 g/dl haemoglobin.

rocytes for 8 min and measured individual auto-fluorescence emission spectra of which ◘ Fig. 16.3C depicts a typical example. The produced fluorescent entity displayed an emission spectrum that peaked at around 510 nm, exactly around the peak of the Indo-1 and Fura-2 emission and thus renders simultaneous measurements of intracellular calcium with such indicators in erythrocytes virtually impossible.

Due to the known biochemical decomposition of haemoglobin[13] in conjunction with the spectral appearance of the recorded fluorescence (◘ Fig. 16.3C) in comparison with published data[14]

it is very likely that the induced autofluorescence originates from bilirubin isomers[15].

The transient nature of this fluorescence during continuous illumination might hint to the fact that bilirubin itself might be photochemically altered to produce less or not fluorescent products. We thus conclude that irradiation of living erythrocytes with UV light (here 360 nm) leads to a time-dependent production of bilirubin which fluorescence largely overlaps with the emission spectra of both, Fura-2 and Indo-1. Although this effect will be less pronounced in cuvette measurements because of the mixture of illuminated and none-illuminated

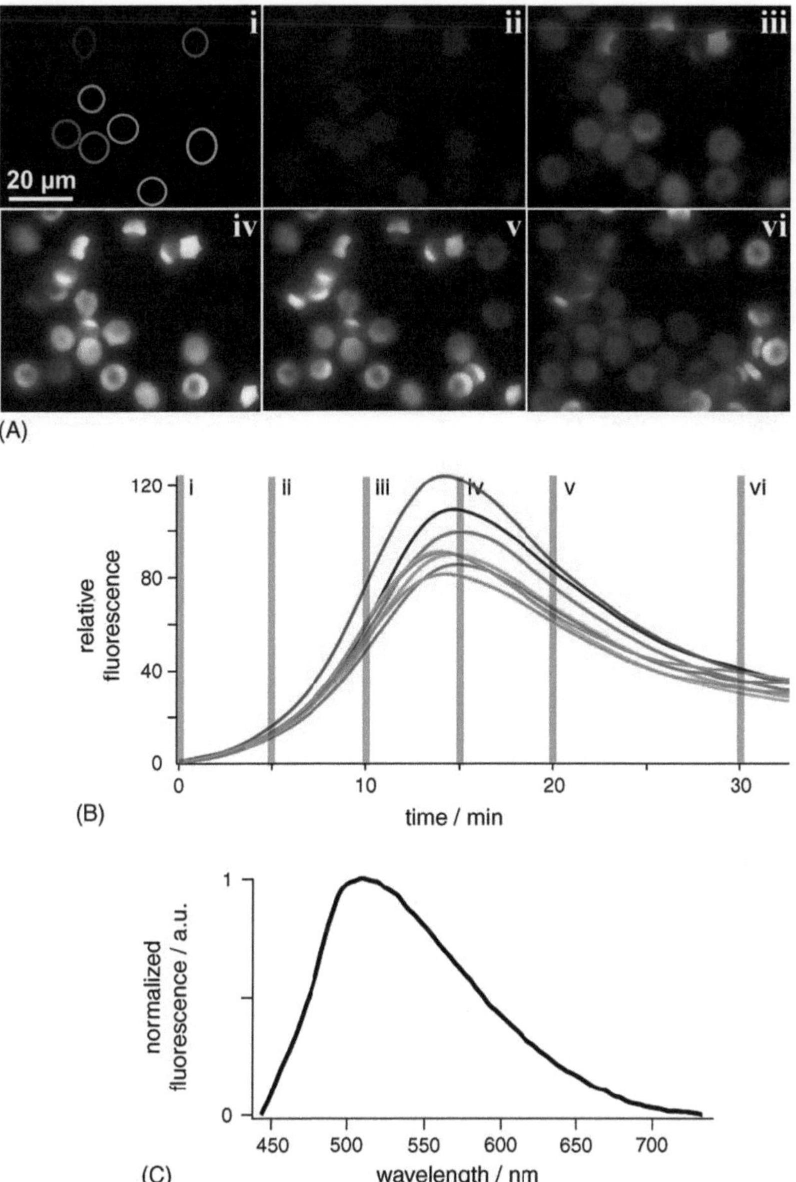

(A)

(B)

(C)

◻ Fig. 16.3 Effects upon UV-A (360 nm) irradiation of human erythrocytes. Part A shows fluorescence images of an erythrocyte population before, during and after irradiation with UV-A (360 nm). The fluorescence images i, ii, iii, iv, v and vi correspond to exposure times of 0, 5, 10, 15, 20 and 30 min, respectively. The light power out of the objective was approximately 35 µW. Warm colours correspond to a high fluorescence intensity and cold colours to a weak fluorescence. There is virtually no fluorescence before the irradiation (image i). The erythrocyte population is a representative example, and the result was verified with erythrocytes of three different donors. Part B shows fluorescence intensities vs. exposure time. The colour-codes of the marked erythrocytes in image i (part A) correspond to the colour-codes of the curves in part B. Although the fluorescence intensities and time curves are slightly different, all erythrocytes show a clear response. The curves show the background corrected self-ratio. The fluorescence spectrum of the erythrocytes corresponding to image ii (part A) is presented in part C. The spectrum suggests the UV-A induced formation of bilirubin or bilirubin-type products as the source of the fluorescence. Upon irradiation with 340 and 380 nm the increase of the fluorescence were qualitatively the same (data not shown). (For interpretation of the references to color in this figure legend, the reader is referred to the web version of the original article.)

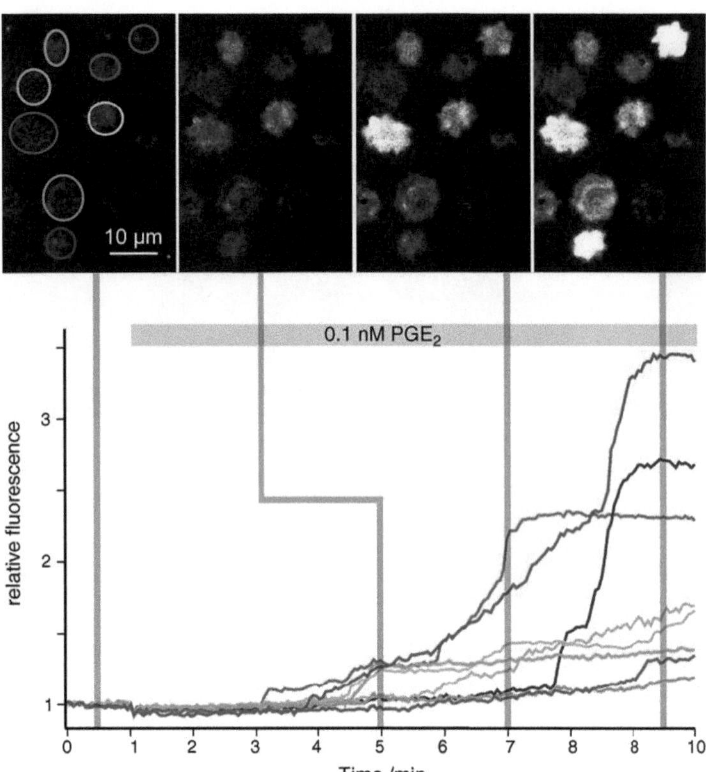

▣ Fig. 16.4 Fluorescence-imaging of human erythrocytes treated with PGE$_2$ using the calcium fluorophor Fluo-4. Erythrocytes of healthy donors were loaded with the esterified dye Fluo-4 and placed on a cover slip. Fluorescence imaging was realised by excitation using the wavelength of 480 nm. In the first fluorescence image seven erythrocytes are marked as regions of interest. The fluorescence intensity of these regions of interest over time is plotted in the graph, which is a background corrected self-ratio. The time point of the addition of 0.1 nM PGE$_2$ is marked. Furthermore, the fluorescence images on selected time points are provided. On the graphs as well as on the images one can see that six out of seven marked erythrocytes react significantly on the addition of PGE$_2$. The increase of the fluorescence intensity must originate from a calcium entry.

erythrocytes, over time this effect will also prevent reliable calcium measurements in cell populations. In contrast, irradiation (at 480 nm) of similar prepared erythrocytes did not result in changes of the autofluorescence, even for high intensities and extended periods of time (data not shown).

16.3.3 Fluo-4 fluorescence reveals PGE$_2$-induced increases in intracellular calcium in individual erythrocytes

▣ Fig. 16.4 depicts experimental evidence of PGE$_2$-induced elevations of intracellular free calcium in individual erythrocytes. Despite the finding that PGE$_2$ responses were quite variable between cells, both in amplitude and time-course, the fact that non-responding cells did not exhibit any changes in their basal fluorescence clearly supported our notion that even extended periods of irradiation

with excitation light at around 480 nm as used here did not induce any changes in autofluorescence. These results were similar to the ones we have reported recently[11]. In all experiments around 44.5% of all erythrocytes responded ($n = 577$ from 10 blood samples). When we applied PGE$_2$ in the absence of extracellular calcium ions (buffered with EDTA) no calcium elevations could be measured providing additional evidence that the PGE$_2$-induced calcium increase measured with Fluo-4 was not an artefact from changes in the photophysical properties of either Fluo-4 or haemoglobin (data not shown). Furthermore, experiments in which we permeabilised the plasma membrane of erythrocytes showed a robust increase in intracellular calcium when extracellular calcium was present. This rise was abolished in the absence or upon removal of extracellular calcium (data not shown). Taken together these results clearly demonstrated that PGE$_2$-stimulation induced an increase in the Fluo-4 fluorescence that was due to a rise in cal-

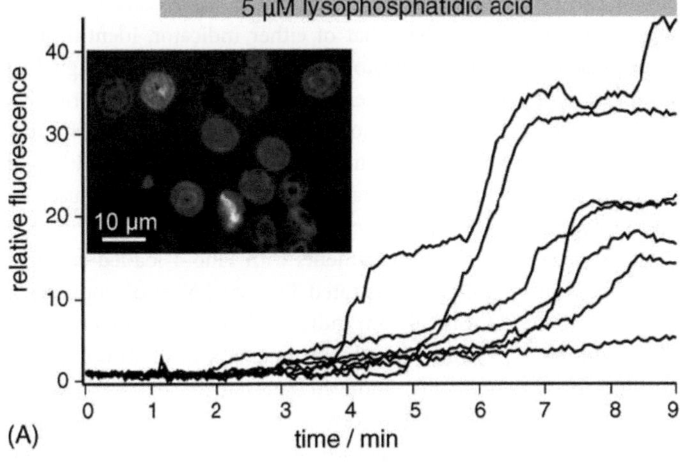

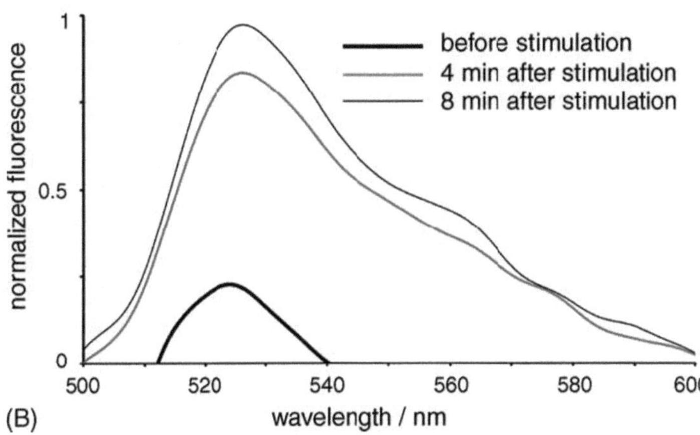

Fig. 16.5 (A) Fluorescence-imaging of human erythrocytes treated with LPA using the calcium fluorophor Fluo-4. Erythrocytes of healthy donors were loaded with the esterified dye Fluo-4 and placed on a cover slip. Fluorescence imaging was realised by excitation using the wavelength of 480 nm. The fluorescence intensity of seven cells over time is plotted in the graph, which is a background corrected self-ratio. The time point of the addition of 5 µM LPA is marked. The inset shows the fluorescence image 8 min after beginning of the stimulation. (B) Time-course of the fluorescence spectrum of a single erythrocyte loaded with Fluo-4 and exposed to 5 µM LPA. The spectra are representative examples (n = 7) and the curves are smoothed.

cium rather than altered properties of the indicator itself. Thus, Fluo-4 indeed appeared to be a much better candidate for following calcium changes in erythrocytes than other indicators that require UV-excitation.

In addition to PGE_2-stimulation, also application of LPA was thought to result in increases of intracellular calcium[16]. Therefore, we loaded erythrocytes with Fluo-4 and stimulated calcium rises with 5 µM LPA as depicted in ◘ Fig. 16.5A. Shortly after administration of the stimulus, intracellular calcium levels in all ($n = 48$) erythrocytes started to elevate. In order to confirm the nature of the fluorescence increase, we repeated these experiments on the set-up that allowed for the recording of single cell emission spectra. We acquired emission spectra at three different time points during the stimulation process as shown in ◘ Fig. 16.5B. The shape of the spectra and their peak at around 525 nm strongly suggested that this emission originated from Fluo-4 excitation.

16.4 Discussion

In the current manuscript we investigated the spectral properties of three popular calcium indicators and their applicability for the measurement of intracellular free calcium concentrations in individual erythrocytes. Indo-1 and Fura-2 require excitation in the UV range while Fluo-4 can be excited in the blue wavelength range (approximately 480 nm). We found that in the presence of haemoglobin all of these indicators underwent significant emission and/or excitation quenching in a concentration dependent manner. In addition, both Fura-2 and Indo-1 also displayed dramatic alterations in their calcium-dependent spectral properties, which rendered these indicators virtually unusable for measurements of intracellular free calcium in erythrocytes.

The overall quenching effect of haemoglobin on all three indicators most likely results from the strong absorbance of the protein in the relevant spectral regions as depicted in ◘ Fig. 16.1. The absorbance spectrum of haemoglobin is dominated by the soret-band but also shows quite a few neighbouring peaks. With indicators that require either dual excitation such as Fura-2 or dual emission recordings, the effect of haemoglobin is spectrally heterogeneous and alters the relationship between the recorded ratio and the calcium concentration in a rather complex way (◘ Fig. 16.2). Surprisingly this quenching appears to be also dependent on the haemoglobin concentration (◘ Fig. 16.2). These findings most likely explain former problems with calcium measurements in erythrocytes and would certainly foster our conclusion not to use these indicators anymore in erythrocyte calcium measurements. In contrast to that, Fluo-4 despite the enormous quenching by haemoglobin did not display any spectral shifts in its emission spectra for all calcium and haemoglobin concentrations tested (◘ Fig. 16.2). We thus conclude that although detection of Fluo-4 emission will certainly be more difficult in the presence of haemoglobin, changes in intracellular calcium concentrations can be measured readily and with the assumption of a resting calcium concentration, it might even be possible to obtain good estimations of absolute calcium concentrations[17].

In addition to these spectral interaction between the calcium indicators and haemoglobin, sole UV-irradiation of erythrocytes results in generation of a highly fluorescent specimen, probably a bilirubin isomer. Unfortunately, the emission spectrum of this bilirubin isomer largely overlaps with the emission spectra of both Indo-1 and Fura-2. Since the bilirubin fluorescence is much larger than that of either indicator, identification of the indicator fluorescence in a large signal of bilirubin fluorescence is rather tedious and renders the use of Indo-1 or Fura-2 even more unfavourable. Similar illumination of erythrocytes with light of Fluo-4 excitation wavelength did not induce any auto-fluorescence whatsoever.

Our experiments with Fluo-4-loaded erythrocytes demonstrated the possibility of application of this calcium indicator in the study of single cell erythrocyte calcium imaging. Both stimuli, PGE_2 and LPA, induced substantial increases in the Fluo-4 fluorescence, indicating a robust increase in intracellular calcium. Since both agonists did not cause increases in all cells, the non-responding erythrocytes confirmed our earlier finding that illumination of these cells with the excitation wavelength for Fluo-4 did not induce any increases in autofluorescence at the Fluo-4 emission wavelengths.

In conclusion, we have demonstrated here that the use of Indo-1 and Fura-2 for calcium recordings in erythrocytes are rather limited and ought not to be continued and that Fluo-4 as a representative of calcium indicators with visible light excitation are much more suitable for calcium imaging in individual erythrocytes. Another possible way out of this dilemma might be the application of fluorescence lifetime measurements, a method that does not discriminate fluorescence entities based on their colour and/or fluorescence intensity but instead is based on different and/or changing lifetimes[18].

Thus, the results of this report shall foster future research in calcium signalling of individual erythrocytes with calcium indicators that work in the visible spectral range, such as Fluo-4.

16.5 References

[1] T. Tiffert, R.M. Bookchin and V.L. Lew, Calcium homeostasis in normal and abnormal human red cells, I. Bernhardt, J.C. Ellory, Editors, Red Cell Membrane Transport in Health and Disease, Springer Verlag, Berlin (2003), pp. 373–405.

[2] V.L. Lew, Z. Etzion, R.M. Bookchin, R. daCosta, H. Vaananen, M. Sassaroli and J. Eisinger, The distribution

of intracellular calcium chelator (fura-2) in a population of intact human red cells. Biochim. Biophys. Acta, 1148 (1993), pp. 152–156.

[3] M.J. Morris, M.D. Dufilho and M.A. Devynck, In vivo and in vitro effects of isradipine on cytosolic Ca^{2+} concentration in erythrocytes from spontaneously hypertensive rats. Am. J. Hypertens., 5 (1992), pp. 887–891.

[4] M. David-Dufilho, C. Astarie, M.G. Pernollet, M. Del Pino, J. Levenson, A. Simon and M.A. Devynck, Control of the erythrocyte free Ca^{2+} concentration in essential hypertension. Hypertension, 19 (1992), pp. 167–174.

[5] A.M. Blackwood, G.A. Sagnella, N.D. Markandu and G.A. MacGregor, Problems associated with using Fura-2 to measure free intracellular calcium concentrations in human red blood cells. J. Hum. Hypertens., 11 (1997), pp. 601–604.

[6] L. Kaestner, C. Bollensdorff and I. Bernhardt, Non-selective voltage-activated cation channel in the human red blood cell membrane. Biochim. Biophys. Acta, 1417 (1999), pp. 9–15.

[7] L. Kaestner, P. Christophersen, I. Bernhardt and P. Bennekou, The non-selective voltage-activated cation channel in the human red blood cell membrane: reconciliation between two conflicting reports and further characterisation. Bioelectrochemistry, 52 (2000), pp. 117–125.

[8] L. Kaestner and I. Bernhardt, Ion channels in the human red blood cell membrane: their further investigation and physiological relevance. Bioelectrochemistry, 55 (2002), pp. 71–74.

[9] L. Soldati, C. Lombardi, D. Adamo, A. Terranegra, C. Bianchin, G. Bianchi and G. Vezzoli, Arachidonic acid increases intracellular calcium in erythrocytes. Biochem. Biophys. Res. Commun., 293 (2002), pp. 974–978.

[10] Y.V. Kucherenko, E. Weiss and I. Bernhardt, Effect of the ionic strength and prostaglandin E_2 on the free Ca^{2+} concentration and the Ca^{2+} influx in human red blood cells. Bioelectrochemistry, 62 (2004), pp. 127–133.

[11] L. Kaestner, W. Tabellion, P. Lipp and I. Bernhardt, Prostaglandin E_2 activates channel-mediated calcium entry in human erythrocytes: an indication for a blood clot formation supporting process. Thromb. Haemost., 92 (2004), pp. 1269–1272.

[12] J.S. Wiley and K.E. McCulloch, Calcium ions, drug action and the red cell membrane. Pharmacol. Ther., 18 (1982), pp. 271–292.

[13] C.K. Mathews and K.E.v. Holde, Degradation of heme in animals, K.E.v. Holde, C.K. Mathews, Editors , Biochemistry, The Benjamin/Cummings Publishing Company, Redwood City (1993), pp. 734–736.

[14] A. Cu, G.G. Bellah and D.A. Lightner, On the fluorescence of bilirubin. J. Am. Chem. Soc., 97 (1975), pp. 2579–2580.

[15] L. Kaestner, A. Juzeniene and J. Moan, Erythrocytes—the 'house elves' of photodynamic therapy. Photochem. Photobiol. Sci., 3 (2004), pp. 981–989.

[16] L. Yang, D.A. Andrews and P.S. Low, Lysophosphatidic acid opens a Ca(++) channel in human erythrocytes. Blood, 95 (2000), pp. 2420–2425.

[17] D. Thomas, S.C. Tovey, T.J. Collins, M.D. Bootman, M.J. Berridge and P. Lipp, A comparison of fluorescent Ca^{2+} indicator properties and their use in measuring elementary and global Ca^{2+} signals. Cell Calcium, 28 (2000), pp. 213–223.

[18] W. Becker, A. Bergmann, M.A. Hink, K. Konig, K. Benndorf and C. Biskup, Fluorescence lifetime imaging by time-correlated single-photon counting. Microsc. Res. Tech., 63 (2004), pp. 58–66.

Calcium Signalling in Cardiac Myocytes

Reduced Cardiac L-Type Ca^{2+} Current in Ca$_V$ß$_2$$^{-/-}$ Embryos Impairs Cardiac Development and Contraction With Secondary Defects in Vascular Maturation

Petra Weissgerber, Brigitte Held, Wilhelm Bloch, Lars Kaestner, Kenneth R. Chien, Bernd K. Fleischmann, Peter Lipp, Veit Flockerzi, Marc Freichel

Reprint from Circulation Research (2006) **99**, 749-757.

▪ Abstract

Cardiac myocyte contraction depends on transmembrane L-type Ca^{2+} currents and the ensuing release of Ca^{2+} from the sarcoplasmic reticulum. Here we show that these L-type Ca^{2+} currents are essential for cardiac pump function in the mouse at developmental stages where the functional significance of the heart becomes imperative to blood flow and to the continuing growth and survival of the embryo. Disruption of the Ca$_V$ß$_2$ gene, which encodes for the predominant ancillary β subunit of cardiac Ca^{2+} channels, resulted in diminished L-type Ca^{2+} currents in cardiomyocytes of embryonic day 9.5 (E9.5). This led to a functionally compromised heart, causing defective remodeling of intra- and extraembryonic blood vessels and embryonic death following E10.5. The defects in vascular remodeling were also observed when the Ca$_V$ß$_2$ gene was selectively targeted in cardiomyocytes, demonstrating that they are secondary to cardiac failure rather than a result of the lack of Ca$_V$ß$_2$ proteins in the vasculature. Partial rescue of the Ca^{2+} channel currents by a Ca^{2+} channel agonist significantly postponed embryonic death in Ca$_V$ß$_2$$^{-/-}$ mice. Taken together, these data strongly support the essential role of L-type Ca^{2+} channel activity in cardiomyocytes for normal heart development and function and that this is a prerequisite for proper maturation of the vasculature.

17.1   Introduction

Depolarisation of the plasma membrane during the cardiac action potential causes the activation of high voltage-gated L-type Ca^{2+} channels.[1] The ensuing Ca^{2+} influx across the plasma membrane triggers Ca^{2+} release from the sarcoplasmic reticulum (SR) via ryanodine receptors through a process called Ca^{2+}-induced Ca^{2+} release (CICR).[2,3] The resulting Ca^{2+} transients initiate myocyte contraction. Although this scenario is well understood in neonatal and adult cardiomyocytes, surprisingly little is known with respect to the developing cardiomyocytes, especially at very early embryonic stages.

High-voltage-activated (HVA) L-type Ca^{2+} channels are complexes of a pore-forming α$_1$ subunit (Ca$_V$1) of ≈190 to 230 kDa associated with auxiliary β subunits (Ca$_V$β), α$_2$δ subunits, and in some cases a γ subunit.[4] In the heart, the principal α$_1$ subunit is encoded by the Ca$_V$1.2 gene.[5] However, expression of this gene is not necessary for spontaneous beating and rhythmic activity of the heart and embryonic development until embryonic day 12.5 (E12.5).[6] Nevertheless, Ca$_V$1.2$^{-/-}$ embryos die before E14.5 for unknown reasons.[6] Ca$_V$1.2$^{-/-}$ cardiomyocytes have been shown to retain L-type Ca^{2+} channels, and it has been assumed that other Ca$_V$1 subunits act as a substitute of Ca$_V$1.2 in spontaneous contraction during cardiac development.[7] A function for Ca$_V$1.1[8] and Ca$_V$1.4 has not been reported in the heart, but Ca$_V$1.3 appears to be important for the generation of spontaneous diastolic depolarisation in sinoatrial node cells of the adult heart,[9,10] although $Ca_V1.3^{-/-}$ mice are viable and fertile.

Molecular cloning has identified altogether 10 $Ca_V\alpha_1$ and 4 $Ca_V\beta$ subunits, and a comparison of the expression pattern of $Ca_V\alpha_1$ and $Ca_V\beta$ subunits indicates that there is certainly not an exclusive association between particular pairs of $Ca_V\alpha_1$ and $Ca_V\beta$ subunits.[4,11–13] The $Ca_V\beta$ subunits bind to the main pore-forming $Ca_V\alpha_1$ subunits and as ancillary cytoplasmic proteins modulate the Ca^{2+} channel function.[14] Among other effects, they shift the voltage dependence of activation in the hyperpolarising direction, so that the channels open at less-depolarised potentials, and they are also required to enhance the number of functional channels at the plasma membrane.[14]

We made use of this latter property of $Ca_V\beta$ proteins to analyse the contribution of L-type Ca^{2+} channel activity independently of the underlying Ca_V1 pore protein to early murine cardiac function and the developing embryo. By generating mouse lines in which the predominant murine cardiac $Ca_V\beta$ gene, $Ca_V\beta_2$, was disrupted, we efficiently diminished L-type Ca^{2+} channel currents in cardiac myocytes. We deleted the $Ca_V\beta_2$ gene because mice deficient in $Ca_V\beta_1$,[15] $Ca_V\beta_3$,[12,16] and $Ca_V\beta_4$[17] all survived embryonic stages and did not show any significant cardiac phenotype. Our results show that L-type Ca^{2+} channel activity is expressed robustly in the wild-type murine heart as early as E9.5 and plays a dominant role for the progression of cardiac looping, cardiomyocyte contraction, blood flow, and blood vessel maturation. The impact of Ca^{2+} channel activity on blood vessel maturation depends on the cardiac performance because disruption of the $Ca_V\beta_2$ gene in the entire embryo or tissue specific only in cardiomyocytes provoked similar vascular defects.

17.2 Materials and Methods

An expanded Materials and Methods section is available in the online data supplement at http://circres.ahajournals.org/content/99/7/749/suppl/DC1. All animal experiments were carried out in accordance with German legislation on protection of animals.

17.2.1 $Ca_V\beta_2$ Gene Targeting and Generation of Mouse Lines

Cre-loxP-mediated recombination was used to generate mouse lines carrying a null allele ($Ca_V\beta_2^{+/-}$) or a floxed allele ($Ca_V\beta_2^{+/flox}$), respectively. $Ca_V\beta_2^{+/-}$ mice were crossed with tgGFP mice[18] to improve the visualisation of early cardiac morphogenesis. For cardiomyocyte-restricted disruption of the $Ca_V\beta_2$ gene, $Ca_V\beta_2^{+/-}$ mice were crossed with MLC2a-Cre animals (MLC2a$^{tg/0}$).[19] Mice with a cardiomyocyte-restricted inactivation of the $Ca_V\beta_2$ gene ($Ca_V\beta_2^{-/flox}$/MLC2a$^{tg/0}$) were obtained by further crossing with $Ca_V\beta_2^{flox/flox}$ mice. Cardiomyocyte restricted expression of the MLC2a-Cre transgene was demonstrated by crossing with the Z/EG reporter strain.[20]

17.2.2 Antibodies and Western Blots

Western blots were performed using anti-$Ca_V\beta$ antibodies.[11–13] Results were confirmed using additionally generated rabbit polyclonal antibodies 424 and 425 (anti-mouse $Ca_V\beta_2$), 828 (anti-mouse $Ca_V\beta_3$), and 830 (anti-mouse $Ca_V\beta_4$). We confirmed specificity of antibodies using microsomal membrane protein fractions from wild-type mice and mice deficient in $Ca_V\beta_2$ (this study), $Ca_V\beta_3$,[12,13] and $Ca_V\beta_4$.[17]

17.2.3 Histology and Whole-Mount Immunostaining

Paraformaldehyde-fixed embryos and yolk sacs were stained with an endothelial cell-specific marker platelet endothelial cell adhesion molecule (PECAM) (rat anti-mouse PECAM, Pharmingen) and a goat anti-rabbit IgG conjugated with biotin.

17.2.4 Electrophysiology and Ca^{2+} Imaging

Ca^{2+} channel currents using Ca^{2+} or Ba^{2+} as charge carrier were recorded using the whole-cell configuration of the patch-clamp technique.[21] Ca^{2+}

transients in E10.5 cardiomyocytes were recorded after loading with 2 μmol/l Fura-2/acetoxymethyl ester (Fura-2 AM) with a fluorescence video imaging configuration (TILL Photonics).

17.3   Results

17.3.1 Ca$_V$β$_2$ Gene Expression Is Essential for Embryonic Survival, Cardiac Rhythmic Activity, and Contractility and Blood Vessel Maturation

Ca$_V$β$_2$ proteins are expressed in the developing mouse heart,[22] and their expression in the developing heart tube begins at approximately E8.5 (◼ Fig. 17.1A). By a Cre-loxP-dependent strategy (Figure I in the online data supplement), we deleted a substantial part of the SH3 domain of Ca$_V$β$_2$ as well as the guanylate kinase domain (supplemental Figure VI) and obtained Ca$_V$β$_2$-deficient embryos (Ca$_V$β$_2^{-/-}$), which died following E10.5 (◼ Fig. 17.1B) and were already reabsorbed on E11.5. Following E10.5, only viable wild-type and heterozygous embryos were identified. The Ca$_V$β$_2^{+/-}$ mice were born healthy and did not exhibit any obvious abnormalities during adulthood. Western blot analysis with various anti-Ca$_V$β antibodies demonstrated that Ca$_V$β$_2$ expression was reduced in Ca$_V$β$_2^{+/-}$ hearts and confirmed that Ca$_V$β$_2^{-/-}$ E10.5 hearts no longer expressed the Ca$_V$β$_2$ protein (◼ Fig. 17.1C), whereas Ca$_V$β$_3$ expression was unchanged and Ca$_V$β$_1$ and Ca$_V$β$_4$ are not detectable (not shown).

Microscopic analysis of 127 Ca$_V$β$_2^{-/-}$ embryos harboring a green fluorescent protein (GFP) transgene expressed in cardiomyocytes (Ca$_V$β$_2^{-/-}$/tgGFP) for improved visualisation of cardiac development displayed no abnormalities compared with control Ca$_V$β$_2^{+/+}$/tgGFP littermates until E9.5 (◼ Fig. 17.1D). The size of the wild-type embryos was more than doubled between E9.5 and 10.5 (◼ Fig. 17.1D), whereas significant retardation of Ca$_V$β$_2^{-/-}$ embryos including poor growth of the facial and brain primordial tissue was observed at E10.5 (◼ Fig. 17.1D). In addition, a massive pericardial effusion frequently associated

with hemorrhages develops at E10.5 (◼ Fig. 17.1D). Closer inspection of Ca$_V$β$_2^{-/-}$ embryos at E10.5 (◼ Fig. 17.1E) revealed abnormalities in the progression of looping architecture with amorphous heart tubes lying distorted within the pericardial cavity, and a stenosis was frequently seen between the emerging right ventricle and the outflow tract. Ultrastructural analysis revealed no obvious differences in myofibrillar structures and cell-cell contacts (◼ Fig. 17.1F). However, a decrease in myocardial wall thickness was observed in most Ca$_V$β$_2^{-/-}$ tissue sections with numerous disruptions of the epicardial and endocardial layers (◼ Fig. 17.1F, bottom).

In embryonic heart, Ca$_V$β$_2$ is homogeneously expressed within the atrial and ventricular myocardium, including the atrioventricular canal and the outflow tract.[22] The phenotype of the Ca$_V$β$_2^{-/-}$ embryos indicates that Ca$_V$β$_2$ is required for normal cardiac and embryonic development, which depends increasingly on cardiac pump function and circulatory delivery of oxygen, rather than diffusion at these stages. We, therefore, studied the cardiac performance of Ca$_V$β$_2^{-/-}$ E10.5 embryos *in vivo* by video microscopy. Wild-type embryos exhibited coordinated fillings of the cardiac inflow tract and subsequent peristaltic ejection of blood into the dorsal aorta and the circulation (see supplemental Movie E10-5_Wt). In contrast, Ca$_V$β$_2^{-/-}$ hearts show only barely noticeable tremors without any noticeable ejection volume (see supplemental Movie E10-5_Cacnb2-ko_01/02). Moreover, the frequencies of spontaneous heart beats (◼ Fig. 17.1G) were drastically reduced in explanted E10.5 Ca$_V$β$_2^{-/-}$ hearts (63±2.88 bpm; n=9) compared with controls (120±3.15 bpm; n=11), whereas no difference was observed at E9.5. Although inhibition of L-type Ca^{2+} currents can lead to negative chronotropic effects, the reduced heart rate of E10.5 Ca$_V$β$_2^{-/-}$ hearts might also be secondary to the morphological aberrations present at this developmental stage (◼ Fig. 17.1E).

The impairment of cardiac pump function during development is frequently associated with impaired angiogenesis, which coincides with the onset of blood flow and is influenced by thereby evoked fluid forces.[23] In yolk sacs, where

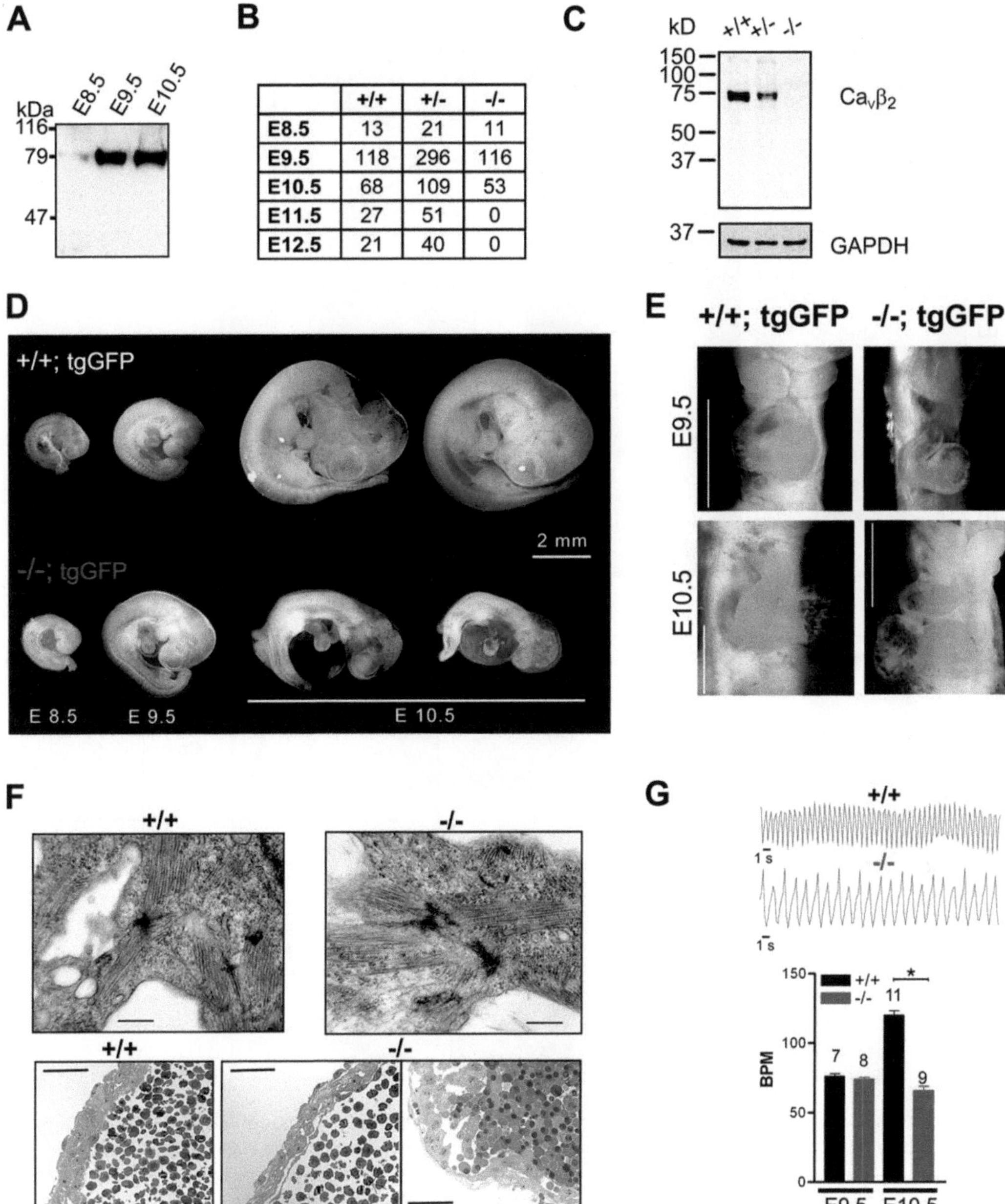

◘ Fig. 17.1 $Ca_v\beta_2$ is essential for cardiac morphogenesis and embryonic development. A, Western blots of protein extracts from embryonic (E8.5 to E10.5) hearts using $Ca_v\beta_2$-specific antibody 425. B, Numbers of surviving wild-type (+/+), heterozygous (+/−), and $Ca_v\beta_2^{-/-}$ embryos (−/−) at E8.5 to E12.5. Time point of embryonic death in $Ca_v\beta_2^{-/-}$ embryos was independent on the expression of the GFP transgene. C, Western blots of protein extracts (50 µg per lane) from E10.5 wild-type, $Ca_v\beta_2^{+/-}$, and $Ca_v\beta_2^{-/-}$ mouse embryos using the $Ca_v\beta_2$-specific antibody 425 and a GAPDH antibody, for control. D, Development of wild-type and $Ca_v\beta_2^{-/-}$ embryos. Embryonic hearts show GFP fluorescence attributable to the cardiac-restricted expression of the GFP transgene (tgGFP). E, Cardiac development at E9.5 and E10.5 by front-view microscopy. F, Ultrastructural (top, bar=50 nm) and histological analysis (bottom, bar=50 µm) of E10.5 wild-type and $Ca_v\beta_2^{-/-}$ myocardium. G, Representative records of spontaneous beatings of explanted E10.5 wild-type and $Ca_v\beta_2^{-/-}$ hearts (top) and averaged results from E9.5 and E10.5 hearts (bottom). *P<0.001. HF indicates heart frequency; BPM, beats per minute; n, number of hearts.

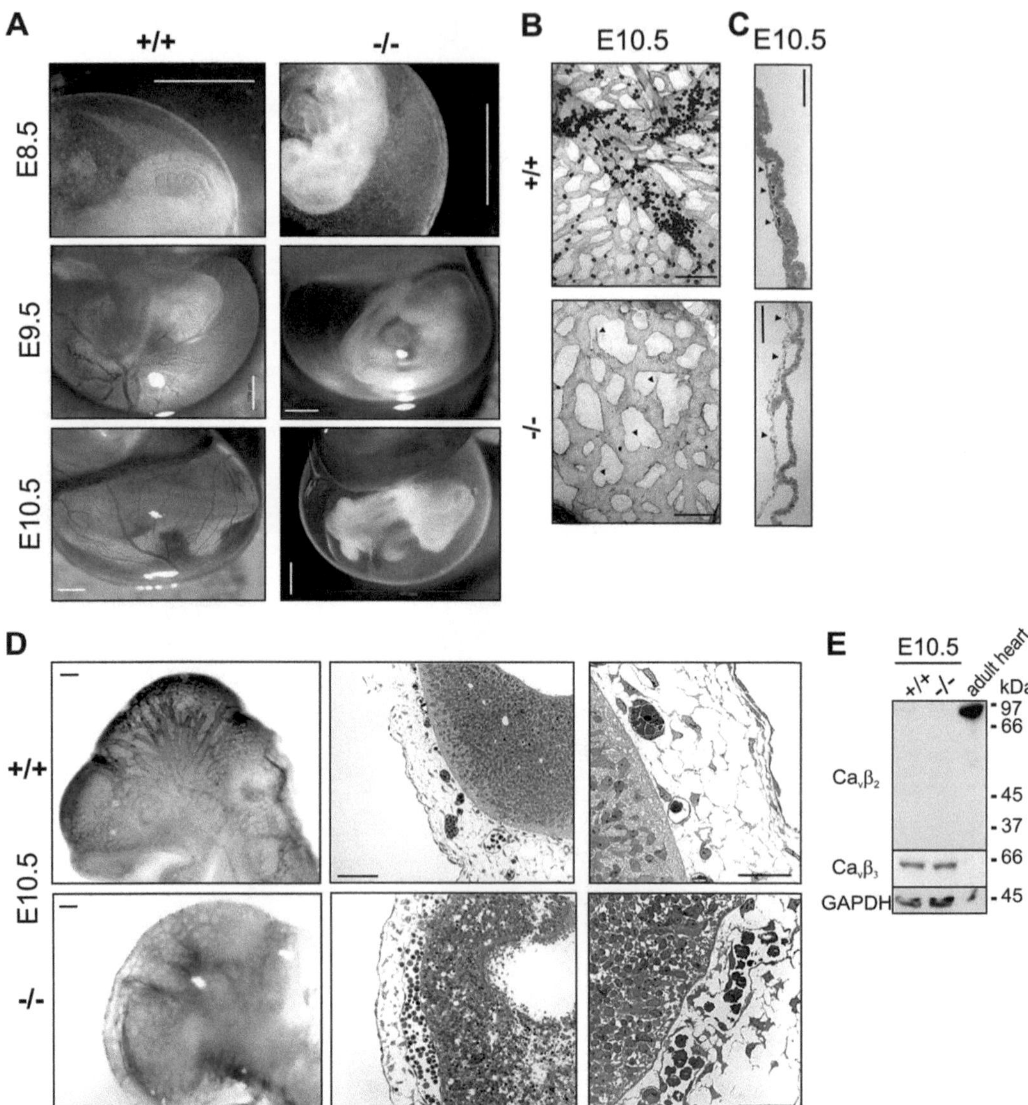

■ **Fig. 17.2** Ca$_v$β$_2$ is essential for blood vessel maturation. A, Vascular formation in the yolk sac of wild-type (+/+) and Ca$_v$β$_2$$^{-/-}$ (−/−) embryos at E8.5, E9.5, and E10.5 (bar=1 mm). B and C, Whole-mount immunostaining for PECAM (B) and methylene blue-stained cross-sections (C) of E10.5 yolk sacs (bar=100 μm). Arrowheads indicate initiation of capillary sprouting (B) and enlarged blood vessel diameter (C) in Ca$_v$β$_2$$^{-/-}$ tissues. D, Whole-mount anti-PECAM staining of the head (bar=100 μm) (left panels), intraembryonic peripheral blood vessels in cross-sections through regions of the head (bar=100 μm) (middle panels), and the back (bar=50 μm) (right panels). E, Western blots of protein extracts (40 μg per lane) from E10.5 wild-type (+/+) and Ca$_v$β$_2$$^{-/-}$ (−/−) yolk sacs using Ca$_v$β$_2$- andCa$_v$β$_3$-specific antibodies. Microsomal membrane proteins from adult wild-type heart served as a control for the Ca$_v$β$_2$-specific antibody.

blood vessel formation is observed first during embryonic development, a delineation of blood vessels occurred in Ca$_V$β$_2$$^{+/+}$ (■ Fig. 17.2A) and Ca$_V$β$_2$$^{+/-}$ littermate controls (data not shown) starting on E9.5 and blood vessels showed a hierarchical arborisation with diameters tapering from large to small as expected (■ Fig. 17.2A and ■ Fig. 17.2B). In contrast, no blood vessels were detectable in $Ca_V\beta_2^{-/-}$ embryos using stereomicroscopic analysis (◻ Fig. 17.2A), but whole-mount immunostaining (◻ Fig. 17.2B) revealed that vasculogenesis had occurred and the vascular plexus was formed also in $Ca_V\beta_2^{-/-}$ tissues. Initiation of capillary sprouting had also taken place in $Ca_V\beta_2^{-/-}$ yolk sacs (◻ Fig. 17.2B, arrowheads), but remodeling from primary capillary plexus into treelike mature blood vessels had been arrested, and a honeycomb-like network of tubules with enlarged and uniform diameters had occurred in both the extra- and intraembryonic vasculature (◻ Fig. 17.2B through ◻ Fig. 17.2D). Furthermore, vessels in $Ca_V\beta_2^{-/-}$ yolk sacs were barely filled with hemangioblasts (◻ Fig. 17.2B), consistent with poor cardiac pump function, and the formation of vitelline vessels could not be detected. However, $Ca_V\beta_2$ transcripts were only barely detectable in E10.5 wild-type yolk sacs (supplemental Figure IIA), and no $Ca_V\beta_2$ proteins were identified (◻ Fig. 17.2E). Because we could not identify expression of $Ca_V\beta_2$ proteins in the yolk sac, we hypothesised that the disturbed vascular maturation in $Ca_V\beta_2^{-/-}$ embryos might be a consequence of impaired cardiac pump function, rather than attributable to the lack of $Ca_V\beta_2$ proteins in the vasculature itself.

17.3.2 L-type Ca^{2+} Channel Activity and CICR in Very Early Embryonic Cardiomyocytes

At least in cells from neonatal or adult hearts, the cardiac performance essentially relies on the operation of voltage-activated Ca^{2+} channels and the CICR mechanism. At E9.5 both low-voltage-activated (LVA) and HVA Ca^{2+} channels are already functional. LVA currents evoked by our protocol did not differ significantly among cells of the 3 genotypes (◻ Fig. 17.3A and supplemental Figure IIIA). HVA currents could be readily detected in E9.5 cardiomyocytes either with Ca^{2+} (1.8 mmol/l; supplemental Table I) or Ba^{2+} (10 mmol/l) as charge carrier (at 0 mV: I_{Ca}, -6.27 ± 0.7 pA/pF [n=21]; I_{Ba}, -29.3 ± 2.6 pA/pF [n=16]). These currents did not differ significantly in wild-type and $Ca_V\beta_2^{+/-}$ cardiomyocytes

(supplemental Figure IIIA and IIIB; I_{Ca}, -5.5 ± 0.51 pA/pF [n=34]; I_{Ba}, -33.8 ± 5.8 pA/pF [n=12]) but were significantly larger by ≈1.4-fold in E10.5 cells compared with E9.5 cells (see supplemental Table I).

In $Ca_V\beta_2^{-/-}$ cells, the density of HVA Ca^{2+} currents was reduced to approximately one-fourth to one-third (◻ Fig. 17.3B and supplemental Figure IIIB), whereas mean cell capacitance was the same for wild-type and mutant cells (wild type, 29.8 ± 1.2 pF [n=106]; $Ca_V\beta_2^{+/-}$, 27.4 ± 1.1 pF [n=101]; $Ca_V\beta_2^{-/-}$, 31.4 ± 1.8 pF [n=87]; P>0.05). In addition, the peak of the whole-cell current was typically at 0 mV in control and slightly shifted to more positive voltages in $Ca_V\beta_2^{-/-}$ cells. Accordingly, the activation midpoints in the normalised curves of ◻ Fig. 17.3C and of supplemental Figure IIIC, and the potentials for half-activation fitted with a Boltzmann equation[24] were significantly shifted to more depolarised potentials. A similar depolarising shift in the voltage dependence of Ca^{2+} channel activation was observed in sensory neurones where the $Ca_V\beta_3^{-/-}$ gene was inactivated[12] or when HVA $Ca_V\alpha_1$ subunits were expressed without any $Ca_V\beta$ subunit.[14]

HVA Ca^{2+} currents in embryonic cardiomyocytes are sensitive to dihydropyridines. Application of 10 µmol/l (±)-isradipine blocked the currents in cardiomyocytes from E10.5 wild-type (◻ Fig. 17.3D; I_{Ca} density, -9.9 ± 1.2 pA/pF in the absence and -0.3 ± 0.2 pA/pF in the presence of isradipine; n=10) and $Ca_V\beta_2^{-/-}$ (◻ Fig. 17.3D; I_{Ca} density, -2.0 ± 0.5 pA/pF in the absence and 0.1 ± 0.1 pA/pF in the presence of isradipine; n=8) mice. Similar results were obtained using cardiomyocytes isolated at E9.5.

We analysed electrically induced Ca^{2+} transients in E10.5 cardiomyocytes (◻ Fig. 17.3E through 17.3J). In $Ca_V\beta_2^{-/-}$ cells, the amplitude of electrically induced Ca^{2+} transients was reduced by more than 80% (◻ Fig. 17.3E and ◻ Fig. 17.3F). In contrast, caffeine-evoked Ca^{2+} responses were unchanged in both amplitude and decay time constant τ (◻ Fig. 17.3G through ◻ Fig. 17.3I), indicating that the Ca^{2+} content of the SR and Ca^{2+} extrusion from the cytosol is unchanged in $Ca_V\beta_2^{-/-}$ cardiomyocytes. ◻ Fig. 17.3J illustrates that the overall Ca^{2+} transients comprised components from both

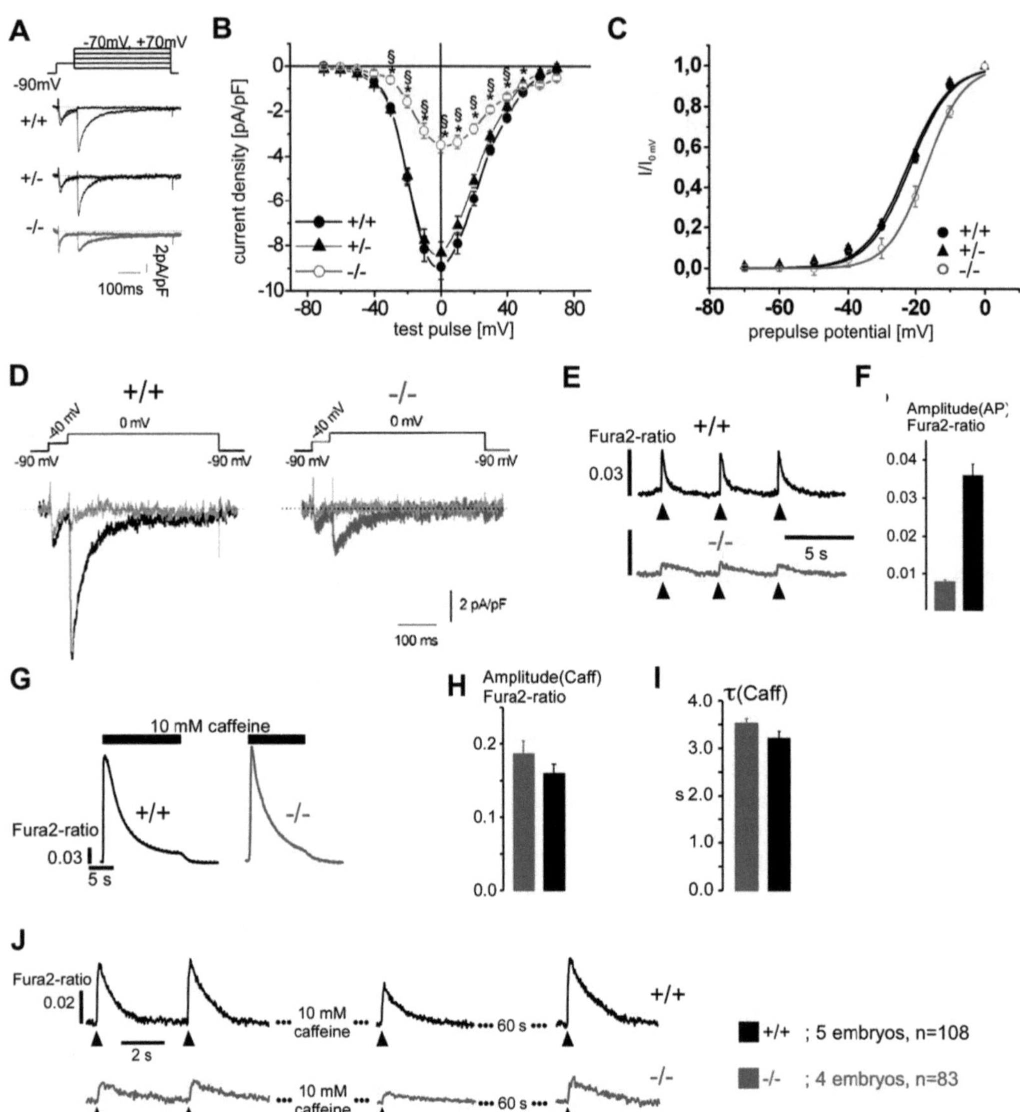

◘ Fig. 17.3 Impaired L-type Ca^{2+} channel activity and Ca^{2+}-induced Ca^{2+} release in E10.5 Ca$_v$ß$_2$$^{-/-}$ cardiomyocytes. A, Representative I_{Ca} traces recorded in wild-type, Ca$_v$ß$_2$$^{+/-}$, and Ca$_v$ß$_2$$^{-/-}$ cardiomyocytes during a 400-ms depolarisation to −70 mV and 0 mV after a 50-ms depolarisation to −40 mV to record T-type currents, which did not differ significantly among cells of the 3 genotypes (peak current density at −40 mV: wild type, −6.6±0.8 pA/pF [n=45]; Ca$_v$ß$_2$$^{+/-}$, −8.3±1.2 pA/pF [n=38]; Ca$_v$ß$_2$$^{-/-}$, −5.9±1.1 pA/pF [n=26]; P>0.05). B, Average current/voltage relationships of I_{Ca} from cardiomyocytes: wild type (n=45 cells), Ca$_v$ß$_2$$^{+/-}$ (n=38 cells), and Ca$_v$ß$_2$$^{-/-}$ (n=26 cells). *Ca$_v$ß$_2$$^{-/-}$ vs wild type, §Ca$_v$ß$_2$$^{-/-}$ vs Ca$_v$ß$_2$$^{+/-}$ (P<0.05). C, Average activation curve (normalised to the peak current at 0 mV) of the L-type I_{Ca}. The continuous lines are a Boltzmann fit to the data points from each genotype, respectively. The activation midpoints in the normalised curves are as follows: wild type (n=45 cells; V$_{1/2}$, −21.6 mV), Ca$_v$ß$_2$$^{+/-}$ (n=38 cells; V$_{1/2}$, −22.5 mV), and Ca$_v$ß$_2$$^{-/-}$ (n=26 cells; V$_{1/2}$, −17.0 mV). D, I_{Ca} traces under control conditions (wild type, black; Ca$_v$ß$_2$$^{-/-}$, red) and in the presence of 10 μmol/l (±)-isradipine (gray). Dotted lines indicate 0 current. Original traces (E) and averaged amplitude (F) of intracellular Ca^{2+} transients in wild-type (108 cells from 5 embryos) and Ca$_v$ß$_2$$^{-/-}$ (83 cells from 4 embryos) cardiomyocytes during electrical pacing (0.25 Hz). G through I, Original traces (G), averaged amplitude (H), and averaged decay time constant (I) of caffeine-induced Ca^{2+} release from intracellular Ca^{2+} stores. J, Representative recordings of electrically induced Ca^{2+} transients. Reloading of the SR was achieved by repetitive electrical field stimulation after caffeine washout.

transmembrane Ca^{2+} influx and Ca^{2+} release from the SR. For this, electrically induced Ca^{2+} transients were recorded before and immediately after caffeine application. We found a marked reduction in the Ca^{2+} transient amplitude probably attributable to the reduced or absent contribution of SR Ca^{2+} release in the presence of caffeine. The rightmost traces in ◘ Fig. 17.3J illustrate electrically induced Ca^{2+} transients after washout of caffeine. During this period, the SR has reloaded to the control steady-state Ca^{2+} levels for both wild-type and $Ca_V\beta_2^{-/-}$ myocytes. These data strongly suggest that the SR Ca^{2+} uptake function was not significantly impaired by the disruption of the $Ca_V\beta_2$ gene and support the conclusion that changes of overall Ca^{2+} transients can mainly be ascribed to the reduced voltage-dependent transmembrane Ca^{2+} influx. A decreased Ca^{2+} influx, then, resulted in markedly reduced overall Ca^{2+} transient and, consecutively, contraction of the myocytes, and pumping of the entire heart was diminished.

17.3.3 Vascularisation Defects Are Secondary to Cardiac Failure

The inactivation of the $Ca_V\beta_2$ gene in tissues other than heart could still contribute to the observed lethal phenotype. To test this possibility, the $Ca_V\beta_2$ gene was inactivated in cardiomyocytes by using the transgenic mouse line MLC2a$^{tg/0}$, which carries the Cre recombinase gene under the myosin light chain 2a promoter.[19] Recombination induced by the MLC2a-Cre transgene was demonstrated as early as E8.5, when formation of the linear heart tube is initiated, and was restricted to atrial and ventricular cardiomyocytes (supplemental Figure IV). Embryos with a cardiac-restricted disruption of the $Ca_V\beta_2$ gene ($Ca_V\beta_2^{-/flox}/MLC2a^{tg/0}$) could be identified until E13.5 (◘ Fig. 17.4A and ◘ Fig. 17.3B), but a large number of E11.5 and E12.5 $Ca_V\beta_2^{-/flox}/MLC2a^{tg/0}$ embryos already showed striking morphological destruction incompatible with survival (not shown). Additionally, in $Ca_V\beta_2^{-/flox}/MLC2a^{tg/0}$ embryos extra- and intraembryonic blood vessels seemed to be arrested in a premature stage with a uniform and enlarged diameter and lacked hier-

archical branching (◘ Fig. 17.3C and ◘ Fig. 17.3D), which is strikingly similar to the defects observed in $Ca_V\beta_2^{-/-}$ embryos (◘ Fig. 17.2).

The $Ca_V\beta_2$ protein was not detectable in the E13.5 $Ca_V\beta_2^{-/flox}/MLC2a^{tg/0}$ hearts (◘ Fig. 17.3E), whereas it was present in the hearts from controls, which carry at least 1 functional $Ca_V\beta_2$ allele. In addition, the $Ca_V\beta_3$ protein but no $Ca_V\beta_1$ and $Ca_V\beta_4$ proteins were detectable in the heart at E13.5. Although Cre-mediated excision induced by the MLC2a-Cre transgene is highly effective and is initiated at least from E8.5 onward (supplemental Figure IV), low amounts of $Ca_V\beta_2$ transcripts were occasionally detectable by RT-PCR in 1 of 4 hearts isolated from E13.5 $Ca_V\beta_2^{-/flox}/MLC2a^{tg/0}$ embryos (supplemental Figure IIB and IIC). Most probably, these $Ca_V\beta_2$ transcripts are encoded by alleles that have already been transcribed before Cre-mediated excision or arise from noncardiomyocytes present in the heart, such as vascular smooth muscle cells, neurons, endothelial cells, or fibroblasts. The translated protein levels are, however, too low to be detectable by our antibodies but sufficient to postpone somewhat the $Ca_V\beta_2^{-/-}$ phenotype. Similar to these results, Wettschureck et al.[19] also reported an incomplete penetrance of the MLC2a-Cre transgene, resulting in a variable phenotype most likely attributable to individual differences in the time course of Cre expression and/or Cre-mediated recombination. The low level of $Ca_V\beta_2$ transcript expression in yolk sac (supplemental Figure IIA) was not affected by the cardiac-specific targeting (supplemental Figure IIB and IIC).

Whole-cell voltage-activated Ca^{2+} channel currents (I_{Ca}) were recorded in $Ca_V\beta_2^{-/flox}/MLC2a^{tg/0}$ E13.5 cardiomyocytes and in the various control cells (◘ Fig. 17.3F). As in the $Ca_V\beta_2^{-/-}$ cells, the current density (-3.0 ± 0.2 pA/pF, n=52) was significantly reduced in $Ca_V\beta_2^{-/flox}/MLC2a^{tg/0}$ cardiomyocytes compared with controls, and the potentials for half-activation were shifted to depolarised potentials ($Ca_V\beta_2^{-/flox}/MLC2a^{tg/0}$ cells, -12.6 ± 0.6 mV [n=52]; for example, $Ca_V\beta_2^{+/flox}/MLC2a^{tg/0}$ control cells, -16.9 ± 0.7 mV [n=36]). Also, the beating frequency of explanted $Ca_V\beta_2^{-/flox}/MLC2a^{tg/0}$ E13.5 hearts was reduced significantly (◘ Fig. 17.3G).

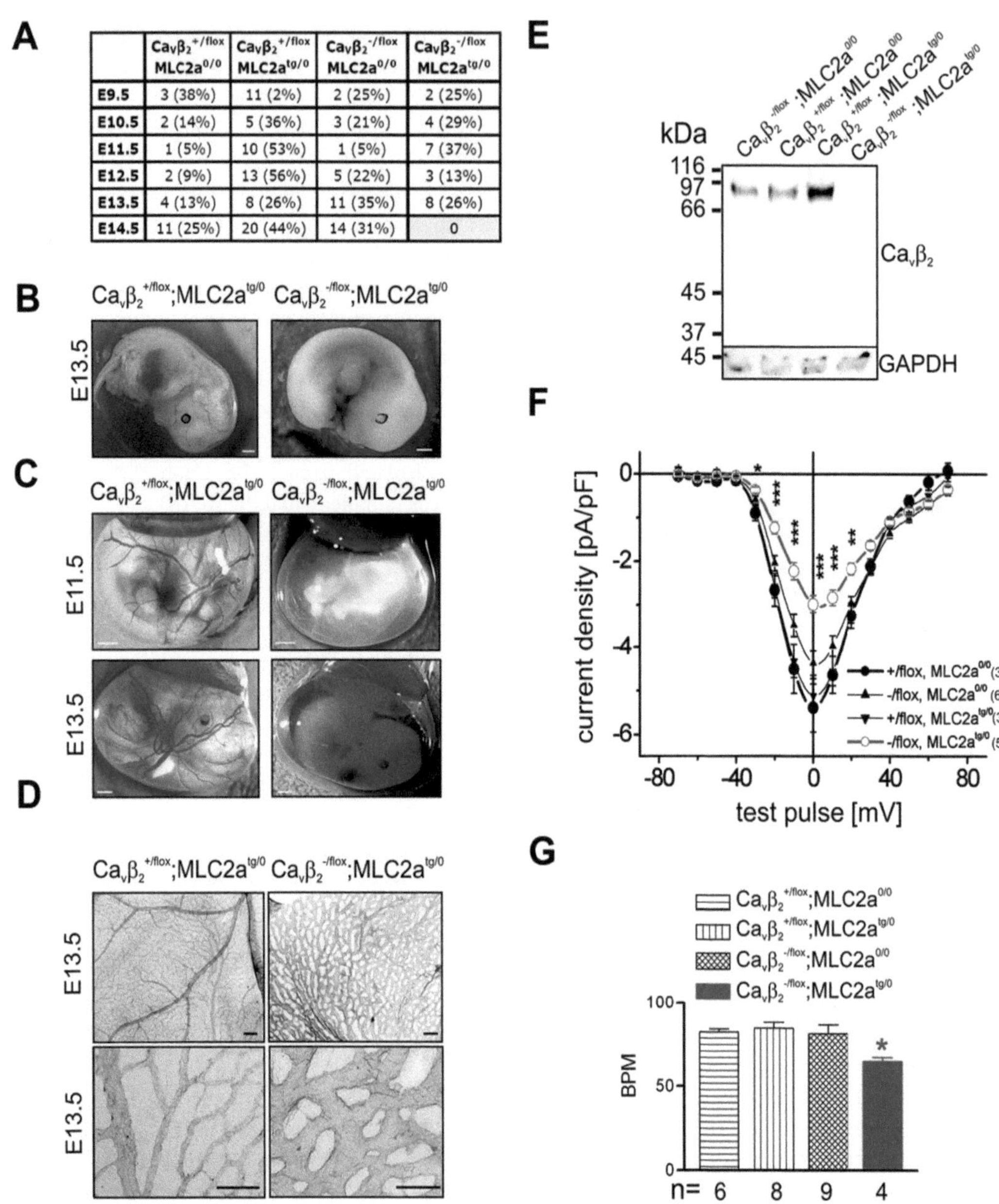

Fig. 17.4 Cardiomyocyte-restricted inactivation of the Ca$_v$β$_2$ gene. A, Numbers of surviving embryos from controls carrying at least 1 functional Ca$_v$β$_2$ allele (columns 1 to 3) and from Ca$_v$β$_2^{-/flox/MLC2atg/0}$ (column 4). B, Microscopic analysis of Ca$_v$β$_2^{-/flox/MLC2atg/0}$ embryos compared with controls (Ca$_v$β$_2^{+/flox; MLC2atg/0}$) at E13.5 (bar=1 mm). C, Microscopic assessment of the yolk sac and its vasculature in Ca$_v$β$_2^{-/flox/MLC2atg/0}$ embryos and controls (Ca$_v$β$_2^{+/flox/MLC2atg/0}$; bar=1 mm). D, Anti-PECAM staining of E13.5 yolk sacs (bar=100 μm). E, Western blot (50 μg of protein extracts per lane) using a Ca$_v$β$_2$ and a GAPDH antibody, for control: the Ca$_v$β$_2$ protein is expressed in hearts of E13.5 controls (Ca$_v$β$_2^{-/flox/MLC2a0/0}$, Ca$_v$β$_2^{+/flox/MLC2atg/0}$, and Ca$_v$β$_2^{+/flox/MLC2a0/0}$) but not in hearts of E13.5 Ca$_v$β$_2^{-/flox/MLC2atg/0}$ embryos. F, Averaged L-type I$_{Ca}$ current/voltage relationships of E13.5 cardiomyocytes from Ca$_v$β$_2^{-/flox/MLC2atg/0}$ embryos and controls. Number of cells are given. *Ca$_v$β$_2^{+/flox/MLC2a0/0}$ vs. Ca$_v$β$_2^{-/flox/MLC2atg/0}$ (*P<0.05, **P<0.01, ***P<0.001). G, Reduced spontaneous beatings of Ca$_v$β$_2^{-/flox/MLC2atg/0}$ E13.5 hearts. *P<0.05. BPM indicates beats per minute; n, number of explanted hearts.

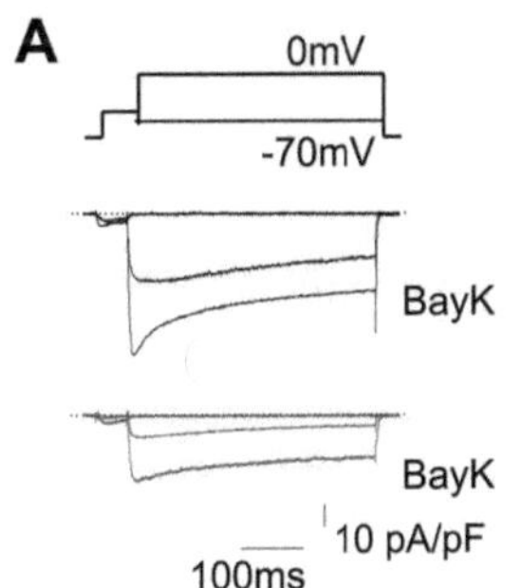

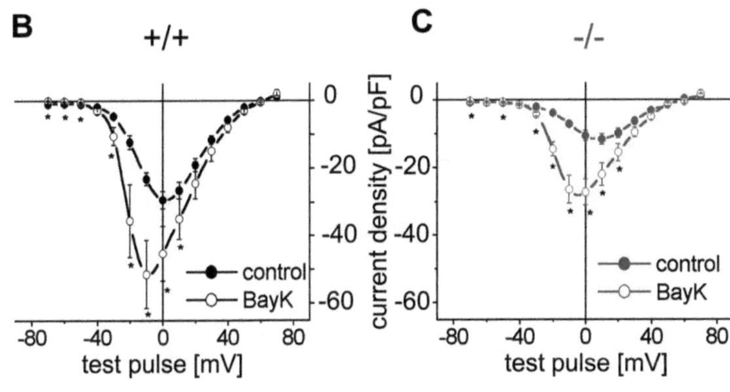

	+/+	+/-	-/-
E10.5	4 (20%)	11 (55%)	5 (25%)
E11.5	7 (19%)	19 (53%)	10 (27%)
E12.5	5 (42%)	7 (58%)	0

B +/+

C -/-

■ **Fig. 17.5** Partial pharmacological rescue of $Ca_V\beta_2^{-/-}$ embryos by (−)-BayK 8644. A, Voltage protocol (top) and representative I_{Ba} traces during a 400-ms depolarisation to −70 mV and to 0 mV in E9.5 wild-type (middle, black) and $Ca_V\beta_2^{-/-}$ (bottom, red) cardiomyocytes in the absence and presence of (−)-BayK 8644 (BayK). B and C, Average L-type I_{Ba} current/voltage relationships for wild-type (B) and $Ca_V\beta_2^{-/-}$ (C) E9.5 cardiomyocytes in the absence (wild type, n=17 cells; $Ca_V\beta_2^{-/-}$, n=14 cells) and presence of 1 µmol/l (−)-BayK 8644 (open symbols: wild type, n=8 cells; $Ca_V\beta_2^{-/-}$, n=9 cells). D, Numbers of viable wild-type, $Ca_V\beta_2^{+/-}$, and $Ca_V\beta_2^{-/-}$ embryos at E10.5, E11.5, and E12.5 after (−)-BayK 8644 treatment.

17.3.4 Partial Rescue of Embryonic Lethality by L-Type Ca^{2+} Channel Agonist

The residual Ca^{2+} channel current was still dihydropyridine sensitive in $Ca_V\beta_2^{-/-}$ cells (■ Fig. 17.3D). In the presence of (−)-BayK 8644 (1 µmol/l), the I_{Ba} was increased ≈1.8-fold from −29.3±2.6 pA/pF at 0 mV (n=17) to −51.4±10.1 pA/pF at −10 mV (n=8) in wild-type cells (■ Fig. 17.5A and ■ Fig. 17.5B) and ≈2.2-fold from −11.7±1.6 pA/pF at +10 mV (n=14) to −27.3±3.8 pA/pF at 0 mV (n=9) in $Ca_V\beta_2$-deficient cells (■ Fig. 17.5A and ■ Fig. 17.5C). The peaks of the whole-cell currents were typically shifted to less-positive voltages. This increase in current density could be sufficient to rescue the lethal phenotype and also to alleviate the detrimental consequences of cardiac failure in $Ca_V\beta_2^{-/-}$ embryos. Therefore, (−)-BayK 8644 was applied to the pregnant mice of heterozygous intercrosses. Inspection of E10.5, E11.5, and E12.5 embryos revealed that living $Ca_V\beta_2^{-/-}$ embryos could now be identified until E11.5 at the anticipated Mendelian ratio (■ Fig. 17.5D). Without treatment, a survival of $Ca_V\beta_2^{-/-}$ embryos at E11.5 has never been observed (■ Fig. 17.1B). Thus, the survival of $Ca_V\beta_2^{-/-}$ embryos can be prolonged in the presence of a Ca^{2+} channel agonist, supporting the overall conclusion that the described $Ca_V\beta_2^{-/-}$ phenotype was caused by a reduced L-type current density in cardiomyocytes resulting from the lack of $Ca_V\beta_2$. Unfortunately, the Ca^{2+} channel agonist (−)-BayK 8644 induces lethal self-biting behavior,[25] which limited considerably the applicable doses and the duration of treatment in mice.

17.4 Discussion

Little information is available on the developmental expression and the biophysical and pharmacological properties of cardiac L-type Ca^{2+} channels during the course of early embryonic development, especially at E9.5 and E10.5, when the heart starts to beat regularly. Therefore, we first characterised L-type Ca^{2+} currents in E9.5 and

E10.5 embryonic cardiomyocytes (■ Fig. 17.3 and supplemental Figure III). The L-type currents were sensitive to dihydropyridines, but, in agreement with previous studies,[26,27] we did not detect sensitivity to catecholamine or cAMP stimulation (supplemental Figure V). The L-type current density increased from E9.5 to E10.5, probably because of the augmentation of the L-type Ca^{2+} channel Ca$_V$1.1, Ca$_V$1.2, and Ca$_V$1.3 subunits, which are present throughout the fetal murine heart development.[7] The transmembrane Ca^{2+} influx through the L-type channels contributes to overall Ca^{2+} transients (■ Fig. 17.3), indicating that CICR is functioning in E10.5 cardiomyocytes and underlies contraction of these cells.

The role of HVA Ca^{2+} channels in the development of cardiac morphology and function is unclear. Here we show that in early embryonic stages L-type Ca^{2+} channel activity is an essential determinant of cardiomyocyte contraction and cardiac pump function, which in turn is required for cardiac morphogenesis, oxygen supply to peripheral organs, and blood vessel maturation. By disrupting the gene of the ancillary Ca^{2+} channel subunit Ca$_V$β$_2$, a significant reduction of L-type current density was observed in cardiac myocytes as early as E9.5. Concomitantly, the Ca^{2+} transient was diminished as was cardiac contractility and heart beat frequency. The function of the SR was, however, not changed in Ca$_V$β$_2^{-/-}$ cardiomyocytes, indicating that the observed effects were mainly confined to the voltage-activated Ca^{2+} influx pathway without additional alterations of other Ca^{2+} transport mechanisms, such as Na$^+$/Ca^{2+} exchange or Ca^{2+} release from the SR.

In mammals, embryonic survival and tissue integrity are largely dependent on cardiac pump function and blood flow, and our results demonstrate that L-type channel activity is an essential determinant of these processes. The reduction of L-type Ca^{2+} channel activity by disruption of the Ca$_V$β$_2$ gene led to embryonic heart failure associated with pericardial effusion and, finally, embryonic death after E10.5. Consistent with our results, inactivation of the Ca$_V$β$_2$ gene was reported to lead to preterm lethality, but embryogenesis and L-type Ca^{2+} channels were not analysed at all in that study.[28] Ineffective Ca^{2+} entry into myocytes

might perturb cell growth and integrity at the transcriptional level but the size of cardiomyocytes, the myofibrillar structure, and cell-cell contacts appeared unchanged indicating a normal morphology. Considering the reduction in heart beat frequency and the virtual absence of coordinated blood ejection, the embryonic death can be traced to cardiac pump dysfunction. Compared with the normal myofibrillar structure in Ca$_V$β$_2^{-/-}$ cardiomyocytes, myofibrillogenesis is strongly affected in cardiomyocytes lacking either the Ca^{2+}-binding protein calreticulin, in which bradykinin-induced elevations of intracellular Ca^{2+} concentration is almost completely absent,[29] or the sodium-calcium exchanger (NCX1), which lack both spontaneous Ca^{2+} transients and contraction.[30] In these examples, a role for Ca^{2+} as a trophic factor regulating gene expression and cardiomyocyte differentiation was suggested, but given the intact morphology of Ca$_V$β$_2^{-/-}$ cardiomyocytes, there is no indication that such mechanisms might be of major relevance when L-type Ca^{2+} channel-mediated Ca^{2+} entry is reduced in the absence of Ca$_V$β$_2$ proteins.

As a consequence of impaired cardiac contraction, the hemodynamic forces are altered, which cause additional functional disturbances including impaired maturation of extra- and intraembryonic vasculature and reduced oxygen supply in Ca$_V$β$_2^{-/-}$ embryos. One could expect expression of Ca$_V$β$_2$ in embryonic blood vessels and the observed vascular defects being a result of Ca$_V$β$_2$ gene disruption in this tissue. However, Ca$_V$β$_2$ transcripts were only barely discernible in wild-type yolk sac, and we could not detect expression of the Ca$_V$β$_2$ protein in this tissue (■ Fig. 17.2E). In addition, mouse embryos that died after cardiomyocyte-restricted inactivation of the Ca$_V$β$_2$ gene share strikingly similar cardiac and vascular aberrations with embryos that died after ubiquitous inactivation of the Ca$_V$β$_2$ gene. Although initial vascular plexus had been formed and capillary formation exists in wild-type and mutated mouse lines, the development into a mature vascular network was lacking in both mutant mouse lines. These results agree with the finding that the capability to form vessel tubes and sprouts is a cell autonomous property of vascular progenitor cells, whereas vascular maturation and remodeling depends on blood flow evoked by a cardiac contraction-induced pressure gradient. Defects in vascular maturation associated with cardiac failure have been described,[30] but the causative role of embryonic heart malfunction on the developing vasculature by cell specific inactivation of the gene in cardiomyocytes as described in this study has not yet been reported.

Final support for the conclusion that L-type Ca^{2+} channel activity is essential for embryonic development and survival resulted from the partial rescue of the observed phenotype by $(-)$-BayK 8644 acting as an agonist on Ca^{2+} channel activity. The residual L-type Ca^{2+} channel activity in $Ca_V\beta_2^{-/-}$ cardiomyocytes was sensitive to dihydropyridines, and application of $(-)$-BayK 8644 to pregnant mice protracts death in $Ca_V\beta_2^{-/-}$ embryos. Taken together, our results indicate a dominant role of L-type Ca^{2+} channel activity in cardiomyocytes for CICR, cardiac contraction, and heart pump function and that the functioning early embryonic heart depends on transmembrane Ca^{2+} influx via HVA L-type Ca^{2+} channels.

17.5 References

[1] Reuter H. Calcium channel modulation by neurotransmitters, enzymes and drugs. Nature. 1983; 301: 569–574.

[2] Bers DM. Cardiac excitation-contraction coupling. Nature. 2002; 415: 198–205.

[3] Fabiato A, Fabiato F. Calcium and cardiac excitation-contraction coupling. Annu Rev Physiol. 1979; 41: 473–484.

[4] Catterall WA. Structure and regulation of voltage-gated Ca^{2+} channels. Annu Rev Cell Dev Biol. 2000; 16: 521–555.

[5] Mikami A, Imoto K, Tanabe T, Niidome T, Mori Y, Takeshima H, Narumiya S, Numa S. Primary structure and functional expression of the cardiac dihydropyridine-sensitive calcium channel. Nature. 1989; 340: 230–233.

[6] Seisenberger C, Specht V, Welling A, Platzer J, Pfeifer A, Kuhbandner S, Striessnig J, Klugbauer N, Feil R, Hofmann F. Functional embryonic cardiomyocytes after disruption of the L-type alpha1C (Cav1.2) calcium channel gene in the mouse. J Biol Chem. 2000; 275: 39193–39199.

[7] Xu M, Welling A, Paparisto S, Hofmann F, Klugbauer N. Enhanced expression of L-type Cav1.3 calcium channels in murine embryonic hearts from Cav1.2-deficient mice. J Biol Chem. 2003; 278: 40837–40841.

[8] Chaudhari N. A single nucleotide deletion in the skeletal muscle-specific calcium channel transcript of muscular dysgenesis (mdg) mice. J Biol Chem. 1992; 267: 25636–25639.

[9] Mangoni ME, Couette B, Bourinet E, Platzer J, Reimer D, Striessnig J, Nargeot J. Functional role of L-type Cav1.3 Ca2+ channels in cardiac pacemaker activity. Proc Natl Acad Sci U S A. 2003; 100: 5543–5548.

[10] Zhang Z, Xu Y, Song H, Rodriguez J, Tuteja D, Namkung Y, Shin HS, Chiamvimonvat N. Functional Roles of Ca(v)1.3 (alpha(1D)) calcium channel in sinoatrial nodes: insight gained using gene-targeted null mutant mice. Circ Res. 2002; 90: 981–987.

[11] Ludwig A, Flockerzi V, Hofmann F. Regional expression and cellular localization of the alpha1 and beta subunit of high voltage-activated calcium channels in rat brain. J Neurosci. 1997; 17: 1339–1349.

[12] Murakami M, Fleischmann B, De Felipe C, Freichel M, Trost C, Ludwig A, Wissenbach U, Schwegler H, Hofmann F, Hescheler J, Flockerzi V, Cavalie A. Pain perception in mice lacking the beta3 subunit of voltage-activated calcium channels. J Biol Chem. 2002; 277: 40342–40351.

[13] Berggren PO, Yang SN, Murakami M, Efanov AM, Uhles S, Kohler M, Moede T, Fernstrom A, Appelskog IB, Aspinwall CA, Zaitsev SV, Larsson O, de Vargas LM, Fecher-Trost C, Weissgerber P, Ludwig A, Leibiger B, Juntti-Berggren L, Barker CJ, Gromada J, Freichel M, Leibiger IB, Flockerzi V. Removal of Ca^{2+} channel beta3 subunit enhances Ca^{2+} oscillation frequency and insulin exocytosis. Cell. 2004; 119: 273–284.

[14] Richards MW, Butcher AJ, Dolphin AC. Ca^{2+} channel beta-subunits: structural insights AID our understanding. Trends Pharmacol Sci. 2004; 25: 626–632.

[15] Gregg RG, Messing A, Strube C, Beurg M, Moss R, Behan M, Sukhareva M, Haynes S, Powell JA, Coronado R, Powers PA. Absence of the beta subunit (cchb1) of the skeletal muscle dihydropyridine receptor alters expression of the alpha 1 subunit and eliminates excitation-contraction coupling. Proc Natl Acad Sci U S A. 1996; 93: 13961–13966.

[16] Namkung Y, Smith SM, Lee SB, Skrypnyk NV, Kim HL, Chin H, Scheller RH, Tsien RW, Shin HS. Targeted disruption of the Ca^{2+} channel beta3 subunit reduces N- and L-type Ca^{2+} channel activity and alters the voltage-dependent activation of P/Q-type Ca^{2+} channels in neurons. Proc Natl Acad Sci U S A. 1998; 95: 12010–12015.

[17] Burgess DL, Jones JM, Meisler MH, Noebels JL. Mutation of the Ca^{2+} channel beta subunit gene Cchb4 is associated with ataxia and seizures in the lethargic (lh) mouse. Cell. 1997; 88: 385–392.

[18] Fleischmann M, Bloch W, Kolossov E, Andressen C, Muller M, Brem G, Hescheler J, Addicks K, Fleischmann BK. Cardiac specific expression of the green fluorescent protein during early murine embryonic development. FEBS Lett. 1998; 440: 370–376.

[19] Wettschureck N, Rutten H, Zywietz A, Gehring D, Wilkie TM, Chen J, Chien KR, Offermanns S. Absence of pressure overload induced myocardial hypertrophy after conditional inactivation of Galphaq/Galpha11 in cardiomyocytes. Nat Med. 2001; 7: 1236–1240.

[20] Novak A, Guo C, Yang W, Nagy A, Lobe CG. Z/EG, a double reporter mouse line that expresses enhanced green fluorescent protein upon Cre-mediated excision. Genesis. 2000; 28: 147–155.

[21] Hamill OP, Marty A, Neher E, Sakmann B, Sigworth FJ. Improved patch-clamp techniques for high-resolution current recording from cells and cell-free membrane patches. Pflugers Arch. 1981; 391: 85–100.

[22] Acosta L, Haase H, Morano I, Moorman AF, Franco D. Regional expression of L-type calcium channel subunits during cardiac development. Dev Dyn. 2004; 230: 131–136.

[23] Carmeliet P. Mechanisms of angiogenesis and arteriogenesis. Nat Med. 2000; 6: 389–395.

[24] Pearson HA, Dolphin AC. Inhibition of omega-conotoxin-sensitive Ca^{2+} channel currents by internal Mg^{2+} in cultured rat cerebellar granule neurones. Pflugers Arch. 1993; 425: 518–527.

[25] Jinnah HA, Yitta S, Drew T, Kim BS, Visser JE, Rothstein JD. Calcium channel activation and self-biting in mice. Proc Natl Acad Sci U S A. 1999; 96: 15228–15232.

[26] An RH, Davies MP, Doevendans PA, Kubalak SW, Bangalore R, Chien KR, Kass RS. Developmental changes in beta-adrenergic modulation of L-type Ca^{2+} channels in embryonic mouse heart. Circ Res. 1996; 78: 371–378.

[27] Davies MP, An RH, Doevendans P, Kubalak S, Chien KR, Kass RS. Developmental changes in ionic channel activity in the embryonic murine heart. Circ Res. 1996; 78: 15–25.

[28] Ball SL, Powers PA, Shin HS, Morgans CW, Peachey NS, Gregg RG. Role of the beta(2) subunit of voltage-dependent calcium channels in the retinal outer plexiform layer. Invest Ophthalmol Vis Sci. 2002; 43: 1595–1603.

[29] Li J, Puceat M, Perez-Terzic C, Mery A, Nakamura K, Michalak M, Krause KH, Jaconi ME. Calreticulin reveals a critical Ca(2+) checkpoint in cardiac myofibrillogenesis. J Cell Biol. 2002; 158: 103–113.

[30] Koushik SV, Wang J, Rogers R, Moskophidis D, Lambert NA, Creazzo TL, Conway SJ. Targeted inactivation of the sodium-calcium exchanger (Ncx1) results in the lack of a heartbeat and abnormal myofibrillar organization. FASEB J. 2001; 15: 1209–1211.

Overexpression of junctin causes adaptive changes in cardiac myocyte Ca^{2+} signaling

Uwe Kirchhefer, Gabriela Hanske, Larry R. Jones, Isabel Justus, Lars Kaestner, Peter Lipp, Wilhelm Schmitz, Joachim Neumann

Reprint from Cell Calcium (2006) **39**, 131-142.

■ **Abstract**

In cardiac muscle, junctin forms a quaternary protein complex with the ryanodine receptor (RyR), calsequestrin, and triadin 1 at the luminal face of the junctional sarcoplasmic reticulum (jSR). By binding directly the RyR and calsequestrin, junctin may mediate the Ca^{2+}-dependent regulatory interactions between both proteins. To gain more insight into the underlying mechanisms of impaired contractile relaxation in transgenic mice with cardiac-specific overexpression of junctin (TG), we studied cellular Ca^{2+} handling in these mice. We found that the SR Ca^{2+} load was reduced by 22% in cardiomyocytes from TG mice. Consistent with this, the frequency of Ca^{2+} sparks was diminished by 32%. The decay of spontaneous Ca^{2+} sparks was prolonged by 117% in TG. This finding was associated with a lower Na^+-Ca^{2+} exchanger (NCX) protein expression (by 67%) and a higher basal RyR phosphorylation at Ser^{2809} (by 64%) in TG. The shortening- and $\Delta[Ca]_i$-frequency relationships (0.5–4 Hz) were flat in TG compared to wild-type (WT) which exhibited a positive staircase for both parameters. Furthermore, increasing stimulation frequencies hastened the time of relaxation and the decay of $[Ca]_i$ by a higher percentage in TG. We conclude that the impaired relaxation in TG may result from a reduced NCX expression and/or a higher SR Ca^{2+} leak. The altered shortening-frequency relationship in TG seems to be a consequence of an impaired excitation–contraction coupling with depressed SR Ca^{2+} release at higher rates of stimulation. Our data suggest that the more prominent frequency-dependent hastening of relaxation in TG results from a stimulation of SR Ca^{2+} transport reflected by corresponding changes of $[Ca]_i$.

18.1 Introduction

The Ca^{2+} release units of the jSR are formed by stable uniform complexes of four proteins consisting of the Ca^{2+} release channel (i.e., ryanodine receptor, RyR), calsequestrin (CSQ), triadin 1 (TRD), and junctin (JCN)[1,2]. The RyR contains a large cytoplasmic domain that reaches into the space between jSR and the T-tubules and a transmembrane domain that forms the conduction pore of the Ca^{2+} release channel[3,4]. CSQ is a Ca^{2+} storage protein in the lumen of the jSR[5]. It forms an electron-dense matrix in the jSR lumen which was ascribed to the ability of CSQ to sequester Ca^{2+} near the Ca^{2+} release sites[6-8]. CSQ may act as a Ca^{2+} sensor that inhibits the Ca^{2+} release channel at low luminal calcium[9] and is anchored to the RyR either directly or via two auxiliary proteins, TRD and JCN. TRD and JCN are structurally similar transmembrane proteins constituted of short N-terminal cytoplasmic domains and highly charged C-terminal domains projecting into the jSR lumen[10,11]. TRD and JCN may be involved in SR luminal Ca^{2+} sensing by mediating the inhibitory interactions between CSQ and the RyR[2]. In addition, a direct modulatory role of TRD has been proposed by inhibiting the activity of the skeletal muscle RyR in reconstitutional systems[12].

JCN like TRD has a jSR luminal domain which accounts for the binding of CSQ and to the RyR. The interaction of JCN with CSQ is diminished at high Ca^{2+} concentrations[2]. Furthermore, JCN can also bind to TRD independently of different Ca^{2+} concentrations and to itself[2]. Junctin runs as a 26-kDa protein in SDS-PAGE and is highly enriched in the jSR in both cardiac and skeletal muscle[11]. Ultrastructural studies have shown that JCN is mainly localised at the peripheral (i.e., junctional) SR[11,13,14]. It has been suggested that JCN (and TRD) may play more direct roles in Ca^{2+} release beyond simply anchoring CSQ to the RyR. In line with this hypothesis, the application of JCN to the luminal side of purified cardiac RyRs increased the open probability of the channels[9]. To shed light on the functional role of JCN in the regulation of SR Ca^{2+} release in the heart, we developed a transgenic mouse model (TG) with cardiac-specific overexpression of JCN[15]. The 10-fold overexpression of JCN was associated with a tighter packing of CSQ in proximity of the jSR membrane[14]. Moreover, TG mice exhibited a mild hypertrophy and a reduced expression of TRD and the RyR. The relaxation was prolonged in papillary muscles and isolated cardiomyocytes of TG mice. Consistently, the decay of Ca^{2+} transients was prolonged in TG[16].

Here, we have extended these initial observations by studying in depth the physiological mechanisms accounting for the impaired relaxation in TG. Overexpression of JCN slowed the decay of spontaneous Ca^{2+} sparks. This effect was associated with a dramatically down-regulated NCX protein expression in TG and an enhanced basal phosphorylation of the RyR at Ser^{2809}. Moreover, relaxation was more hastened at increasing stimulation frequencies in TG. Thus, our data suggest that the prolonged relaxation in TG is the result of an impaired global cellular Ca^{2+} homeostasis which is caused by local structural and functional alterations of the sarcolemmal and jSR Ca^{2+} cycling.

18.2   Methods

18.2.1   Experimental animals

TG mice with cardiac-specific overexpression of canine junctin were generated and identified as previously described[15]. All experiments were performed on mice of 4–5 months of age. Mice were handled and maintained according to protocols approved by the animal welfare committees of the University of Münster, Germany, and Indiana University, USA, which also conform to the NIH Guidelines for the Care and Use of Laboratory Animals.

18.2.2   Ca²⁺ATPase assay

SR vesicles were prepared from individual ventricles of TG and WT mice according to a modified protocol as described[17]. Ca^{2+}ATPase in SR vesicles was determined by measuring colorimetrically the release of inorganic phosphate (P_i) from ATP at different $CaCl_2$ concentrations[18,19].

18.2.3   Measurement of Ca²⁺ transients, cell shortening, and SR Ca²⁺ load

Cardiomyocytes were enzymatically isolated from TG and WT hearts by a standard procedure as reported before[17]. Intracellular Ca^{2+} transients were determined at 0.5 Hz and 2 mM extracellular Ca^{2+} with Indo-1/AM (Sigma, St. Louis, MO, USA). The fluorescence indicator was excited at 365 nm. The emitted fluorescence was detected at 405 and 495 nm. The cytosolic free Ca^{2+} concentration was estimated by calculating the ratio of Indo-1 signals (405/495 nm). The maximum Ca^{2+} concentration, the minimum Ca^{2+} concentration, and their difference were set as the systolic ratio, diastolic ratio, and $\Delta[Ca]_i$, respectively. Where indicated, stimulation frequency was increased stepwise from 0.5 to 4 Hz. At each frequency steady state was reached after 5 min. The shortening of cardiomyocytes was recorded simultaneously using a video edge detection system[16]. Moreover, the SR Ca^{2+} content was estimated by measurement of Ca^{2+} transients in the presence of 10 mM caffeine and 5 mM Ni^{2+}, blocking effectively the $I_{Na/Ca}$ current[20].

18.2.4   Detection of Ca²⁺ sparks by confocal microscopy

Cardiomyocytes of either TG or WT animals were placed on poly-L-lysin coated coverslips and loaded with 2 µM of the indicator fluo-4/AM (Molecular Probes, Leiden, The Netherlands) for 30 min followed by a period of 10 min for de-esterification of the internalised dye. Excitation of fluo-4 was performed with the 488 nm line of an Argon- or Argon-Krypton laser. The Ca^{2+} spark frequency and the characteristics of individual Ca^{2+} sparks were measured with laser-scanning confocal microscopes. We used a Nipkow-disk-based confocal microscope (QLC-100, VisiTech International, Sunderland, UK) to estimate the Ca^{2+} spark frequency in quiescent cardiomyocytes during 27 s long recording periods at a constant frame rate of 19 Hz. Spatiotemporal fine details of individual Ca^{2+} sparks were recorded on an ultra-fast confocal microscope (VTeye; Visi-Tech International, Sunderland, UK). For this we acquired 2D-data from resting cardiomyocytes at a constant recording frequency of 120 or 360 Hz. The resulting images were analysed using ImageJ (Wayne Rasband, NIH, Bethesda, USA) and Igor-Pro 4.0 software (WaveMetrics, Lake Oswego, USA). The amplitude of individual Ca^{2+} sparks was analysed by the following procedure. A box of 5 × 5 pixels was drawn around each maximal increased fluo-4 fluorescence signal (F), which had to be at least 25% above the minimal resting level (F_0). The background was subtracted from both parameters and ratio F/F_0 was calculated.

18.2.5 Heart perfusion

To determine the phosphorylation of PLN, hearts from TG and WT mice were perfused in the antero-grade mode using a working-heart set-up[17]. After a 30-min stabilisation period, either buffer alone or buffer supplemented with isoproterenol (Iso, 0.1 µM) was applied to the perfusion system for further 10 min. For measurement of basal RyR phospho-rylation, hearts were perfused using the procedure described above with 1 mM NaF in the perfusate[21].

18.2.6 Immunoblotting

Aliquots (50 µg) of prepared membrane vesicles from ventricles of TG and WT mice were solu-bilised in 5% SDS-stop solution[22], subjected to 8% SDS-PAGE[23], and transferred to nitrocellulose. The nitrocellulose was incubated with the mouse monoclonal antibody C2C12 raised to purified canine cardiac NCX (ABR, Golden, CO, USA)[20] and with a rabbit antibody to canine cardiac CSQ[24]. Antibody-reacting bands were detected by an anti-mouse antibody conjugated with alkaline phosphatase (Sigma) and by ^{125}I-labeled protein A (Perkin-Elmer, Boston, MA, USA). Signals were quantified using a PhosphorImager (Bio Rad, Hercules, CA, USA). In addition, perfused hearts from TG and WT mice were homogenised in 10 mM $NaHCO_3$, 50 mM NaF, and 5 mM NaP_2O_7 (pH 7.4) at 4 °C for 90 s. Homogenates were centrifuged at $10,000 \times g$ for 20 min at 4 °C. Aliquots of superna-tants were incubated at 30 °C with equal volumes of loading buffer (7.5% SDS, 62.5 mM Tris–HCl, pH 6.8, 20% glycerol, 40 mM dithiothreitol, and a trace of bromphenol blue). Supernatant protein was sep-arated on 5% or 10% SDS-PAGE[23] and transferred to nitrocellulose membranes which were probed separately with antibodies against PLN phospho-serine 16 (Upstate Biotechnology, Lake Placid, NY, USA), phosphothreonine 17 (Badrilla, Leeds, UK), and the non-phosphorylation-specific A1 monoclonal antibody (Upstate Biotechnology)[25] or against RyR phosphoserine 2809 (Badrilla) and the non-phosphorylation-specific RyR 1E9 monoclo-nal antibody[2]. Antibody binding was detected by secondary antibodies and then quantified using a PhosphorImager (Bio-Rad).

18.2.7 Statistical analysis

Data are reported as mean ± S.E.M. Statistical dif-ferences between the different types of mice were calculated by ANOVA followed by Bonferroni's t-test. $P < 0.05$ was considered significant.

18.3 Results

18.3.1 Unchanged SERCA2a function and enrichment of CSQ in SR vesicles in TG

Here we found that a decreased SR Ca^{2+} uptake by SERCA2a cannot account for the slower decline of $[Ca]_i$ in TG[16]. The maximal velocity of the Ca^{2+}

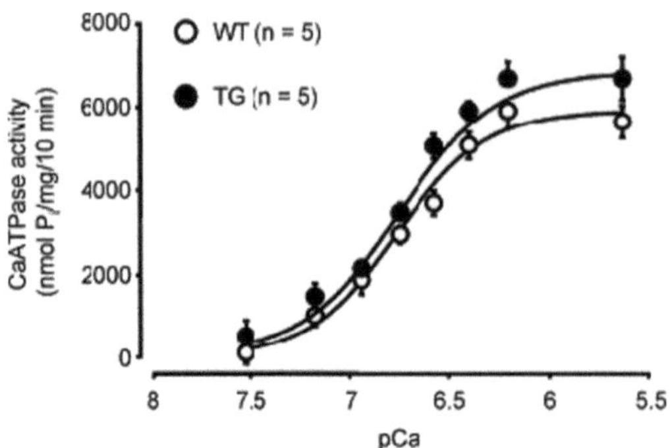

◘ Fig. 18.1 Ca²⁺ ATPase activity in membrane vesicles. Ca²⁺ ATPase activity was determined in membrane vesicles of TG and WT hearts. The reaction was incubated for 10 min and the Ca²⁺-dependent ATPase activity was determined with the malachite green method.

ATPase activity (V_{max}) was unchanged between TG and WT hearts (◘ Fig. 18.1) when measured at saturating ionised Ca²⁺ concentrations (≥1 µM). Moreover, K_{Ca}, indicating the Ca²⁺ concentration required for half-maximal activation of SERCA2a, was comparable between TG and WT hearts (◘ Fig. 18.1). Thus, Ca²⁺ transport kinetics between TG and WT appeared identical. In addition, CSQ protein expression was increased by 23% in SR vesicles from TG compared to WT hearts ($n = 6$–7, $P < 0.05$).

18.3.2 Reduced expression of the NCX and lower SR Ca²⁺ load in TG

Interestingly, caffeine-induced Ca²⁺ transients were prolonged in cardiomyocytes from TG hearts in comparison to WT[16] suggesting an impaired function of the NCX. Indeed, we found that the protein expression of the NCX was reduced by 67% in TG taking together the sum of both molecular mass forms at 120 and 160 kDa (◘ Fig. 18.2A and B). This finding was supported by experimental results that depicted a greater reduction in the decline of Ca²⁺ transients in WT in comparison to TG cardiomyocytes when inhibiting NCX function with 5 mM Ni²⁺ (◘ Fig. 18.2C). Since an impaired NCX function may affect the Ca²⁺ extrusion during the rise of [Ca]$_i$[26], resulting

in an increased amplitude of the Ca²⁺ transient, we analyzed Δ[Ca]$_i$ in the presence and absence of Ni²⁺ and found that the amplitude reduction of caffeine-induced Ca²⁺ transient that we have found in TG cardiomyocytes[16] was unchanged by the presence of Ni²⁺ (◘ Fig. 18.2D).

18.3.3 Altered Ca²⁺ spark parameters in TG

It has been proposed that the Ca²⁺ spark frequency might directly depend on the loading state of the SR[27]. Thus, we recorded spontaneous Ca²⁺ sparks and estimated their frequency in TG and WT cardiomyocytes using confocal microscopy and found that the frequency of Ca²⁺ sparks was reduced by 32% in TG (◘ Fig. 18.3A and B). Moreover, the Ca²⁺ spark amplitude (F/F₀) was slightly reduced in TG compared to WT (from 1.51 ± 0.02 to 1.44 ± 0.01, $n = 13$–22, $P < 0.05$). In addition, the spatial spread of Ca²⁺ sparks, measured as the full width at half-maximal fluorescence intensity (FWHM), was increased by 73% in TG (◘ Fig. 18.3C–E). In parallel to the increased spatial spread we also found an increased relaxation phase of the TG Ca²⁺ sparks as depicted in ◘ Fig. 18.3F. The time constant of decay of Ca²⁺ sparks was enhanced by 117% in TG (◘ Fig. 18.3G). The distribution of Ca²⁺ spark amplitudes revealed one population both in TG and WT (◘ Fig. 18.4A). Furthermore, only one population of Ca²⁺ sparks was also observed for FWHM (◘ Fig. 18.4B) and the time constant of decay (◘ Fig. 18.4C) both in TG and WT.

18.3.4 Hyperphosphorylation of the RyR in TG

The diminished SR Ca²⁺ load in heart failure has been attributed to a hyperphosphorylation of the RyR by cAMP-dependent protein kinase (PKA), which may result in »leaky« SR Ca²⁺ release channels[28]. Therefore, we tested the basal phosphorylation state of the RyR (◘ Fig. 18.5A). We found that the basal phosphorylation of the RyR at Ser²⁸⁰⁹, representing the amino acid residue which is phos-

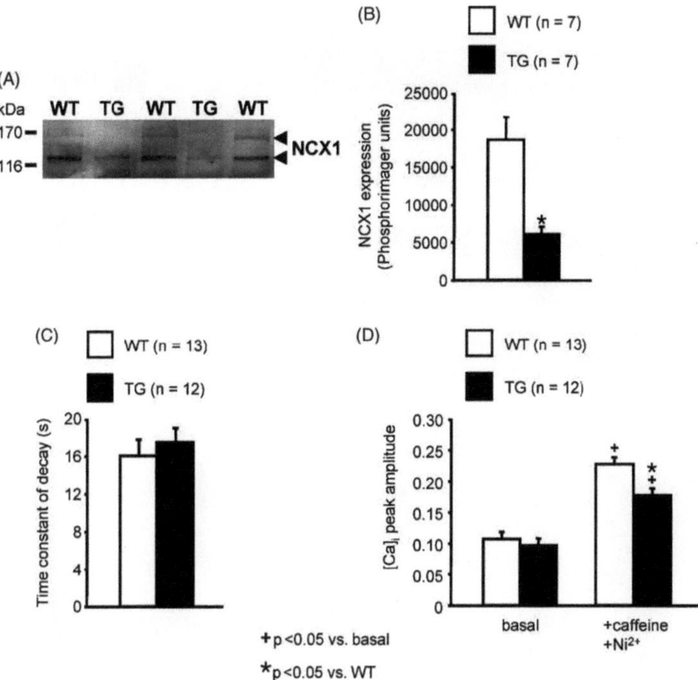

Fig. 18.2 Expression of the NCX and effects of caffeine and Ni^{2+} on Ca^{2+} transients. The protein expression of the NCX was measured in membrane vesicles of TG and WT hearts (A). Identical amounts of protein (50 μg) were loaded per lane. Blots were probed with an antibody against cardiac NCX as described under Section 18.2. Shown is a representative immunoblot. The indicated 120-kDa and 160-kDa molecular mass forms of NCX (arrowheads) were analysed and summarised (B). The time constant of decay of Ca^{2+} transients $[Ca^{2+}]_i$ was determined in the presence of 10 mM caffeine and 5 mM Ni^{2+} in unstimulated TG and WT cardiomyocytes (C). The amplitude of the Ca^{2+} transients ($\Delta[Ca]_i$) was measured at basal conditions in electrically stimulated cardiomyocytes and in response to 10 mM caffeine and 5 mM Ni^{2+} in quiescent TG and WT cardiomyocytes (D).

phorylated by Ca^{2+}/calmodulin-dependent protein kinase II (CaMKII) and PKA, was increased by 64% (i.e., hyperphosphorylated) in TG compared to WT hearts (**Fig. 18.5B**)[21].

18.3.5 Increased frequency-dependent acceleration of relaxation in TG

Fig. 18.6A and B show representative shortening-frequency relationships in cardiomyocytes of WT and TG mice, respectively. Frequency was increased stepwise from 0.5 to 4 Hz in both groups. The shortening-frequency relationship was positive in WT compared to TG where shortening-frequency relationship was flat (**Fig. 18.6C**). The ratio of shortening at 4 Hz/0.5 Hz was used as an index of inotropy mediated by increasing frequency. A ratio over 1.0 represent a positive inotropy, whereas a ratio less than 1.0 a negative inotropy[29]. The ratio was increased by 122% in WT. In contrast, there was no significant difference of this ratio in TG (**Fig. 18.6D**). Increasing stimulation frequencies accelerated the half-time

of relengthening in WT, but to a higher extent in TG (**Fig. 18.6E**). This is because of a prolonged relaxation in TG at the basal stimulation frequency of 0.5 Hz. The ratio of $t_{50\%}$ at 4 Hz/0.5 Hz was used as an index of the frequency-dependent acceleration of relaxation (FDAR)[29]. FDAR decreased $t_{50\%}$ by 36% in WT, whereas in TG the effect of FDAR was even 52% (**Fig. 18.6F**). To test whether the frequency-dependent effects on cardiomyocyte shortening and relaxation are paralleled by alterations in Ca^{2+} transients, $\Delta[Ca]_i$ and decay of $[Ca]_i$ was measured simultaneously. Typical traces of $[Ca]_i$ measurements are shown for WT (**Fig. 18.7A**) and TG (**Fig. 18.7B**). **Fig. 18.7C** shows that the relationship between $\Delta[Ca]_i$ and the frequency was positive in WT cardiomyocytes, whereas this relationship was nearly flat in TG. The $\Delta[Ca]_i$ was reduced from 2 to 4 Hz in TG. However, the ratio of $\Delta[Ca]_i$ at 4 Hz/0.5 Hz was comparable between TG and WT cardiomyocytes (**Fig. 18.7D**). FDAR was clearly apparent both in TG and WT (**Fig. 18.7E**). The time to 50% decay of $[Ca]_i$ was prolonged at 0.5 Hz in TG. The increase of the stimulation frequency up to

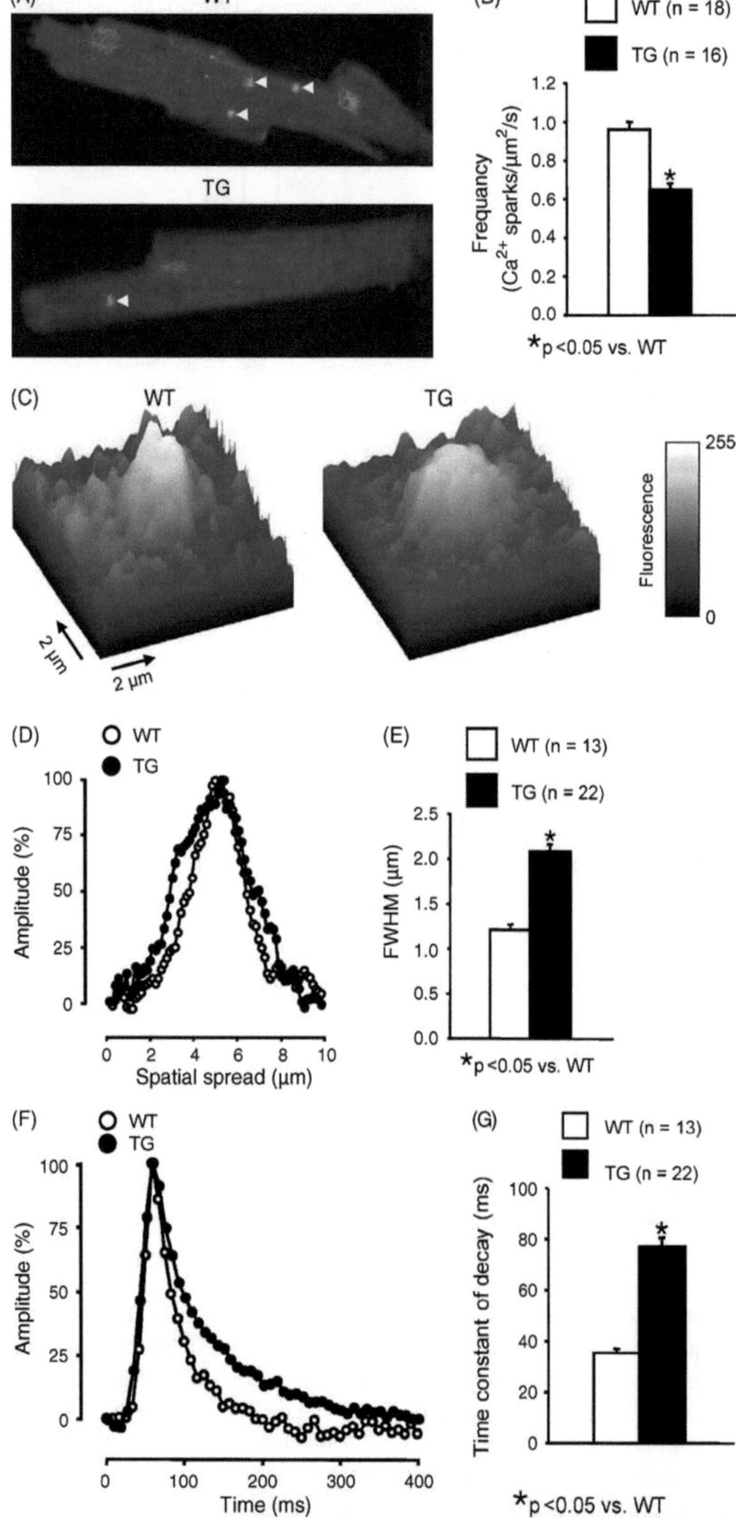

☐ **Fig. 18.3** Characteristics of spontaneous Ca^{2+} sparks. Shown are representative Fluo-4-loaded cardiomyocytes from WT and TG mice (A). Spontaneous Ca^{2+} release events are marked by arrowheads. The summarised data for Ca^{2+} spark frequency are given (B). In addition, shown are representative Fluo-4 fluorescence signals depicting individual spontaneous Ca^{2+} sparks in resting WT and TG cardiomyocytes (C). The spatial spread was averaged on 20 individual Ca^{2+} sparks of both groups (D). The full width at half-maximal fluorescence intensity (FWHM) was increased in TG cardiomyocytes (E). The upstroke and decay kinetics were averaged on 20 individual Ca^{2+} sparks of TG and WT cardiomyocytes (F). Note the prolonged decline of the Fluo-4 fluorescence signal in the TG cardiomyocytes. The summarised data are given (G).

18

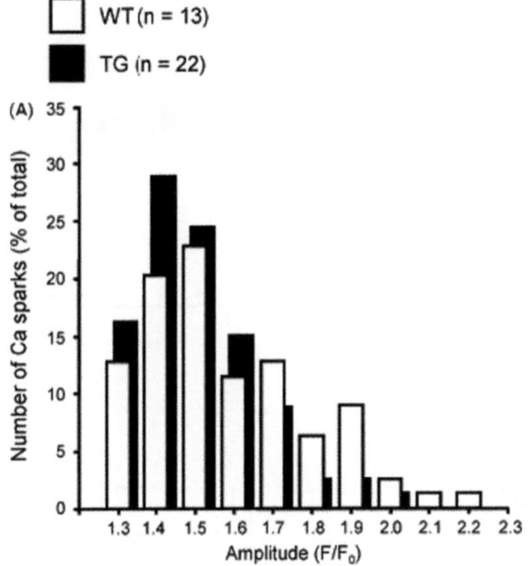

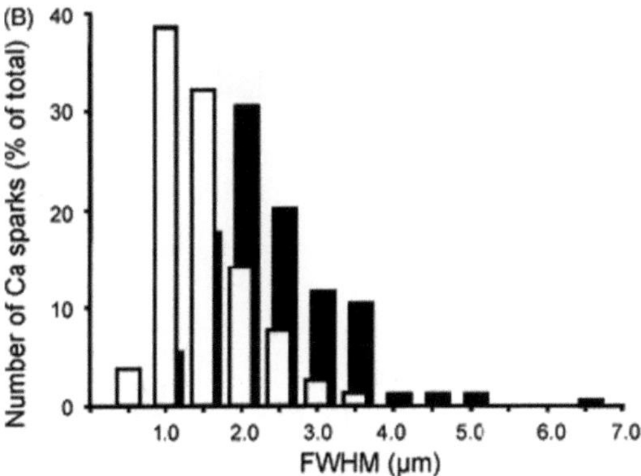

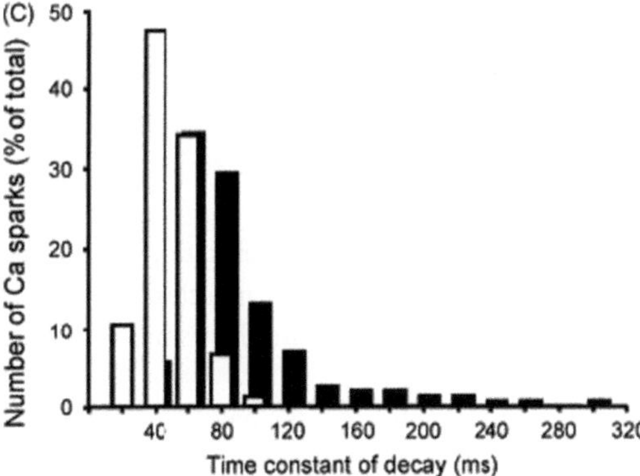

◘ **Fig. 18.4** Distribution of Ca^{2+} spark parameters. Shown are histograms of the number of events for the Ca^{2+} spark amplitude (A), the full width at half-maximal fluorescence intensity (B) and the time constant of decay (C) in TG and WT cardiomyocytes.

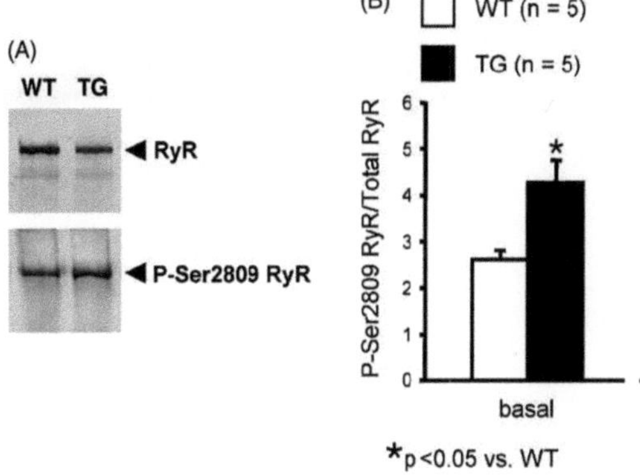

◘ **Fig. 18.5** Phosphorylation state of the RyR. Proteins in homogenates of isolated perfused WT and TG hearts were separated by SDS-PAGE and transferred to nitrocellulose membranes. RyR phosphorylated at Ser^{2809} and total RyR protein were detected with specific antibodies and representative immunoblots are given (A). Antibody-reacting bands for phosphorylated RyR at Ser^{2809} and non-phosphorylated RyR were quantified under basal conditions and the ratio is given (B).

4 Hz resulted in a comparable half-time of $[Ca]_i$ decay between both groups. Thus, FDAR reduced $t_{50\%}$ of $[Ca]_i$ decay (4 Hz/0.5 Hz) by 79% in TG and by 64% in WT (◘ Fig. 18.7F). This represents a significant change of FDAR in TG for both time to 50% relaxation and 50% $[Ca]_i$ decay.

18.3.6 Higher PLN phosphorylation at Thr^{17} under Iso stimulation

To further examine the cellular mechanisms of an impaired relaxation in TG compared to WT, we measured the phosphorylation state of PLN in both groups (◘ Fig. 18.8A). The phosphorylation state of PLN at Ser^{16} (◘ Fig. 18.8B) and Thr^{17} (◘ Fig. 18.8C) was comparable under basal conditions in TG and WT hearts. In addition, the application of Iso was associated with an unchanged phosphorylation state of PLN at Ser^{16} between TG and WT (◘ Fig. 18.8B). In contrast, the overexpression of JCN resulted in a 1.6-fold increased phosphorylation state of PLN at Thr^{17} under β-adrenergic receptor (β-AR) stimulation (◘ Fig. 18.8C).

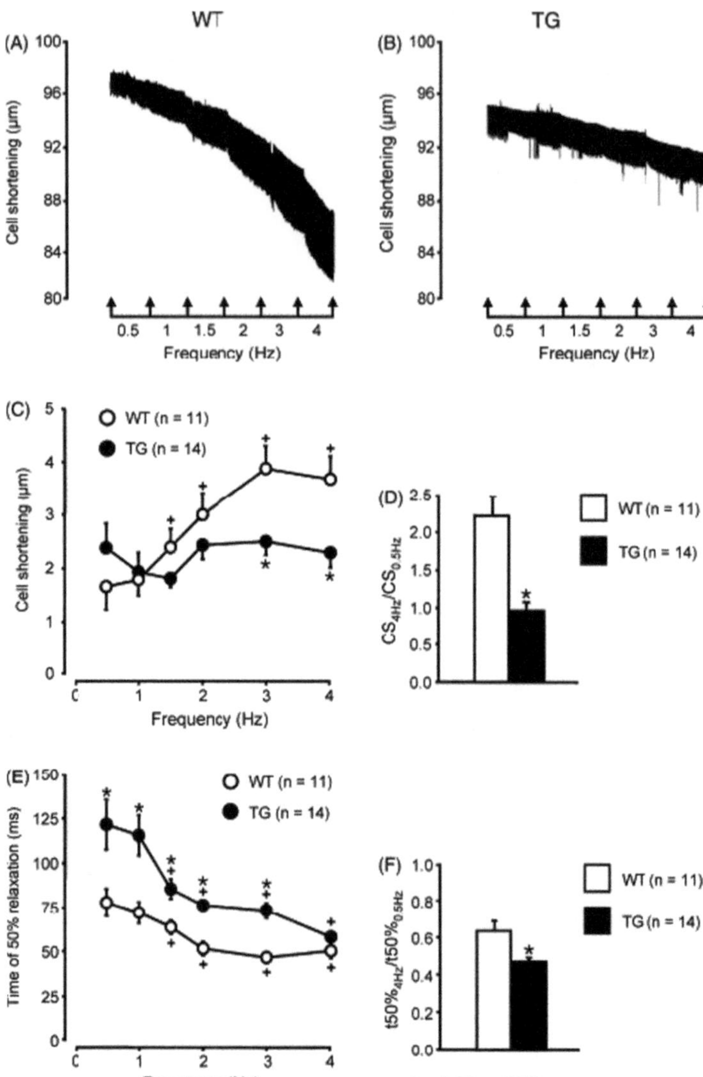

◩ Fig. 18.6 Frequency-dependent changes of cell shortening and relaxation. Shown are representative traces of edge detection measurements recorded on cardiomyocytes of WT (A) and TG mice (B). Frequency was stepwise increased every 5 min from 0.5 to 4 Hz. Average data of experiments are given for cell shortening (C). The frequency-dependent inotropic effect was determined by the ratio of cell shortening (CS) at 4 Hz/0.5 Hz (D). Cardiomyocytes of both groups exhibited an acceleration of relaxation under increasing frequencies as measured by the time to 50% relengthening from 0.5 to 4 Hz (E). However, those of TG mice showed a higher reduction of the FDAR (frequency-dependent acceleration of relaxation) index (F, the ratio of t$_{50\%}$ at 4 Hz/0.5 Hz).

18.4   Discussion

The SR Ca^{2+} release is maintained by structural units, which are organised into quaternary protein complexes consisting of the RyR, CSQ, TRD, and JCN[2]. Biochemical studies revealed that JCN enables the Ca^{2+}-dependent tethering of CSQ to the RyR[2,9]. The use of transgenic models with cardiac-specific overexpression of JCN provided new insight into the dynamic processes of SR Ca^{2+} release[15,30]. The 10-fold JCN overexpression was associated with an impaired relaxation in isolated cardiomyocytes and papillary muscles[16]. However, the underlying cellular mechanisms remained undefined.

The dominant process that determines the decay of [Ca]$_i$ is thought to be Ca^{2+} transport by the SR Ca^{2+}ATPase. The fraction of Ca^{2+} transported out of the cytosol by SERCA2a is ~90% in mouse heart[31]. If the impaired relaxation in TG is caused by an abolished removal of Ca^{2+} into the SR then the expression and/or activ-

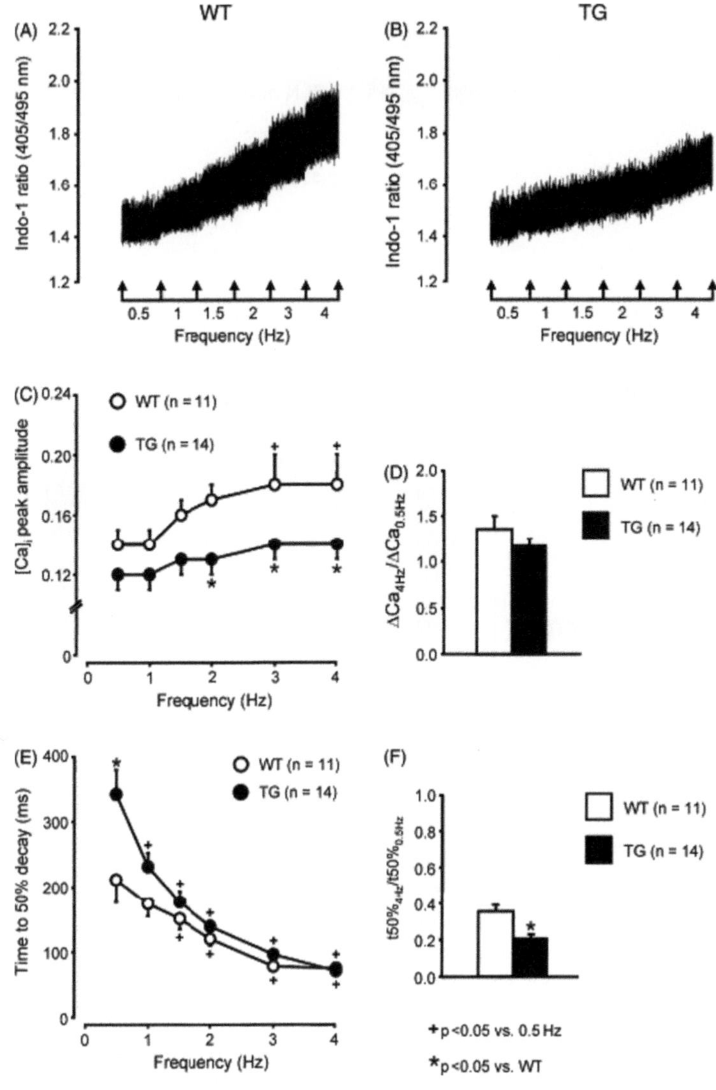

Fig. 18.7 Frequency-dependent effects on Ca^{2+} transients. Representative Ca^{2+} transient recordings are shown for WT (A) and TG (B) cardiomyocytes. Cardiomyocytes were stimulated for 5 min at 0.5, 1, 1.5, 2, 3 and 4 Hz as in Fig. 18.6. WT cardiomyocytes showed a significant increase of $\Delta[Ca]_i$, whereas in TG an increment of the stimulation frequency had no effect on $\Delta[Ca]_i$ (C). The ratio of $\Delta[Ca]_i$ at 4 Hz/0.5 Hz was similar in TG and WT cardiomyocytes (D). The time to 50% decay of $[Ca]_i$ was hastened at increasing frequencies in TG and WT cardiomyocytes (E). The FDAR (frequency-dependent acceleration of relaxation) index (the ratio of $t_{50\%}$ of $[Ca]_i$ decay at 4 Hz/0.5 Hz) was more reduced in TG than in WT cardiomyocytes (F).

ity of SERCA2a should be diminished. However, we found that the maximal velocity of the Ca^{2+}ATPase activity (V_{max}) was similar in TG and WT, which is consistent with the unchanged SERCA2a expression between both groups[16]. Moreover, the apparent Ca^{2+} affinity of SERCA2a (K_{Ca}) and the basal phosphorylation of PLN at Ser[16] and Thr[17] were not found to be significantly different between TG and WT. Normally, the removal of Ca^{2+} from the cytosol during relaxation by the NCX plays only a minor role in mouse heart (~9% of total activated Ca^{2+})[31].

However, the expression of the NCX was reduced by 67% in TG. These data suggest that the NCX is slower in extruding Ca^{2+} from the cytosol in TG. Consistently, we have previously reported a prolonged decay of caffeine-induced $[Ca]_i$ in TG[16]. The additional application of Ni^{2+} (i.e., blocking $I_{Na/Ca}$) resulted in a comparable decline of caffeine-triggered $[Ca]_i$ in TG and WT. Thus, the prolonged decay of $[Ca]_i$ in TG, resulting in an impaired mechanical relaxation, is caused, at least in part, by a decreased activity of the NCX besides the reduced expression of NCX.

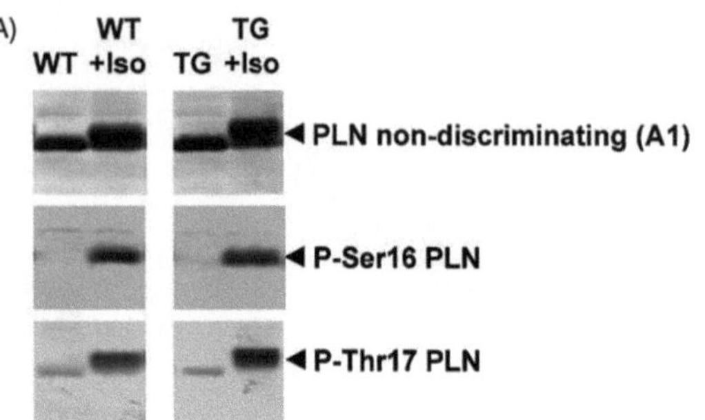

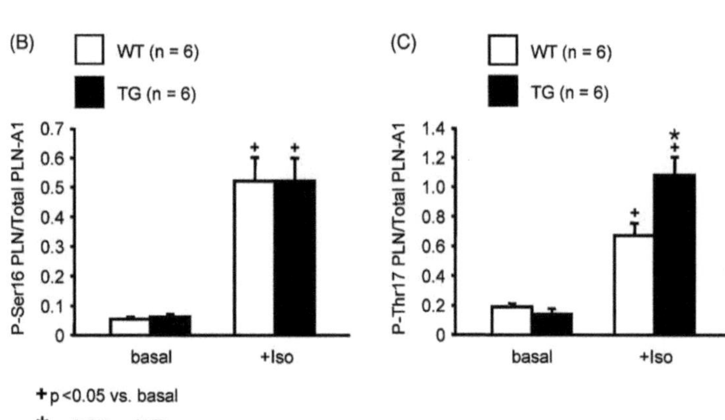

◪ Fig. 18.8 Phosphorylation state of PLN. Proteins in homogenates of isolated perfused WT and TG hearts were separated by SDS-PAGE and transferred to nitrocellulose membranes. Hearts were perfused in the absence or presence of 0.1 µM isoproterenol (+Iso). Total PLN and PLN phosphorylated at Ser^{16} or at Thr^{17} were identified with specific antibodies as described in Section 18.2. Representative immunoblots are given (A). Antibody-reacting bands for phosphorylated PLN at Ser^{16} (B) and at Thr^{17} (C) were quantified under basal and stimulated (+Iso) conditions using a Phosphorimager.

Our results on the measurement of Ca^{2+} transients in the presence of caffeine and Ni^{2+} depicted a reduction of $\Delta[\text{Ca}]_i$ by 22% in TG (i.e., a reduced SR Ca^{2+} load). The lower SR Ca^{2+} load in TG was consistent with the lower Ca^{2+} spark frequency in TG. Both parameters correlate well in resting ventricular cardiomyocytes of different species[27,32,33]. These data suggest that overexpression of junctin can regulate the SR Ca^{2+} content and thereby the SR Ca^{2+} release. This is supported by the observation that CSQ was more enriched in SR vesicles of TG hearts (by 23%). In addition, we found in electron microscopic studies that the overexpression of JCN was associated with a tighter packing of CSQ in proximity of the SR membrane[14]. Thus, Ca^{2+} may be displaced from its binding sites on CSQ by an increased occupation of these sites by JCN in TG resulting in a lower Ca^{2+} binding capacity or an altered Ca^{2+} sensitivity of CSQ or an impaired gating of the RyR. More evidence for a direct functional role of JCN in the regulation of the SR Ca^{2+} release process came from a study with ablation of JCN by gene targeting. JCN knockout mice exhibited an enhanced cardiac function in echocardiography and in isolated cardiomyocytes, which was paralleled by a higher $\Delta[\text{Ca}]_i$ and an increased SR Ca^{2+} load[34]. The data of the JCN knockout and overexpression studies suggest that JCN may act as a »placeholder« for Ca^{2+} which competes, together with the SR luminal Ca^{2+}, for the free binding sites on CSQ.

Interestingly, the decay of $[\text{Ca}]_i$ was not changed in cardiomyocytes of JCN knockout mice[34]. This may result from a functional replacement of JCN by TRD in the knockout mice. The decline of Ca^{2+} sparks is determined by Ca^{2+} diffusion and the removal of Ca^{2+} by the SR Ca^{2+}ATPase[35]. The expression and the activity of the SR Ca^{2+} pump is unchanged between TG and WT. Thus, the reduced expression of NCX may lead to a delayed termination of Ca^{2+} sparks in TG[36]. It has been shown that the NCX has access to the subsarcolemmal space or so-called »fuzzy space« into which Ca^{2+} is released from the jSR[37]. Thus, a lower amount of Ca^{2+} can be transported into the extracellular space by the NCX in TG resulting in a prolonged decay of Ca^{2+} sparks. Moreover, an enhanced diffusion space around the jSR Ca^{2+} release units, which is caused by the mild hypertrophy in TG, can account for the prolonged decline of Ca^{2+} sparks. Alternatively, PKA hyperphosphorylation of the RyR at Ser^{2809} in TG may lead to a higher Ca^{2+} sensitivity of the RyR and a depletion of FKBP12.6. This results in a reduced coupled gating between individual RyRs. The longer opening of the »leaky« Ca^{2+} release channels in resting cardiomyocytes may then contribute to the depletion of SR Ca^{2+} stores[38]. However, the contribution of a PKA hyperphosphorylation of the RyR to the prolonged decay of Ca^{2+} sparks and $[Ca]_i$ remains controversial[21,28,39,40]. The small reduction of the Ca^{2+} spark amplitude in TG may be attributable to the diminished SR Ca^{2+} load or the reduced protein expression of the RyR rather than to an altered activity. Moreover, the higher spread width of Ca^{2+} sparks (FWHM) in TG suggests a higher recruitment of single RyRs in each Ca^{2+} release event. This conclusion is supported by the fact that the protein expression of the RyR is reduced by 32% in TG[16]. Alternatively, the lower NCX expression may reduce the apparent Ca^{2+} diffusion coefficient.

Frequency is a powerful modulator of excitation–contraction coupling influencing the amplitude and relaxation of cardiac contraction. The frequency-dependent effects depend, in part, on the SR Ca^{2+} load at the initial frequency and on the recovery of the RyR from inactivation at the following frequencies. The basal stimulation at 0.5 Hz was associated with a similar shortening and $\Delta[Ca]_i$ between both groups despite a reduced SR Ca^{2+} load in TG. A higher maximal-stimulated fractional SR Ca^{2+} release could have contributed to that effect in TG. We suggest that the loss of I_{Ca} at higher frequencies, which is related to slow recovery of the L-type Ca^{2+} channel from inactivation[41], results in a lower fractional SR Ca^{2+} release in TG and consequently in a flat staircase. Indeed, the inactivation kinetics of I_{Ca} were prolonged at basal conditions in TG[16]. A lower phosphorylation of PLN at Ser^{16} may also account for the lower shortening and $\Delta[Ca]_i$ in TG at higher frequencies[42]. Increasing frequen-

cies affect not only the amplitude of contraction but also relaxation. Frequency-dependent acceleration of relaxation (FDAR) occurs in mammalian ventricular preparations independently of a positive or negative staircase[43-45]. Here we measured a higher effect of increasing frequencies on the relengthening of TG cardiomyocytes paralleled by changes of the decay of $[Ca]_i$. There is controversy in the literature as to whether the SR Ca^{2+} load contributes to FDAR. On one side, it has been shown that higher frequencies elevate intracellular $[Ca]_i$ leading to an increased activation of CaMKII[29,46], phosphorylation of PLN at Thr^{17} and higher SR Ca^{2+} load[47,48]. This frequency-dependent phosphorylation of PLN at Thr^{17} is closely correlated with a decrease in the relaxation time ($t_{50\%}$) of rat ventricular cardiomyocytes[48]. On the other side, a dependence of FDAR on CaMKII activation[45,49,50] and a higher PLN phosphorylation after an increase of the stimulation frequency[49,50] were not observed. Moreover, FDAR also occurred in PLN knockout mice and was inhibited by the potent CaMKII inhibitor KN-93 suggesting that FDAR depends on CaMKII-dependent stimulation of SR Ca^{2+} release but does not require PLN phosphorylation[29]. Thus, a potential involvement of CaMKII-dependent phosphorylation of PLN at Thr^{17} during FDAR remains controversial. Here, we did not detect the PLN phosphorylation status or the CaMKII activity at different stimulation frequencies. It is conceivable that a higher phosphorylation of SERCA2a by CaMKII[51] may contribute to an acceleration of SR Ca^{2+} transport and improved relaxation at higher frequencies in TG. However, the study by Valverde and coworkers[50] found no evidence for an association of FDAR with changes in the CaMKII-dependent phosphorylation of SERCA2a. These authors suggested that high intracellular Ca^{2+} by itself, without CaMKII activation, may effect the interaction between PLN and SERCA2a, as demonstrated by *in vitro* studies[52]. The enhanced phosphorylation of PLN at Thr^{17} under β-AR stimulation in beating TG hearts does not indicate that CaMKII-dependent mechanisms are involved in FDAR. It has been shown by use of transgenic models with mutation of Ser^{16} and Thr^{17} to alanine that the phosphorylation of Ser^{16} in PLB by PKA is sufficient in mediating its maximal contractile response to β-AR stimulation[53,54]. Thus, the phosphorylation of PLN at Ser^{16} might be a prerequisite for Thr^{17} phosphorylation of PLN. Moreover, it has been shown more recently that the increase in PLN phosphorylation at Ser^{16} after Iso application was associated with a relaxant effect, which was further enhanced when phosphorylation of PLN at Thr^{17} reached significant levels[50]. This suggests that FDAR and isoproterenol-dependent acceleration of relaxation do not share common underlying mechanisms. Thus, our results lead us conclude that the higher phosphorylation of PLN at Thr^{17} in Iso-stimulated TG hearts may contribute to the hastened relaxation after β-AR stimulation[16].

In summary, here we report novel findings on this model: (i) the lower SR Ca^{2+} load is associated with a reduced Ca^{2+} spark frequency; (ii) the prolonged decay of spontaneous Ca^{2+} sparks in TG is accompanied by a down-regulation of the NCX protein expression and an enhanced phosphorylation of the RyR at Ser^{2809}; and (iii) the overexpression of JCN is paralleled by a higher frequency-dependent acceleration of relaxation and a faster decay of [Ca]$_i$. Thus, we suggest that the impaired relaxation in TG may result from both a lower sarcolemmal Ca^{2+} extrusion and/or a higher SR Ca^{2+} leak.

18.5   References

[1]   W. Guo and K.P. Campbell, Association of triadin with the ryanodine receptor and calsequestrin in the lumen of the sarcoplasmic reticulum. J. Biol. Chem., 270 (1995), pp. 9027–9030.

[2]   L. Zhang, J. Kelly, G. Schmeisser, Y.M. Kobayashi and L.R. Jones, Complex formation between junctin, triadin, calsequestrin, and the ryanodine receptor. J. Biol. Chem., 272 (1997), pp. 23389–23397.

[3]   H. Takeshima, S. Nishimura and T. Matsumoto, et al. Primary structure and expression from complementary DNA of skeletal muscle ryanodine receptor. Nature, 339 (1989), pp. 439–445.

[4]   M.B. Bhat, J. Zhao, H. Takeshima and J. Ma, Functional calcium release channel formed by the carboxyl-terminal portion of ryanodine receptor. Biophys. J., 73 (1997), pp. 1329–1336.

[5]   K.P. Campbell, D.H. MacLennan, A.O. Jorgensen and M.C. Mintzer, Purification and characterization of calsequestrin from canine cardiac sarcoplasmic reticulum and identification of the 53,000 dalton glycoprotein. J. Biol. Chem., 258 (1983), pp. 1197–1204.

[6]   C. Franzini-Armstrong, L.J. Kenney and E. Varriano-Marston, The structure of calsequestrin in triads of vertebrate skeletal muscle: a deep-etch study. J. Cell. Biol., 105 (1987), pp. 49–56.

[7]   Y. Sato, D.G. Ferguson and H. Sako, et al. Cardiac-specific overexpression of mouse cardiac calsequestrin is associated with depressed cardiovascular function and hypertrophy in transgenic mice. J. Biol. Chem., 273 (1998), pp. 28470–28477.

[8]   L.R. Jones, Y.J. Suzuki and W. Wang, et al. Regulation of Ca^{2+} signaling in transgenic mouse cardiac myocytes overexpressing calsequestrin. J. Clin. Invest., 101 (1998), pp. 1385–1393.

[9]   I. Györke, N. Hester, L.R. Jones and S. Györke, The role of calsequestrin, triadin, and junctin in conferring cardiac ryanodine receptor responsiveness to luminal calcium. Biophys. J., 86 (2004), pp. 2121–2128.

[10] W. Guo, A.O. Jorgensen, L.R. Jones and K.P. Campbell, Biochemical characterization and molecular cloning of cardiac triadin. J. Biol. Chem., 271 (1996), pp. 458–465.

[11] L.R. Jones, L. Zhang, K. Sanborn, A.O. Jorgensen and J. Kelley, Purification, primary structure, and immunological characterization of the 26-kDa calsequestrin binding protein (junctin) from cardiac junctional sarcoplasmic reticulum. J. Biol. Chem., 270 (1995), pp. 30787–30796.

[12] M. Ohkura, K.I. Furukawa and H. Fujimori, et al. Dual regulation of the skeletal muscle ryanodine receptor by triadin and calsequestrin. Biochemistry, 37 (1998), pp. 12987–12993.

[13] C. Thompson, W. Arnold, A.C.-Y. Shen, L.R. Jones and A.O. Jorgensen, Subcellular distribution of calsequestrin (CSQ) binding proteins in cardiac sarcoplasmic reticulum. Biophys. J., 72 (1997), pp. 378–386.

[14] P. Tijskens, L.R. Jones and C. Franzini-Armstrong, Junctin and calsequestrin overexpression in cardiac muscle: the role of junctin and the synthetic and delivery pathways for the two proteins. J. Mol. Cell. Cardiol., 35 (2003), pp. 961–974.

[15] L. Zhang, C. Franzini-Armstrong, V. Ramesh and L.R. Jones, Structural alterations in cardiac calcium release units resulting from overexpression of junctin. J. Mol. Cell. Cardiol., 33 (2001), pp. 233–247.

[16] U. Kirchhefer, J. Neumann and D.M. Bers, et al. Impaired relaxation in transgenic mice overexpressing junctin. Cardiovasc. Res., 59 (2003), pp. 369–379.

[17] U. Kirchhefer, J. Neumann and H.A. Baba, et al. Cardiac hypertrophy and impaired relaxation in transgenic mice overexpressing triadin 1. J. Biol. Chem., 276 (2001), pp. 4142–4149.

[18] L.R. Jones and S.E. Cala, Biochemical evidence for functional heterogeneity of cardiac sarcoplasmic reticulum vesicles. J. Biol. Chem., 256 (1981), pp. 11809–11818.

[19] F.N. Briggs, K.F. Lee, A.W. Wechsler and L.R. Jones, Phospholamban expressed in slow-twitch and chronically stimulated fast-twitch muscles minimally affects calcium affinity of sarcoplasmic reticulum Ca^{2+}-ATPase. J. Biol. Chem., 267 (1992), pp. 26056–26061.

[20] U. Kirchhefer, L.R. Jones and F. Begrow, et al. Transgenic triadin 1 overexpression alters SR Ca^{2+} handling and leads to a blunted contractile response to β-adrenergic agonists. Cardiovasc. Res., 62 (2004), pp. 122–134.

[21] S. Reiken, M. Gaburjakova and S. Guatimosim, et al. Protein kinase A phosphorylation of the cardiac calcium release channel (ryanodine receptor) in normal and failing hearts. J. Biol. Chem., 278 (2003), pp. 444–453.

[22] J. Neumann, P. Boknik and A.A. DePaoli-Roach, et al. Targeted overexpression of phospholamban to mouse atrium depresses Ca^{2+} transport and contractility. J. Mol. Cell. Cardiol., 30 (1998), pp. 1991–2002.

[23] M.A. Porzio and A.M. Pearson, Improved resolution of myofibrillar proteins with sodium dodecyl sulfate-polyacrylamide gel electrophoresis. Biochim. Biophys. Acta, 490 (1977), pp. 27–34.

[24] L. Mahony and L.R. Jones, Developmental changes in cardiac sarcoplasmic reticulum in sheep. J. Biol. Chem., 261 (1986), pp. 15257–15265.

[25] T. Suzuki and J.H. Wang, Stimulation of bovine cardiac sarcoplasmic reticulum Ca^{2+} pump and blocking of phospholamban phosphorylation and dephosphorylation by a phospholamban monoclonal antibody. J. Biol. Chem., 261 (1986), pp. 7018–7023.

[26] J.W.M. Bassani, R.A. Bassani and D.M. Bers, Relaxation in rabbit and rat cardiac cells: species-dependent differences in cellular mechanisms. J. Physiol., 476 (1994), pp. 279–293.

[27] H. Satoh, L.A. Blatter and D.M. Bers, Effects of $[Ca^{2+}]_i$, SR Ca^{2+} load, and rest on Ca^{2+} spark frequency in ventricular myocytes. Am. J. Physiol., 272 (1997), pp. H657–H668.

[28] S.O. Marx, S. Reiken and Y. Hisamatsu, et al. PKA phosphorylation dissociates FKBP12.6 from the calcium release channel (ryanodine receptor): defective regulation in failing hearts. Cell, 101 (2000), pp. 365–376.

[29] J. DeSantiago, L.S. Maier and D.M. Bers, Frequency-dependent acceleration of relaxation in the heart depends on CaMKII, but not phospholamban. J. Mol. Cell. Cardiol., 34 (2002), pp. 975–984.

[30] C.S. Hong, M.C. Cho and Y.G. Kwak, et al. Cardiac remodeling and atrial fibrillation in transgenic mice overexpressing junctin. FASEB J., 16 (2002), pp. 1310–1312.

[31] L. Li, G. Chu, E.G. Kranias and D.M. Bers, Cardiac myocyte calcium transport in phospholamban knockout mouse: relaxation and endogenous CaMKII effects. Am. J. Physiol., 274 (1998), pp. H1335–H1347.

[32] H. Cheng, W.J. Lederer and M.B. Cannell, Calcium sparks: elementary events underlying excitation-contraction coupling in heart muscle. Science, 262 (1993), pp. 740–744.

[33] V. Lukyanenko, S. Viatchenko-Karpinski, A. Smirnov, T.F. Wiesner and S. Györke, Dynamic regulation of sarcoplasmic reticulum Ca^{2+} content and release by luminal Ca^{2+}-sensitive leak in rat ventricular myocytes. Biophys. J., 81 (2001), pp. 785–798.

[34] Q. Yuan, G.-C. Fan and X. Sun, et al. Ablation of junctin by gene targeting results in enhanced cardiac contractility and Ca^{2+} kinetics, . Circulation, 110 Suppl. (2004) III-131.

[35] A.M. Gomez, H. Cheng, W.J. Lederer and D.M. Bers, Ca^{2+} diffusion and sarcoplasmic reticulum transport both contribute to $[Ca^{2+}]_i$ decline during Ca^{2+} sparks in rat ventricular myocytes. J. Physiol., 496 (1996), pp. 575–581.

[36] J.I. Goldhaber, S.T. Lamp, D.O. Walter, A. Garfinkel, G.H. Fukumoto and J.N. Weiss, Local regulation the threshold for calcium sparks in rat ventricular myocytes: role of sodium-calcium exchanger. J. Physiol., 520 (1999), pp. 431–438.

[37] A.W. Trafford, M.E. Diaz, S.C. O'Neill and D.A. Eisner, Comparison of subsarcolemmal and bulk calcium concentration during spontaneous calcium release in rat ventricular myocytes. J. Physiol., 488 (1995), pp. 577–586.

[38] S.O. Marx, J. Gaburjakova, M. Gaburjakova, C. Henrikson, K. Ondrias and A.R. Marks, Coupled gating between cardiac calcium release channels (ryanodine receptors). Circ. Res., 88 (2001), pp. 1151–1158.

[39] A.R. Marks, Cardiac intracellular calcium release channels: role in heart failure. Circ. Res., 87 (2000), pp. 8–11.

[40] Y. Li, E.G. Kranias, G.A. Mignery and D.M. Bers, Protein kinase A phosphorylation of the ryanodine receptor does not affect calcium sparks in mouse ventricular myocytes. Circ. Res., 90 (2002), pp. 309–316.

[41] G. Antoons, K. Mubagwa, I. Nevelsteen and K.R. Sipido, Mechanisms underlying the frequency dependence of contraction and $[Ca^{2+}]_i$ transients in mouse ventricular myocytes. J. Physiol., 543 (2002), pp. 889–898.

[42] K. Brixius, A. Wollmer, B. Bölck, W. Mehlhorn and R.H.G. Schwinger, Ser16-, but not Thr17-phosphorylation of phospholamban influences frequency-dependent force generation in human myocyardium. Pflügers Arch., 447 (2003), pp. 150–157.

[43] W.D. Gao, N.G. Perez and E. Marban, Calcium cycling and contractile activation in intact mouse cardiac muscle. J. Physiol., 507 (1998), pp. 175–184.

[44] B. Pieske, B. Kretschmann and M. Meyer, et al. Alterations in intracellular calcium handling associated with the inverse force-frequency relation in human dilated cardiomyopathy. Circulation, 92 (1995), pp. 1169–1178.

[45] Z. Kassiri, R. Myers, R. Kaprielian, H.S. Banijamali and P.H. Backx, Rate-dependent changes of twitch force duration in rat cardiac trabeculae: a property of the contractile system. J. Physiol., 524 (2000), pp. 221–231.

[46] R.A. Bassani, A. Mattiazzi and D.M. Bers, CaMKII is responsible for activity-dependent acceleration of relaxation in rat ventricular myocytes. Am. J. Physiol., 268 (1995), pp. H703–H712.

[47] V.J.A. Schouten, Interval dependence of force and twitch duration in rat heart explained by Ca pump inactivation in sarcoplasmic reticulum. J. Physiol., 431 (1990), pp. 427–444.

[48] D. Hagemann, M. Kuschel, T. Kuramochi, W. Zhu, H. Cheng and R.P. Xiao, Frequency-encoding Thr17 phospholamban phosphorylation is independent of Ser16 phosphorylation in cardiac myocytes. J. Biol. Chem., 275 (2000), pp. 22532–22536.

[49] M. Hussain, G.A. Drago, J. Colyer and C.H. Orchard, Rate-dependent abbreviation of Ca^{2+} transient in rat heart is independent of phospholamban phosphorylation. Am. J. Physiol., 273 (1997), pp. H695–H706.

[50] C.A. Valverde, C. Mundina-Weilenmann and M. Said, et al. Frequency-dependent acceleration of relaxation in mammalian heart: a property not relying on phospholamban and SERCA2a phosphorylation. J. Physiol., 562 (2005), pp. 801–813.

[51] T. Toyofuku, K. Curotto Kurzydlowski, N. Narayanan and D.H. MacLennan, Identification of Ser38 as the site in cardiac sarcoplasmic reticulum Ca^{2+}-ATPase that is phosphorylated by Ca^{2+}/calmodulin-dependent protein kinase. J. Biol. Chem., 269 (1994), pp. 26492–26496.

[52] M. Asahi, E. McKenna, K. Kurzydlowski, M. Tada and D.H. MacLennan, Physiological interactions between phospholamban and sarco(endo)plasmic reticulum Ca^{2+}-ATPases are dissociated by elevated Ca^{2+}, but not by phospholamban phosphorylation, vanadate, or thapsigargin, and are enhanced by ATP. J. Biol. Chem., 275 (2000), pp. 15034–15038.

[53] W. Luo, G. Chu, Y. Sato, Z. Zhou, V.J. Kadambi and E.G. Kranias, Transgenic approaches to define the functional role of dual site phospholamban phosphorylation. J. Biol. Chem., 273 (1998), pp. 4734–4739.

[54] G. Chu, J.W. Lester, K.B. Young, W. Luo, J. Thai and E.G. Kranias, A single site (Ser16) phosphorylation in phospholamban is sufficient in mediating its maximal cardiac responses to beta-agonists. J. Biol. Chem., 275 (2000), pp. 38938–38943.

Remodelling of Ca^{2+} handling organelles in adult rat ventricular myocytes during longterm culture

Karin Hammer, Sandra Ruppenthal, Cedric Viero, Anke Scholz, Ludwig Edelmann, Lars Kaestner, Peter Lipp

Reprint from JMCC (2010) **49**: 427-437.

■ **Abstract**

It is well known that for cardiomyocytes, isolation and culturing induce largely unknown remodelling processes. We analysed changes in the structure of cell compartments with optical techniques such as confocal microscopy and fluorescence redistribution after photobleaching employing adenoviral-mediated transduction of targeted fluorescent proteins and small molecule dyes. We identified characteristic remodelling processes: the T-tubular membrane system was gradually lost by a process referred to as »sequential pinching off«, in an outward direction. Mitochondria fell in one of three classes, very small (0.9 µm length), medium long (1.8 µm) or extended shape (3.6 µm) organelles. Over the culturing time mitochondria gradually fused. Bleaching of individual mitochondria revealed association between apparently separated mitochondria by »tunnelling« via sub-resolution organelle-tubes. This tunnelling process was increasing over the culturing time. A gradual loss of the cross striation arrangement in the endoplasmic/sarcoplasmic reticulum was visualised. Analysis of large populations of Ca^{2+} sparks by video-rate confocal 2D-scanning revealed significant albeit small changes of these elementary SR-Ca^{2+} release events in adult cardiomyocytes that could be related to changes in SR-Ca^{2+} content rather than resting Ca^{2+} concentration. In conclusion, primary isolated cardiomyocytes from adult hearts undergo a well-defined, but reproducible subcellular remodelling during optimised long term culture.

19.1 Introduction

More than 30 years ago, cell isolation procedures for Ca^{2+} tolerant rat cardiac myocytes have been introduced[1] followed by initial attempts to use them in long term culture[2,3]. Those studies have reported a significant remodelling of cellular and subcellular properties of the myocytes. Together with other groups Eppenberger *et al.* have performed seminal work in characterising this process also referred to as *in vitro* remodelling that occurs in cells maintained in medium containing serum (e.g., [4]). Further investigations revealed that serum free medium was much better in retaining cell morphology. Nevertheless after a few days even under these conditions cell de-differentiation occurred[5]. Newer studies also reported characteristic alterations of cell morphology and/or function[6,7].

Shortly after isolation the regular plasma membrane invaginations, referred to as the T-tubular system, reportedly undergo major remodelling in almost all adult myocytes studied so far. These include guinea-pig[8], rabbit[9] and rat[6]. Here, one of the interesting questions is the fate of the T-tubular membrane (TTM), which is still widely unknown. This is of particular interest since major proteins of cardiac ECC are located predominantly at the TTM and alterations of that have been linked to cardiac pathologies (see e.g., [10,11]). It was shown that the majority of plasma membrane Ca^{2+} current is located at the TTM[12]. A similar distribution has been described for the Na^+/Ca^{2+} exchanger (NCX) based on functional data[13] and immunocytochemistry[14]. Brette *et al.* illustrated that the TTM is the major part of the plasma membrane for Ca^{2+} refilling of the SR[15], the prime Ca^{2+} storing organelle[16]. Not much is known about the structure of the ER/SR in living cardiac muscle cells, most of the structural data is either derived from electron microscopy studies (e.g., [17]) or indirectly from immunocytochemistry of SR proteins such as sarco-/endoplasmic reticulum Ca^{2+}-ATPase (SERCA pumps) or ryanodine receptor (RyR) (e.g., [18,19]). Moreover, little is known about changes of the ER/SR structure in long term culture of living adult cardiac myocytes.

Besides ATP production, mitochondria serve a multitude of other physiological functions and pathophysiological roles[16,20-22]. While in many other cell types, mitochondria display a highly dynamic network structure (e.g., [23]), in cardiac myocytes this organelle is arranged in the form of highly organised »packages« along the longitudinal axis of the cell[17]. Mitochondria appear to be located in close proximity to local Ca^{2+} release sites. Duchen *et al.* have provided seminal work reporting local depolarisation in response to local Ca^{2+} release in cardiac muscle cells[24]. Despite all of this knowledge the morphology and distribution of mitochondria in cultured adult myocytes and changes thereof are largely unknown but important for their physiology.

Since isolated cells often serve as single cell *in vitro* models to study their *in vivo* function it is inevitable to gain insight into remodelling processes occurring after isolation. Here we have taken the rat adult cardiac myocyte and demonstrate that remodelling of the cells' functional and morphological state does indeed occur with respect to organelles involved in Ca^{2+} handling. These changes are highly reproducible and will thus allow the application of such cultured cells as single cell models to study cardiac physiology and pathophysiology.

19.2 Material and methods

19.2.1 Cell isolation and culture

The particular steps for cell isolation and culture were performed as described in great details recently[7,25].

19.2.2 Construction of adenoviruses

For construction of the adenoviruses we used the Transpose-Ad™ adenoviral vector system (MP Biomedicals, USA) as described recently[7,25].

For plasma membrane labelling we used a YFP localised at the N-terminus of a glycosylphosphatidylinositol (GPI) sequence[26]; mitochondrial fluorescence was achieved by fusing a DsRed1 to the C-terminus of the subunit VIII of the human cytochrome *C* oxidase targeting sequence and the ER/SR was marked by fusing a DsRed2 between the C-terminus of the ER localisation sequence of calreticulin and the N-terminus of the KDEL (ER-retrieval) sequence.

19.2.3 Organelle-specific staining with small molecule dyes

In order to achieve organelle specific staining with small molecular dyes we used the membrane potential sensitive dye Di-8-ANEPPS (2.5 µM, 10 min); mitoTracker-Green (MTG) for mitochondria (1 µM, 30 min) and ER-Tracker® (1 µM, 30 min) for the ER/SR. All dyes were purchased from Molecular Probes (Eugene, USA).

19.2.4 Electron microscopy

High pressure freezing (HPF) and freeze-substitution of freshly isolated rat ventricular myocytes were performed according to a procedure described recently[27].

19.2.5 Confocal imaging and bleaching experiments

Confocal imaging was performed on multibeam confocal scanner systems (QLC-100 or VTinfinity-3; both from VisiTech Int., Sunderland, UK). Fluorescence recovery after photobleaching (FRAP) was implemented on the Infinity-3 platform by defining the region for bleaching and rapidly scanning across that region at the durations and laser energies given in the text. Image acquisition was achieved either through a 100× water immersion objective (NA 1.1; NIKON, Japan) when using the QLC-100 or through a 60× oil immersion objective (NA 1.4; NIKON, Japan) on the Infinity-3. All experiments were carried out at room temperature.

19.2.6 Image processing

After data acquisition, they were transferred into ImageJ software (W. Rasband, NIH, USA) for further processing. ImageJ was used to construct the 3D-surface plots used.

For construction of power spectra the fluorescence profile of a line (2–4 pixels wide) along the longitudinal or perpendicular axis of the myocytes was calculated within ImageJ and exported to Igor software (Wavemetrics, USA) in which we performed the power spectral analysis.

3D reconstruction of the mitochondria was executed in Imaris (Bitplane, Switzerland). We used surface rendering to generate solid 3D-models of mitochondria in the ventricular myocytes.

For display of weaker fluorescence signals, we enhanced the contrast. Where necessary we colour coded the greyscale images according to the colour of the fluorescence entity used or as indicated.

19.2.7 Statistics

Statistical analysis was carried out in GraphPad Prism software (GraphPad Inc., USA). Depending on the outcome of testing the data for normal distribution (D'Agostino and Pearson omnibus normality test and Shapiro–Wilk normality test), we either used Mann–Whitney (normally distributed) or Kruskal–Wallis (not normally distributed) tests for analysing the statistical significance of changes; $p < 0.05$ (*), $p < 0.01$ (**) or $p < 0.001$ (***). Bar graphs are displayed as the mean ± SEM.

19.3 Results

19.3.1 GPI-YFP and Di-8-ANEPPS based analysis of T-tubules

After isolation, rat ventricular myocytes (RVMs) start »losing« T-tubular structures[9]. Under our optimised culture conditions[7], we also observed a gradual loss of the T-tubules (see ◗ Fig. 19.1(A), (C) and (E)) coinciding with a gradual decrease of the cross section (◗ Fig. 19.1(D)). We labelled the T-tubules with the voltage-sensitive probe Di-8-ANEPPS (◗ Fig. 19.1(A) and (E)). To analyse the fate of the plasma membrane, we expressed a membrane targeted YFP (GPI-YFP, e.g., [28]). In case the membrane is internalised we could identify internalised membrane residues while if the membrane would be digested or externalised, the YFP-labelling should be virtually lost and co-localise exclusively with Di-8-ANEPPS. ◗ Fig. 19.1(C) shows plasma membrane outlined by membrane targeted YFP as early as 24 h after transfection. At later days *in vitro* (DIVs) the TTM could be observed for longer times when compared to Di-8-ANEPPS staining. To quantify that, we measured the length of the TTM over culture time (◗ Fig. 19.1(E), black squares). Although both parameters showed a steady decrease, the time course of the GPI-YFP tracing appeared delayed by 1–2 days, indicating an inward-out process in the »pinching off« of the plasma membrane. The composite image (◗ Fig. 19.1(B) for DIV1) of a double-labelling experiment (red — Di-8-ANEPPS, green — GPI-YFP) supports this notion and highlights that already at DIV1 not all TTM can be accessed from the extracellular space (i.e., labelled with Di-8-ANEPPS).

19.3.2 Properties of mitochondria

We followed the 3D arrangement of the RVM mitochondria in culture by labelling them with MTG (◗ Fig. 19.2). ◗ Fig. 19.2(Aa) depicts representative 3D reconstructions of the mitochondria morphology at the given DIVs. While at DIV0 and DIV1 mitochondria structure appeared stable at DIV3 and 6 mitochondria inherited a more elongated

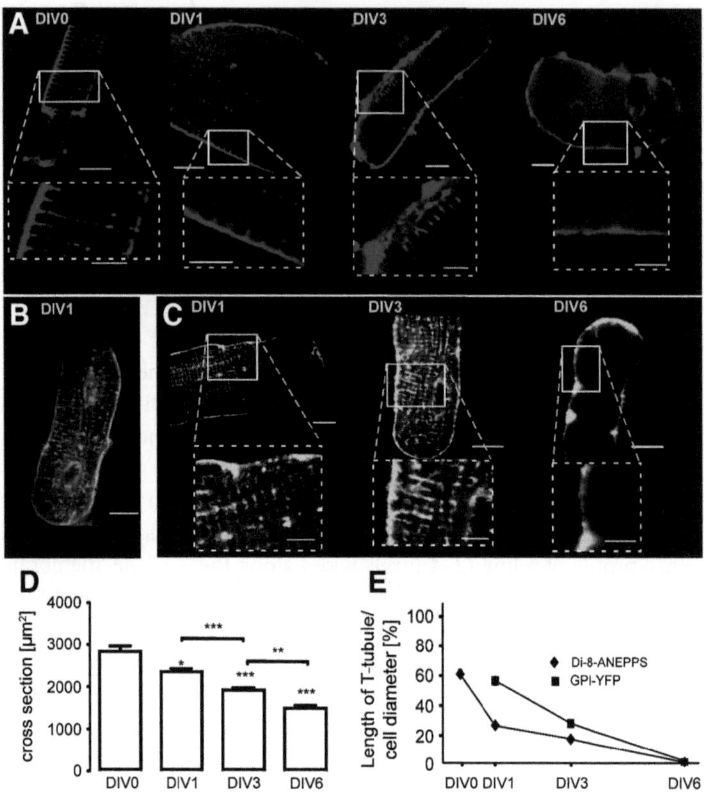

Fig. 19.1 Characterisation of the plasma membrane of RVM. (A) Di-8-ANEPPS labelled RVMs were imaged at the times indicated (from left to right, DIV0, 1, 3 and 6). Insets display enlarged portions (marked by the white rectangle). (B) RVMs were transduced with a GPI-YFP encoding adenovirus at DIV0 and imaged after Di-8-ANEPPS staining. YFP-fluorescence — green, Di-8-ANEPPS — red. (C) RVMs expressing GPI-YFP were imaged at DIV1, 3 and 6 (from left to right). Insets display enlarged portions (marked by the white rectangle). (D) Cross-sectional areas of RVM in confocal sections taken in the middle of the cell depth. Number of cells/hearts was: 72/3 (DIV0), 76/3 (DIV1), 96/3 (DIV3) and 45/2 (DIV6). (E) Fractional length of T-tubules (ratio of T-tubular length divided by the cell width). GPI-YFP staining (filled rectangles) displays apparently longer T-tubules than Di-8-ANEPPS labelling (black filled diamonds) at DIV1 and 3. (For interpretation of the references to color in this figure legend, the reader is referred to the web version of the original article.)

shape, losing some of their cross striation properties. For verification, we expressed mitochondria targeted DsRed (mito-DsRed1, **Fig. 19.2(Ab)**). Due to very low expression levels of the construct recordings at DIV1 were not possible. At DIV3 and 6 mitochondria displayed identical morphological properties when compared to patterns found after labelling with small molecule dyes. For quantification we analysed the spatial properties of the mitochondria along the longitudinal and transversal cell axis (**Fig. 19.2(B)**, yellow lines). For this typical cell, the strong spatial frequency component at 0.5–0.6 μm⁻¹ (spatial wavelength of ~ 1.85 μm) was very close to the expected value for the cross striation (black arrowhead in **Fig. 19.2(Bc)**; see also [7]). To verify such findings with ultra-structural approaches we performed transmission electron microscopy (TEM) of RVM at DIV0. **Fig. 19.2(Ca)** shows typical results for DIV0 depicting that in most cases mitochondria spanned exactly one sarcomere. It almost appeared as if mitochondria were »separated« by the T-tubular invaginations. Surprisingly, beside mitochondria with sarcomeric lengths (marked by white circles), we also identified organelles with almost exactly double the »unitary« length (marked with a rectangle) and very short and round one spanning only half a sarcomeric space (marked with white stars, **Fig. 19.2(Cb)**). Very interestingly, neighbouring mitochondria could be linked together via short and narrow »tube-like« structures (marked with black arrows). A closer look at the power spectra of longitudinal lines (**Fig. 19.2(B)**) indeed also revealed additional, smaller peak components in the spectrum (open arrowheads in **Fig. 19.2(Bc)**) that could be linked to the longer (lower-frequency peak, left) or shorter (higher frequency peak, right) intermitochondrial distance found in the TEM images.

For a more detailed analysis we investigated larger populations of cells for their major frequency components (results displayed in **Fig. 19.2(D)**).

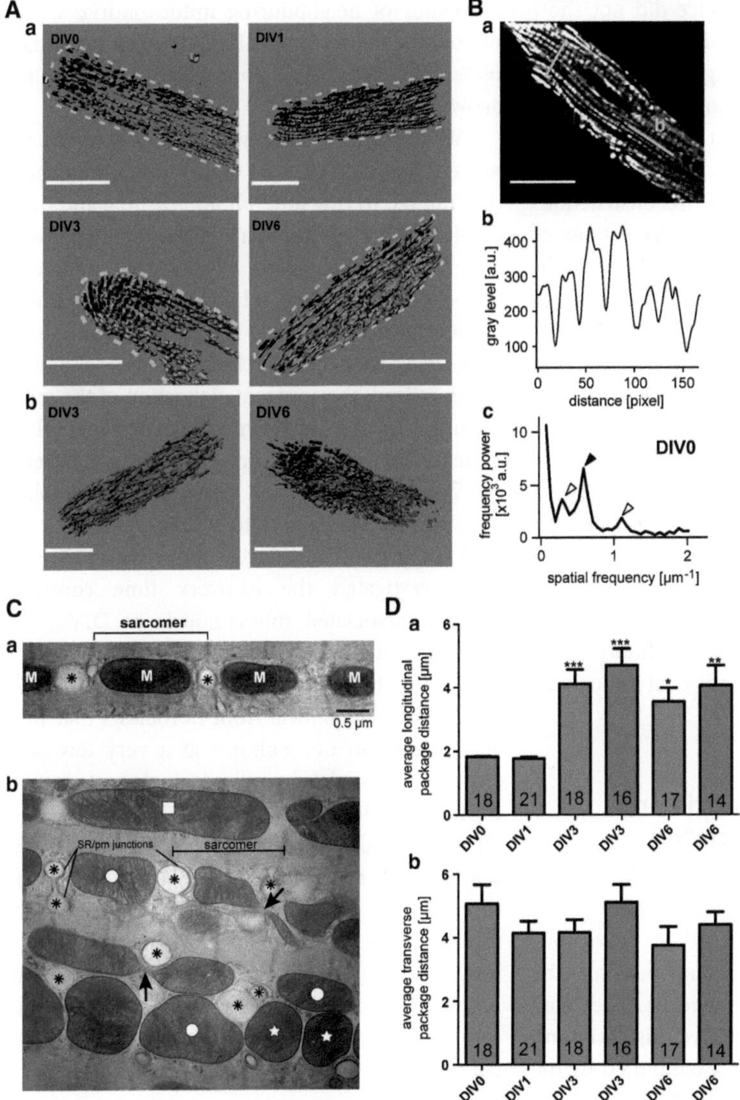

◘ Fig. 19.2 Morphological properties of mitochondria in RVM. (A) RVMs labelled with MTG (Aa) or expressing mito-DsRed1 (Ab). Serial confocal sections (z-steps 0.3 µm) of typical RVMs were acquired and surface rendered at the times indicated. The yellow (Aa) or orange (Ab) dashed lines highlight the cell boundary (scale bars — 10 µm). (B) Power spectral analysis of mitochondrial fluorescence. (Ba) Confocal section through a MTG-labelled RVM at DIV0. Longitudinal (labelled with »b«) and perpendicular lines were analysed. (Bb) Typical longitudinal fluorescence profile depicts regular fluorescence changes while (Bc) illustrates the corresponding power spectrum with its typical large low frequency component (cut off at the top). Filled arrowhead marks the spatial frequency of the sarcomere length, and two open arrowheads point to additional peaks in the spectrum (see text). (C) Electron microscopic properties of mitochondria in RVMs at DIV0. (Ca) Mitochondria labelled with »M«, T-tubules with an asterisk. Regular arrangement of the contractile filaments and the sarcomeric length is indicated. (Cb) Mitochondria in RVM display a variety of different shapes; (i) full sarcomere length (labelled with a white circle), (ii) half sarcomeric length (white star) and (iii) double sarcomeric length (white rectangle). T-tubules marked with asterisks, SR/T-tubular junctions are labelled accordingly. Black arrows point to mitochondrial constrictions at the site of close proximity to T-tubules. (D) Statistical analysis of the longitudinal (Da) and perpendicular (Db) arrangement of the mitochondria in RVM. Numbers given correspond to the numbers of RVMs from 4 rat hearts. Green bars — MTG staining, orange bars — mito-DsRed1 labelling. (For interpretation of the references to colour in this figure legend, the reader is referred to the web version of the original article.)

The transversal packaging distance did not show statistical significances during the culture period (■ Fig. 19.2(Db)), while the longitudinal distance significantly increased (■ Fig. 19.2(Da)). Both observations were independent of the labelling approach. In addition, the detailed analysis of individual mitochondria in the 3D reconstructions (■ Fig. 19.2(A)) or in dedicated x/z-projections of the 3D stacks (data not shown) revealed no additional changes in the mitochondrial morphology over the culturing time. We compared the distances between mitochondria but could not find any culture-dependent change in this parameter (data not shown).

One possible explanation for this finding was partial or complete fusion of »single sarcomere« mitochondria. We thus set out to analyse the »communication« between apparent individual mitochondrial »packages« as visualised by MTG. The basic idea was: if we bleach such individual mitochondria, we could analyse the fluorescence time course in non-bleached adjacent mitochondria.

19.3.3 Cardiac mitochondria are dynamically associated

We thus performed FRAP experiments as shown in ■ Fig. 19.3(A)–(C) for DIV0 RVM mitochondria. Amongst the five mitochondria within the bleached area we found two behaviours as depicted in ■ Fig. 19.3(B). Four of the five displayed no recovery after photobleaching (exemplified by the mitochondrion marked in grey in ■ Fig. 19.3(Abi)), the time course is displayed in matching colours in ■ Fig. 19.3(B). The organelle highlighted with the green arrow (■ Fig. 19.3(Ab)) showed clear recovery (■ Fig. 19.3(B)). Surprisingly the mitochondria directly adjacent to the »green« mitochondrion but localised clearly outside of the bleaching region (marked with the red arrow, ■ Fig. 19.3(Ab)) showed a gradual loss in its fluorescence (see also red tracing in ■ Fig. 19.3(B)). After fitting mono-exponential functions to the red and green traces we found similar absolute time constants for both behaviours (10.2 s for the recovery in green and 9.8 s for the fluorescence loss in red). To accentuate this

behaviour of neighbouring mitochondria, we replotted the fluorescence distribution as 3D-surface plots for selected time points (■ Fig. 19.3(C); black arrows in ■ Fig. 19.3(B)).

We could find these two kinds of behaviour in all cells analysed (n = 54 from 3 rats); mitochondria not recovering after photobleaching and those displaying a clear recovery. These data indicated that some mitochondria were indeed individual mitochondria and a subpopulation was linked without a directly visible connection, the fluorescence was »tunnelling« between the mitochondria.

If, indeed, the apparent elongation of the mitochondria, as seen in ■ Fig. 19.2(Da) was a result of either fusion processes or processes that fostered the apparent connection between mitochondria, then we should be able to monitor this with the FRAP approach introduced above (■ Fig. 19.3(D)). We investigated the recovery time constants between associated mitochondria at DIV0, followed the putative change of those over time and compared the values with those recovery times found in mitochondria from fibroblasts that were also present in our cultures at a very low density and where clearly individual, elongated mitochondria could be identified (■ Fig. 19.3(D)). From DIV0 onwards there was a gradual increase in the recovery speed until at DIV6 this process in the myocytes was almost as fast as that for the mitochondria in fibroblasts (dashed horizontal line). These functional data strongly supported our notion of an increased association and communication between mitochondria during the time of culture.

19.3.4 Changes of the ER/SR over the culture time

To analyse the ER/SR during the culture period, we initially tested small molecule dyes, such as ER-Tracker® blue–white DPX, but the staining was lacking specificity in cardiac myocytes (large unspecific staining of mitochondria; data not shown). We thus constructed an ER/SR-targeted DsRed2. The results after transduction are summarised in ■ Fig. 19.4. Despite the tetrameric nature of the DsRed2 fluorescence protein, we

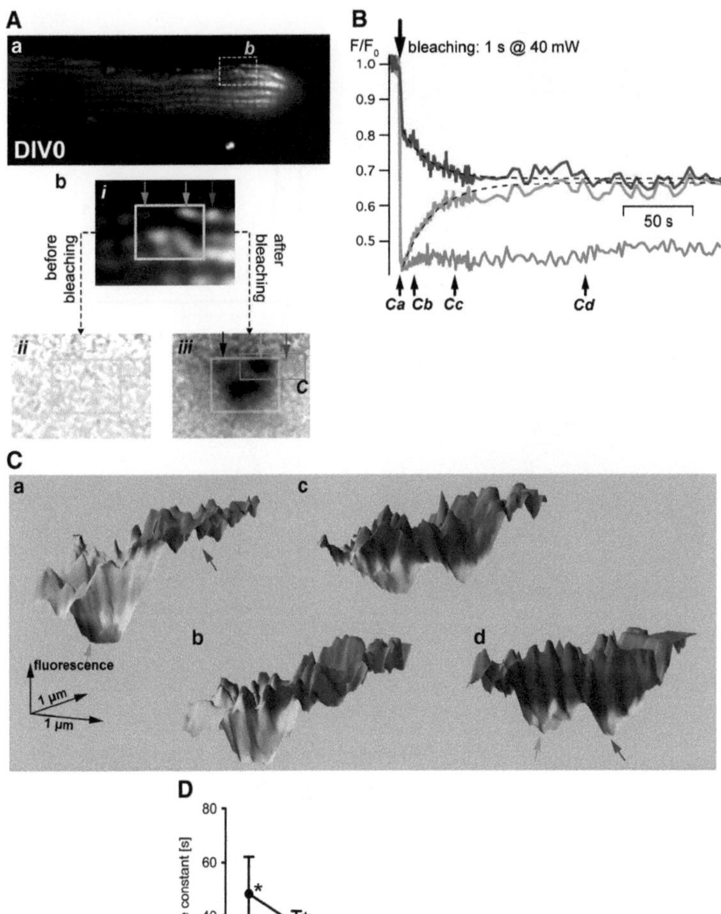

▫ Fig. 19.3 Association of adjacent mitochondria in RVM. (A) FRAP experiments on MTG loaded RVMs at DIV0. (Aa) Confocal section of a DIV0 RVM. Dashed yellow rectangle highlights enlarged area (Ab). (Abi) Enlarged illustration displaying the bleached area (yellow box). Three individual mitochondria have been marked by the coloured arrows. (Abii and iii) depict F/F₀ images. In (Abiii) to (Abii) white corresponds to unchanged and dark for decreased fluorescence (the grey arrow was substituted by the black arrow for illustration reasons). (B) Fluorescence time course averaged at the three locations marked in (Abi) and (Abiii). Black arrow indicates the bleaching event. The arrows beneath the tracings highlight time points for the 3D-surface plots of the fluorescence distribution in (Ci–iv). The two black dashed traces originate from a mono-exponential decay (corresponding red trace) or increase (corresponding green trace). (C) 3D-surface plots of the fluorescence distribution in the orange rectangle (from Abiii). Mitochondrium marked with the green arrow was within the bleached region while the red one was outside. These data were typical of all RVMs analysed (n = 53 from 3 rat hearts). (D) Summarised data obtained from a series of experiments (similar to (A)) taken at DIV0, 1, 3 and 6. We compared the recovery time constant of mitochondria in RVMs with those found in fibroblasts sparsely present in our culture. The dashed horizontal line indicates the mean of all fibroblasts (n = 37 from 3 rat hearts; grey area = SEM). The asterisks close to the data symbols illustrate statistically significant slower recovery in the RVM in comparison to the fibroblasts. The statistical analysis at the bottom indicates that only the DIV6 RVMs displayed a significantly increased recovery speed compared to DIV0. (For interpretation of the references to colour in this figure legend, the reader is referred to the web version of the original article.)

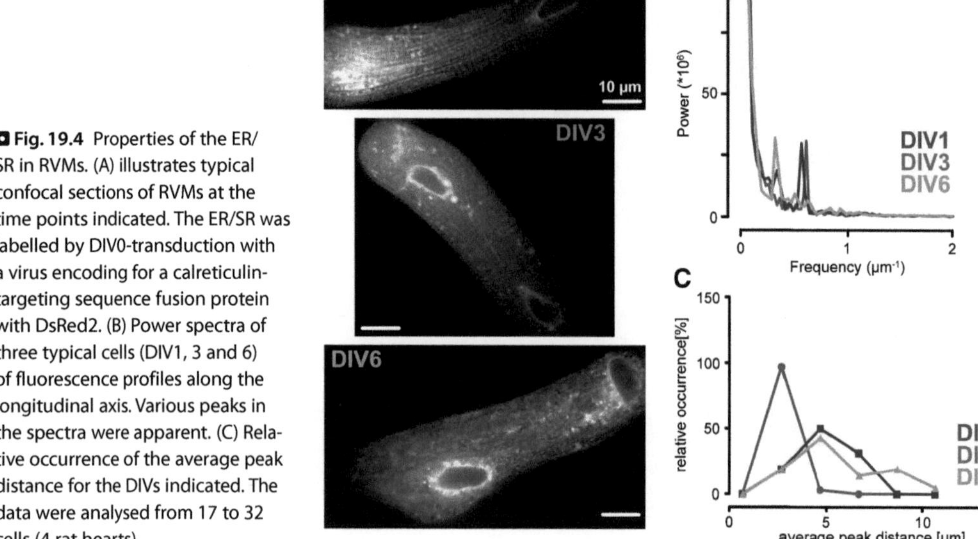

Fig. 19.4 Properties of the ER/SR in RVMs. (A) illustrates typical confocal sections of RVMs at the time points indicated. The ER/SR was labelled by DIV0-transduction with a virus encoding for a calreticulin-targeting sequence fusion protein with DsRed2. (B) Power spectra of three typical cells (DIV1, 3 and 6) of fluorescence profiles along the longitudinal axis. Various peaks in the spectra were apparent. (C) Relative occurrence of the average peak distance for the DIVs indicated. The data were analysed from 17 to 32 cells (4 rat hearts).

were able to detect fluorescence within less than 24 h after transduction and analysed cells at DIV3 and DIV6, **Fig. 19.4(A)**. The regular structure of the SR was well preserved at DIV1 and 3 but at DIV6 it displayed a rather irregular pattern, that we investigated by power spectral analysis (**Fig. 19.4(B)**). While at DIV1 the spectrum displayed a single frequency peak at 0.55 µm⁻¹ (corresponding to 1.8 µm) we were surprised to find two peaks in cells analysed at DIV3 (blue trace in **Fig. 19.4(B)**) and at DIV6 we identified multiple peaks with a major component at lower frequencies (green trace in **Fig. 19.4(B)**). For better quantification we investigated a greater population of cells and constructed histograms of the major spatial frequency component (**Fig. 19.4(C)**). At DIV1 (red graph) there was a sharp peak centred around the sarcomeric length and later broadening and right-shifting of this distribution occurred, indicating a loss of SR-organisation. These data suggest a gradually diminishing ordered SR structure most likely reflecting changes in the TTM (**Fig. 19.1**). We thus wondered whether such a rearrangement of the SR might affect Ca²⁺ signalling in RVMs.

19.3.5 Elementary Ca²⁺ signalling in cultured RVM

For this we analysed elementary Ca²⁺ release signals, Ca²⁺ sparks, which require a highly organised clustering of RyR. Moreover, the properties of Ca²⁺ sparks (e.g., amplitude, decay time constant or spatial spread) are known to be very sensitive to minute changes of the geometry of the cluster and the nanoscopic environment of the RyRs[16,29]. Consequently, Ca²⁺ sparks appeared well suited as a sensitive read-out to study putative changes of cardiac Ca²⁺ signalling morphology during the culture period.

Fig. 19.5(A) depicts exemplified recordings of 3 representative Ca²⁺ spark sites each from typical cardiac myocytes (see insets in **Fig. 19.5(A)**) at the DIVs indicated. Since the properties of Ca²⁺ sparks display an inherent variability, we recorded and analysed large populations of Ca²⁺ sparks. The results of such a large scale analysis are given in **Fig. 19.5(B)** displaying parameters on the basis of individual Ca²⁺ spark sites not individual Ca²⁺ sparks. The parameters we concentrated on in this study were the amplitude

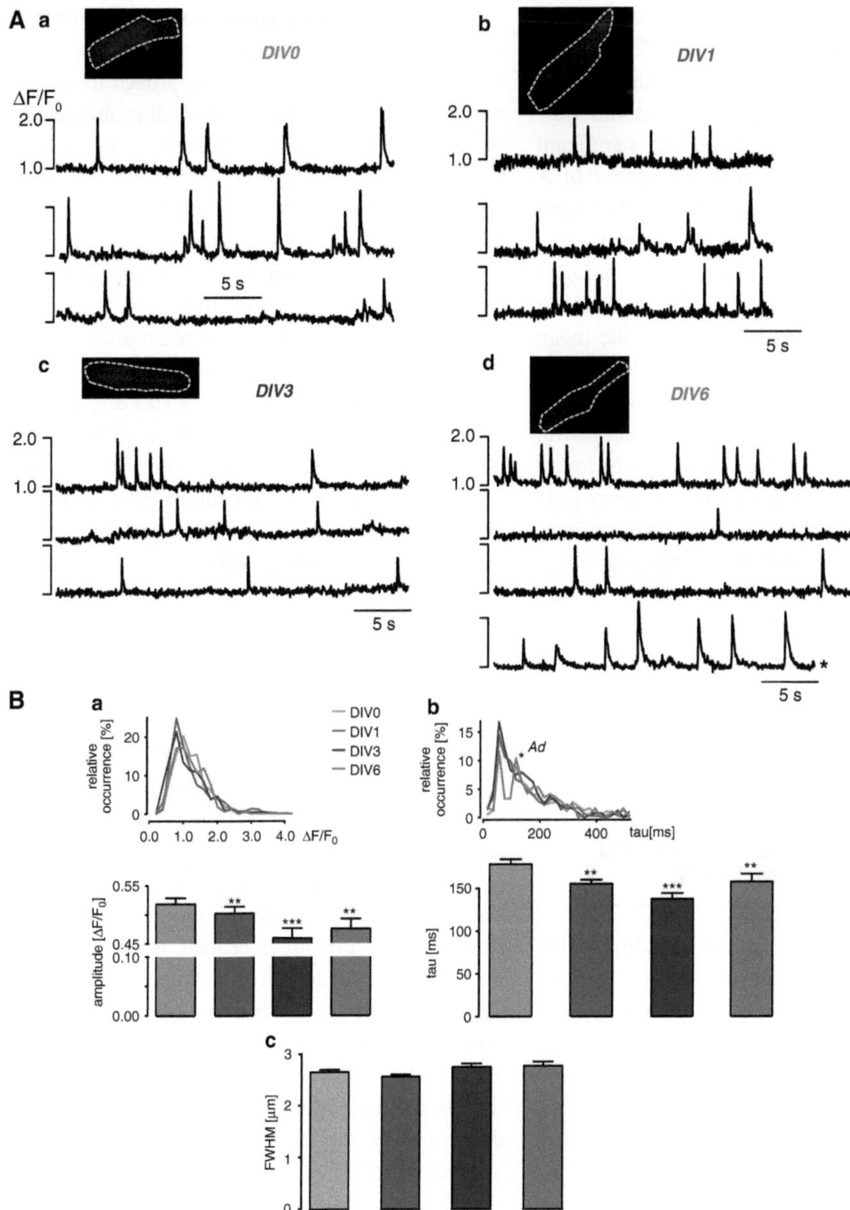

❏ **Fig. 19.5** Ca$_{2+}$-sparks from RVMs in long term culture. (A) RVMs were loaded with Fluo-4 and imaged at 33 frames/ second. Spontaneous Ca^{2+} sparks were recorded at the DIVs indicated (Aa–d). Each panel comprises: confocal cell image (outlined with the yellow dashed line) and typical traces from 3 different Ca^{2+} spark sites. Panel (Ad) depicts an additional trace of a spark site at which longer-lasting Ca^{2+} sparks could be recorded (see below and text). (B) Spark properties over culturing time. (Ba) Analysis of the spark amplitudes as (ΔF/F$_0$) for DIV1–6. Upper panel — distribution of spark amplitudes. Lower panel — bar graphs. Colour code is the same as in the upper panel. Data were obtained from 200 to 350 identified spark sites from 20 to 30 cells taken from 4 rat hearts at each DIV indicated. (Bb) Decay time constants of the Ca^{2+} sparks. Colour coding as in (Ba). The histogram in the upper panel highlights the occurrence of an additional second peak at longer lifetimes, that could also be identified in the original tracings (see Ad, trace with »*«). (Bc) Analysis of the spatial spread (FWHM). (For interpretation of the references to colour in this figure legend, the reader is referred to the web version of the original article.)

(expressed as $\Delta F/F_0$, ■ Fig. 19.5(Ba)), the decay time constant (■ Fig. 19.5(Bb)) and the full width at half maximum (FWHM in μm, ■ Fig. 19.5(Bc)). From these findings it became apparent that the Ca^{2+} spark properties underwent small but significant remodelling, most evident in ■ Fig. 19.5(Ba and b). A more detailed analysis of the amplitude distribution (■ Fig. 19.5(Ba) upper panel) revealed that the distribution width was significantly narrower at DIV1 and 3 than at DIV0 and that a broadening at DIV6 was responsible for the final increase of the mean as depicted in ■ Fig. 19.5(Ba) lower panel. Such findings were not related to out-of-focus sparks since all sparks analysed displayed a time to peak within one frame. A similar behaviour could also be found for the time dependent changes in the decay time constant. While at DIV1 and 3 a progressively sharper distribution resulted in a significant reduction in the time constant (i.e. faster recovery), the occurrence of a second peak at longer decay times (DIV6) caused an increase in the median of the decay times (see green box in the lower panel of ■ Fig. 19.5(Bb)). The spatial spread was unaffected by the culturing time (■ Fig. 19.5(Bc)).

19.3.6 SR-Ca^{2+} content and resting [Ca^{2+}] in cultured RVM

To investigate putative mechanisms leading to the altered spark properties described above, we have tested the SR-Ca^{2+} content and basal [Ca^{2+}]. Both parameters are major determinants of spark properties. As depicted, brief applications of 10 mM caffeine to empty the SR at different times during culture (■ Fig. 19.6(A)) revealed that from DIV0 onwards there was a significant drop in SR-Ca^{2+} content (■ Fig. 19.6(B)). Measurements of the resting fura2-ratio (■ Fig. 19.6(C)) clearly showed the absence of changes in basal Ca^{2+}.

19.4 Discussion

Although we and others have succeeded in introducing culture conditions that allow an extended culture time of adult RVMs with minimised de-differentiation (e.g., [6,7]), approaches to understand and investigate subcellular changes of their morphology and functional consequences thereof are largely lacking. Here we introduce an approach to address this by using a combination of small molecule dyes and targeted expression of fluorescent fusion proteins after viral transduction. This allows us to study their microscopic and functional properties over the time of culture such as organelle–organelle communication and altered elementary Ca^{2+} signalling. For this we aimed our study on three major organelles involved in Ca^{2+} handling: the plasma membrane, the mitochondria and the ER/SR.

Mammalian ventricular myocytes display deep invaginations of the plasma membrane the T-tubular system[30]. Especially these parts of the plasma membrane are believed to contain most of the key proteins of ECC (e.g., [10]). The L-type Ca^{2+} channel, a central protein in early ECC, is predominately located at dyadic junctions between the plasma membrane and the SR membrane (e.g., [31]). The loss of this membrane system either during culture or during experimental interventions has profound effects on myocyte ECC[10]. During long term culture adult ventricular myocytes (AVM) undergo a remodelling of their plasma membrane, the TTM is gradually lost[9]. In the present study we also described this phenomenon but went further to address the fate of the TTM. For this we used a unique feature of the plasma membrane that distinguishes it from other cellular membranes: the formation of lipid rafts. The GPI-anchor is a robust targeting motive to address directed expression of proteins into lipid rafts[28]. In our hands, YFP-GPI expressed quickly (< 24 h) and specifically labelled the plasma membrane (153 cells from 5 rats). Our data strongly suggest that the membrane of the T-tubules is pinched off in an in- to outward direction. More persistent GPI-labelling of such structures in comparison to Di-8-ANEPPS (additional 1–2 days; data not shown) supports the notion that the disconnected membranes stay intact for an extended time period and were not immediately metabolised or disintegrated. Although we have neither addressed the putative mechanism for this nor investigated proteins involved, preliminary experiments stabilising the cytoskeleton appear to suggest a profound role of the actin cytoskeleton in this process (see also [32]).

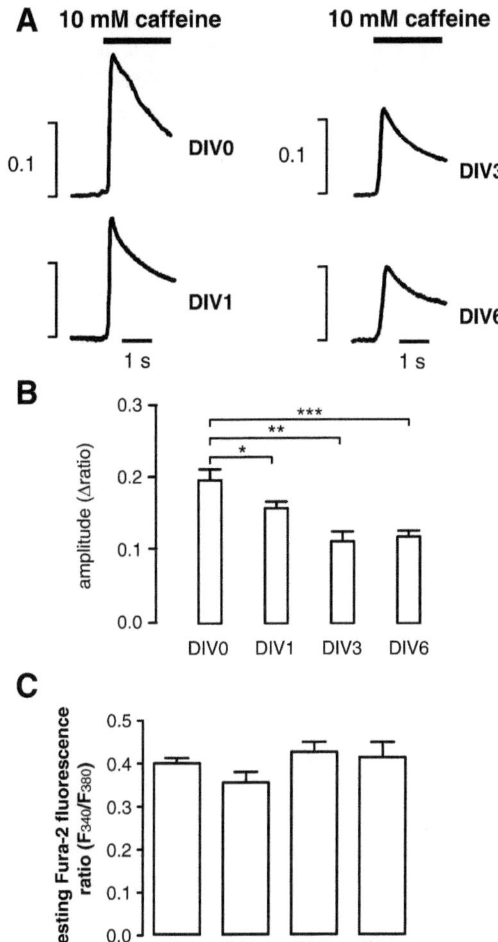

◘ Fig. 19.6 SR-Ca^{2+} content and basal Ca^{2+} in RVM during long term culture. (A) RVMs were loaded with Fura2. Caffeine (10 mM) was applied for the periods indicated. (B) summarises experimental results from larger populations of RVM tested with caffeine. For (C) basal, resting fluorescence ratios were recorded (n = 32 (DIV0), 31 (DIV1), 29 (DIV3) and 28 (DIV6) from 3 rats for B and C).

A possible limitation of our approach is that we rely on the specificity and existence of lipid-raft (or lipid-raft like) membrane domains to be distributed across the entire plasma membrane. In some cells we had been able to perform double-labelling experiments as early as 16–18 h after isolation and transduction. These data show a 100% co-localisation of the two fluorescent entities strongly suggesting that the GPI-anchor is a sufficiently good tool to highlight the plasma membrane (data not shown).

Mitochondria play a profound role for living cells[20]. They are involved in cell differentiation (e.g., [33]), cell death (e.g., [34]) and signalling mechanisms[35]. In cardiac myocytes mitochondria volume takes up more than 30% of the cell volume[30]. In contrast to many cell lines, (e.g., [23]) mitochondria in adult RVMs display a rather stable arrangement and did not show noticeable movement in the time range of up to 20 min.

A detailed analysis of the spatial characteristics of the mitochondria over the culturing period suggested that the distribution of the mitochondria was initially determined by the sarcomeric and/or T-tubule arrangement. These findings nicely correlate to our DIV0 electron microscopy studies and are supportive for electron microscopic data obtained decades ago[36]. Over the culturing time the average mitochondrial length increased along the longitudinal axis while the perpendicular properties remained unchanged. These findings were independent of the method of organelle labelling.

Using FRAP experiments we were able to demonstrate for the first time that in fact the mitochondrial »package« visible on the microscopic level is very often »associated« with neighbouring organelles. This fluorescence exchange was the functional correlate of the fine tubular connections between neighbouring mitochondria visible in all of our electron microscopic images. Since these tubes are not visible in our live-cell experiments one can only speculate about the dynamic behaviour. Mitochondria are well known to be mobile and fusion and fission occur rather frequently, therefore such events are very likely to also occur in living cardiac mitochondria[37]. In FRAP experiments in DIV6 cells we found an increased association between neighbouring mitochondria approaching the exchange rates for elongated mitochondria of cardiac fibroblasts sparsely distributed in our myocyte culture. Thus, the tubular connections between neighbouring mitochondria might consequently widened and allow more ready fluorescence exchange. Nevertheless, as could be depicted from the EM images in ◘ Fig. 19.2, the tubular connections between mitochondria appear rather thin (< 100 nm). This could explain the rather slow recovery time constants observed in our FRAP experiments (◘ Fig. 19.3). Such tubular intra-mitochondrial connections might serve several physiological functions: (i) Ca^{2+} signals occurring locally nearby individual mitochondria might undergo a longer intra-mitochondrial diffusion and will thus influence e.g., ATP production in a larger number of mitochondria, (ii) local mitochondrial depolarisations will be »transmitted« over longer distances and (iii) the coupling produces mitochondria with an increased Ca^{2+} buffer capacity.

We nevertheless concluded that there is no gross change in the morphology and arrangement of the mitochondria during the time of culture. Already at DIV0 a fraction of neighbouring mitochondria displayed inter-organelle communication that pronounced over the culturing time.

ER/SR-targeted expression FPs highlighted the distinctly regular structure of the SR in the living cardiomyocyte. At DIV1 the 3D-structure of the ER/SR appeared to be determined by the sarcomeres and/or T-tubules while at later times partial remodelling of the ladder-like structure of the SR occurred. Interestingly, there were a steady loss of the »sarcomere« frequency peak and a gradual increase in a prominent lower-frequency peak corresponding to double the sarcomeric length. Taken into account the loss of T-tubular structures and remodelling of the mitochondria at later days (especially DIV6) we hypothesise that the distribution of the SR/ER was influenced by the morphology of the mitochondria. In addition the distribution of the peak distances changed from a sharp peak at DIV0 to a very broad distribution of peak distances. These data indicated that the entire SR/ER was undergoing significant remodelling from a well ordered state towards a rather disordered situation at DIV6 because of the loss of T-tubules. Segretain *et al.* described the SR as a structure running alongside the T-tubules[38].

Since the SR is the major Ca^{2+} storing organelle, alterations of the arrangement might translate into changes in Ca^{2+} handling. Recently, we have reported that despite partial remodelling of the SR during the culture time as described in this report, global Ca^{2+} signalling was not altered grossly[7].

Nevertheless this did not exclude changes in Ca^{2+} handling visible on the microscopic level. Ca^{2+} sparks are the elementary Ca^{2+} release events in cardiac myocytes originating from a cluster of RyR in the membrane of the SR[29,39,40]. The properties of Ca^{2+} sparks reflect the integration of a large number of different processes all occurring locally in the vicinity of the RyR. These processes include the particular arrangement of the RyR in the cluster, the cluster size, local Ca^{2+} buffers, SR-Ca^{2+} load, resting $[Ca^{2+}]$, SERCA-pump and NCX activity[29].

Here we analysed the amplitude, spatial spread and the decay time to characterise Ca^{2+} sparks. While at DIV0 the amplitude distribution displayed the expected distribution already described earlier by others (e.g., [29]) a transient decrease of the mean amplitude at DIV1 and 3 occurred which recovered at DIV6. This might reflect variations in the SR-Ca^{2+} content or alterations in the coordination between the RyR in the cluster.

Concomitant to this, we also identified a transient decrease in the decay time constant at DIV1 and 3 while DIV6 displayed a partial recovery to DIV0 values. Interestingly at DIV6 this recovery was mainly due to the occurrence of a second peak in the distribution of the decay time constants rather than a general broadening of the distribution. In fact the leftmost peak at DIV6 depicted an even tighter and narrower distribution. We have tried to relate the Ca^{2+} sparks with a longer lifetime to a particular location in the cells, such as deeper layer, close to the nucleus etc. but they were randomly distributed inside the myocytes (data not shown). Nevertheless, a Ca^{2+} spark site at DIV6 displayed very tight distributions of the decay time constants, indicating that this seemed to be a property of the micro-environment rather than transient variations in the local Ca^{2+} buffering capacity. Both findings, amplitude and decay time variations could be linked to changes of a reduced SR-Ca^{2+} content rather than changes in the resting $[Ca^{2+}]$. This is in line with the strong correlation between SR-Ca^{2+} content and RyR activity described by Györke *et al.*[41]. We speculate that changes in the surface to volume ratio and resulting altered Ca^{2+} fluxes gradually occurring during the loss of T-tubules (◘ Fig. 19.1) might be contributing to the decrease in SR-Ca^{2+} content.

In summary, in this report we have analysed primary isolated RVMs that undergo a distinct but highly reproducible remodelling during the culture time of one week. By using a combination of small molecule dyes and targeted expression of fluorescent proteins we have been able to provide a detailed and quantitative analysis of this process in living cells. Our study was not only limited to morphological studies moreover we have used optical approaches to also study the functional state of some of the organelles such as organelle–organelle association and localised Ca^{2+} release.

In conclusion our study demonstrates the necessity of detailed investigations of the remodelling process that unavoidably occurs in primary RVMs as soon as they are separated from their »natural« micro-environment. For RVMs distinct but highly reproducible processes could be identified that exert measurable but small functional effects. Such investigations represent very important studies to characterise, understand and identify possible limitations of single cell models and also lay the foundation for possible improvements of culture conditions to preserve the morphology and function of primary myocytes in culture. Here, this study is further strongly supportive for using this cell system as a possible high content screening foundation to understand cardiac diseases[42].

19.5　References

[1]　T. Powell and V.W. Twist, A rapid technique for the isolation and purification of adult cardiac muscle cells having respiratory control and a tolerance to calcium. Biochem Biophys Res Commun, 72 1 (1976), pp. 327–333.

[2]　A.C. Nag and M. Cheng, Adult mammalian cardiac muscle cells in culture. Tissue Cell, 13 3 (1981), pp. 515–523.

[3]　T.A. Schwarzfeld and S.L. Jacobson, Isolation and development in cell culture of myocardial cells of the adult rat. J Mol Cell Cardiol, 13 6 (1981), pp. 563–575.

[4]　M.E. Eppenberger, et al. Immunocytochemical analysis of the regeneration of myofibrils in long-term cultures of adult cardiomyocytes of the rat. Dev Biol, 130 1 (1988), pp. 1–15.

[5]　A. Volz, et al. Longevity of adult ventricular rat heart muscle cells in serum-free primary culture. J Mol Cell Cardiol, 23 2 (1991), pp. 161–173.

[6]　T. Banyasz, et al. Transformation of adult rat cardiac myocytes in primary culture. Exp Physiol, 93 3 (2008), pp. 370–382.

[7]　C. Viero, et al. A primary culture system for sustained expression of a calcium sensor in preserved adult rat ventricular myocytes. Cell Calcium, 43 1 (2008), pp. 59–71.

[8]　P. Lipp, et al. Spatially non-uniform Ca^{2+} signals induced by the reduction of transverse tubules in citrate-loaded guinea-pig ventricular myocytes in culture. J Physiol, 497 Pt 3 (1996), pp. 589–597.

[9]　J.S. Mitcheson, J.C. Hancox and A.J. Levi, Action potentials, ion channel currents and transverse tubule density in adult rabbit ventricular myocytes maintained for 6 days in cell culture. Pflugers Arch, 431 6 (1996), pp. 814–827.

[10]　C. Orchard and F. Brette, t-Tubules and sarcoplasmic reticulum function in cardiac ventricular myocytes. Cardiovasc Res, 77 2 (2008), pp. 237–244.

[11]　M.B. Cannell, D.J. Crossman and C. Soeller, Effect of changes in action potential spike configuration, junctional sarcoplasmic reticulum micro-architecture and altered t-tubule structure in human heart failure. J Muscle Res Cell Motil, 27 5–7 (2006), pp. 297–306.

[12]　M. Kawai, M. Hussain and C.H. Orchard, Excitation–contraction coupling in rat ventricular myocytes after formamide-induced detubulation. Am J Physiol, 277 2 Pt 2 (1999), pp. H603–H609.

[13]　M.R. Fowler, et al. Functional consequences of detubulation of isolated rat ventricular myocytes. Cardiovasc Res, 62 3 (2004), pp. 529–537.

[14]　R.S. Kieval, et al. Immunofluorescence localization of the Na–Ca exchanger in heart cells. Am J Physiol, 263 2 Pt 1 (1992), pp. C545–C550.

[15]　F. Brette, et al. Spatiotemporal characteristics of SR Ca(2+) uptake and release in detubulated rat ventricular myocytes. J Mol Cell Cardiol, 39 5 (2005), pp. 804–812.

[16]　D.M. Bers, Cardiac excitation–contraction coupling. Nature, 415 6868 (2002), pp. 198–205.

[17]　E.A. Johnson and J.R. Sommer, A strand of cardiac muscle. Its ultrastructure and the electrophysiological implications of its geometry. J Cell Biol, 33 1 (1967), pp. 103–129.

[18]　D. Rossi, et al. The sarcoplasmic reticulum: an organized patchwork of specialized domains. Traffic, 9 7 (2008), pp. 1044–1049.

[19]　L. Mackenzie, et al. The role of inositol 1, 4, 5-trisphosphate receptors in Ca(2+) signalling and the generation of arrhythmias in rat atrial myocytes. J Physiol, 541 Pt 2 (2002), pp. 395–409.

[20]　H.M. McBride, M. Neuspiel and S. Wasiak, Mitochondria: more than just a powerhouse. Curr Biol, 16 14 (2006), pp. R551–R560.

[21]　H. Tsutsui, S. Kinugawa and S. Matsushima, Mitochondrial oxidative stress and dysfunction in myocardial remodelling. Cardiovasc Res, 81 3 (2009), pp. 449–456.

[22]　E.N. Dedkova and L.A. Blatter, Mitochondrial Ca^{2+} and the heart. Cell Calcium, 44 1 (2008), pp. 77–91.

[23]　T.J. Collins, et al. Mitochondria are morphologically and functionally heterogeneous within cells. EMBO J, 21 7 (2002), pp. 1616–1627.

[24] M.R. Duchen, A. Leyssens and M. Crompton, Transient mitochondrial depolarizations reflect focal sarcoplasmic reticular calcium release in single rat cardiomyocytes. J Cell Biol, 142 4 (1998), pp. 975–988.

[25] Kaestner L. et al., Isolation and genetic manipulation of adult cardiac myocytes for confocal imaging. J Vis Exp 2009. 31.

[26] P. Keller, et al. Multicolour imaging of post-Golgi sorting and trafficking in live cells. Nat Cell Biol, 3 2 (2001), pp. 140–149.

[27] L. Edelmann, et al. High pressure freezing and freeze-substitution of cultured rat myocytes. Microsc Microanal, 13 Suppl. 3 (2007), pp. 146–147.

[28] D.F. Legler, et al. Differential insertion of GPI-anchored GFPs into lipid rafts of live cells. FASEB J, 19 1 (2005), pp. 73–75.

[29] H. Cheng and W.J. Lederer, Calcium sparks. Physiol Rev, 88 4 (2008), pp. 1491–1545.

[30] D.M. Bers, Excitation–contraction coupling and cardiac contractile force, Kluver Academic Publishers, Dordrecht (1991).

[31] D.R. Scriven, P. Dan and E.D. Moore, Distribution of proteins implicated in excitation–contraction coupling in rat ventricular myocytes. Biophys J, 79 5 (2000), pp. 2682–2691.

[32] R.N. Leach, J.C. Desai and C.H. Orchard, Effect of cytoskeleton disruptors on L-type Ca channel distribution in rat ventricular myocytes. Cell Calcium, 38 5 (2005), pp. 515–526.

[33] J. Van Blerkom, Mitochondria in human oogenesis and preimplantation embryogenesis: engines of metabolism, ionic regulation and developmental competence. Reproduction, 128 3 (2004), pp. 269–280.

[34] M. Degli Esposti, Mitochondria in apoptosis: past, present and future. Biochem Soc Trans, 32 Pt3 (2004), pp. 493–495.

[35] V. Soubannier and H.M. McBride, Positioning mitochondrial plasticity within cellular signaling cascades. Biochim Biophys Acta, 1793 1 (2009), pp. 154–170.

[36] P.D. Deshpande, D.D. Hickman and R.W. Von Korff, Morphology of isolated rabbit heart muscle mitochondria and the oxidation of extramitochondrial reduced diphosphopyridine nucleotide. J Biophys Biochem Cytol, 11 (1961), pp. 77–93.

[37] J. Hom and S.S. Sheu, Morphological dynamics of mitochondria—a special emphasis on cardiac muscle cells. J Mol Cell Cardiol, 46 6 (2009), pp. 811–820.

[38] D. Segretain, A. Rambourg and Y. Clermont, Three dimensional arrangement of mitochondria and endoplasmic reticulum in the heart muscle fiber of the rat. Anat Rec, 200 2 (1981), pp. 139–151.

[39] P. Lipp and E. Niggli, A hierarchical concept of cellular and subcellular Ca(2+)-signalling. Prog Biophys Mol Biol, 65 3 (1996), pp. 265–296.

[40] M.J. Berridge, P. Lipp and M.D. Bootman, The versatility and universality of calcium signalling. Nat Rev Mol Cell Biol, 1 1 (2000), pp. 11–21.

[41] S. Gyorke, et al. Regulation of sarcoplasmic reticulum calcium release by luminal calcium in cardiac muscle. Front Biosci, 7 (2002), pp. d1454–d1463.

[42] O. Muller, et al. A system for optical high resolution screening of electrical excitable cells. Cell Calcium, 47 3 (2010), pp. 224–233.

Functional and morphological preservation of adult ventricular myocytes in culture by sub-micromolar cytochalasin D supplement

Qinghai Tian, Sara Pahlavan, Katharina Oleinikow, Jennifer Jung, Sandra Ruppenthal, Anke Scholz, Christian Schumann, Annette Kraegeloh, Martin Oberhofer, Peter Lipp, Lars Kaestner

Reprint from JMCC (2012) **52**: 113-124.

▪ Abstract

In cardiac myocytes, cytochalasin D (CytoD) was reported to act as an actin disruptor and mechanical uncoupler. Using confocal and super-resolution STED microscopy, we show that CytoD preserves the actin filament architecture of adult rat ventricular myocytes in culture. Five hundred nanomolar CytoD was the optimal concentration to achieve both preservation of the T-tubular structure during culture periods of 3 days and conservation of major functional characteristics such as action potentials, calcium transients and, importantly, the contractile properties of single myocytes. Therefore, we conclude that the addition of CytoD to the culture of adult cardiac myocytes can indeed be used to generate a solid single-cell model that preserves both morphology and function of freshly isolated cells. Moreover, we reveal a putative link between cytoskeletal and T-tubular remodeling. In the absence of CytoD, we observed a loss of T-tubules that led to significant dyssynchronous Ca^{2+}-induced Ca^{2+} release (CICR), while in the presence of 0.5 μM CytoD, T-tubules and homogeneous CICR were majorly preserved. Such data suggested a possible link between the actin cytoskeleton, T-tubules and synchronous, reliable excitation–contraction coupling. Thus, T-tubular re-organization in cell culture sheds some additional light onto similar processes found during many cardiac diseases and might link cytoskeletal alterations to changes in subcellular Ca^{2+} signaling revealed under such pathophysiological conditions.

20.1 Introduction

Cytochalasin D (CytoD) is a fungal metabolite that suppresses cytokinesis by blocking the formation of contractile microfilament structures and causes multinucleated cell formation, reversible inhibition of cell movement and the induction of cellular extrusion [1].

In cardiac myocytes, it is believed to act as a F-actin disruptor[2-6]. In such studies, CytoD was used beside diacetyl monoxime (DAM) and butanedione monoxime (BDM), which acts as a mechanical uncoupler on the organ level[5], in tissue[2,3] and with isolated myocytes[6]. However, other reports suggested that in adult ventricular myocytes CytoD stabilises the actin cytoskeleton[7,8]. These studies indicated a positive effect of CytoD on the morphology of cardiomyocytes using 40 μM CytoD as a cell culture supplement.

The aim of the current study was to investigate how CytoD affects the morphology and function of adult cardiomyocytes in culture and whether and how CytoD can be used as a routine supplement in single cell models of cardiomyocytes.

We based our investigation on a previously optimised serum-free culture method of rat ventricular myocytes[9]. The major characteristics were the following: (i) a robust isolation procedure using a well-defined blend of digestion enzymes[10]; (ii) coating of the coverslips with a mixture of extracellular matrix proteins; and (iii) the addition of culture medium supplements including insulin, transferrin and selenite (ITS). We recently characterised this model further and found a gradual loss of T-tubular membranes during a culture period of 6 days *in vitro* (DIV) by a process referred to as »sequential pinching off«, from inside to outside[11]. We speculate that T-tubular remodeling and structural changes in the cytoskeleton might be related to each other. Such a mechanism would be of particular interest because a number of relevant pathologies, particularly the development of heart failure, have recently been associated with T-tubular remodeling[12-14].

20.2   Material and methods

20.2.1   Cell isolation and culture

Ventricular myocytes were isolated from male Wistar rats and cultured as previously described[10]. CytoD (Sigma-Aldrich, Taufkirchen, Germany) was added to the culture medium at the concentrations indicated and was exclusively present when cells were in culture, i.e., during all acute measurements, CytoD was absent.

20.2.2   3D imaging, STED and line scans

For 3D visualization of the plasma membrane, cells were stained with di-8-ANEPPS (Invitrogen, Darmstadt, Germany) and optically sectioned (0.2 µm step size) on a TCS SP5 II (Leica, Mannheim, Germany) followed by deconvolution and 3D rendering as described[15].

Actin was visualised in cells that were chemically fixed at the indicated time points after cell isolation. To visualise the plasma membrane in the fixed cells, cardiomyocytes were transduced with an adenovirus (multiplicity of infection of 20) coding for a GPI-anchor coupled to YFP as described[11]. Actin was stained with phalloidin coupled to ATTO-647N (Atto-Tec GmbH, Siegen, Germany).

Cells were fixed in 4% paraformaldehyde (in PBS) for 10 min, washed in PBS and permeabilised in 0.5% Triton X-100 (in PBS) for 10 min. Samples were then blocked during a 20-min incubation in 5% bovine serum albumin (in PBS). Thereafter,

they were treated with 0.5 µM ATTO-647N phalloidin (2 hours at room temperature), washed with PBS, mounted in ProLong® Gold antifade reagent (Invitrogen) and stored at 4 °C until visualisation.

Super-resolution imaging was performed on a TSC SP5 STED with an HCX Plan APO 100×, 1.4 objective (Leica). For an initial overview of the actin (ATTO-647N phalloidin) and membrane labeling (GPI-YFP), we used the microscope's conventional confocal mode (excitation: 635 nm and 488 nm laser lines, respectively). Emissions were collected in the spectral range of 648–700 nm (ATTO dye) and 495–594 nm (YFP). For the merge of the despeckled images, the individual intensity was adjusted, and a red and green look-up-table was applied in ImageJ (Wayne Rasband, NIH, Bethesda, USA). A rescan of the actin staining at higher magnification was performed in the microscope's STED (stimulated emission depletion) mode. While the probe was excited with a ps pulsed 635 nm laser (PDL800-B, PicoQuant GmbH, Berlin, Germany), stimulated emission was induced by a fs-laser (MaiTai, Newport Corp., Irvine, USA). STED images were deconvolved by custom-made algorithms based on a linear Tikhonov-deconvolution implemented in MatLab (MathWorks, Ismaning, Germany).

Fast line scans were performed on Fluo-4-loaded myocytes as described previously[16]. The effective acquisition speed of the resonant scanner (TCS SP5 II, Leica) was 1000 lines per second, and the analysis of the line scans was performed as described[17].

20.2.3   Functional measurements, power spectral analysis and data statistics

Action potentials (AP) were measured in the whole-cell configuration of the patch-clamp technique as described previously[15]. Ca^{2+} imaging was performed as previously published for mouse cardiomyocytes[18]. We evaluated the contractile performance of the myocytes by sarcomere length measurements as described before[19].

In statistical graphs, if Gaussian distributed, values are given as mean ± SEM, if not Gaussian distributed values are presented in box plots presenting the median and 10% percentile. The number of measurements is given in the format *n = x/y with x = number of cells and y = number of animals.*

The power spectral analysis of the T-tubular membrane system was performed as described previously[9,11]. For the analysis of the actin cytoskeleton, line profiles (5 pixels wide) were generated on parts of the cell that presented fairly straight sections of the actin bundles. These intensity profiles were Fourier transformed and the real component (power) was plotted against the spatial frequency. Finally, these power spectra were normalised to the baseline low frequency to the right of the characteristic peak between 0.5 and 0.7 μm^{-1}, the maximum of which was used for statistical analysis.

Statistical significance was tested in Prism5 (GraphPad Software, La Jolla, USA) with unpaired t-tests. For those data without a Gaussian distribution, a nonparametric t-test (Mann–Whitney) was used. When data were significantly different, they were labeled as follows: $p < 0.05$ (*), $p < 0.01$ (**), $p < 0.001$ (***).

20.3 Results

Based on reports of the beneficial effect of CytoD on the culture of cardiomyocytes[7,8] we applied CytoD in the suggested concentration of 40 μM as a supplement to our optimised culture conditions[9,11]. In an initial series of experiments, we evaluated the CytoD effect on the morphology, basic Ca^{2+} handling and AP characteristics, the results of which are summarised in ◘ Fig. 20.1. Recently, we have reported that T-tubules disappear over a period of 3 days *in vitro* (DIV3) by pinching off from inside to outside[11]. The result of this process, also depicted in ◘ Fig. 20.1Aa (DIV3, Ctrl), was a detectable loss of membrane capacitance (◘ Fig. 20.1Ab, black columns). In the continuous presence of 40 μM CytoD in the culture medium, both membrane capacitance (◘ Fig. 20.1Ab, red columns) and membrane morphology (◘ Fig. 20.1Aa, DIV3, 40 μM CytoD) appeared to be well retained. However, a closer analysis revealed that the T-tubular orientation was altered significantly from a regularly

ordered arrangement at DIV0 to a high density, disordered distribution (cp. ◘ Fig. 20.1A, top row of panels). Moreover, functional parameters such as steady state Ca^{2+} transients (◘ Fig. 20.1B) and action potentials (◘ Fig. 20.1C) were largely altered in the continuous presence of 40 μM CytoD. These data strongly suggested major modulatory and unfavorable effects of CytoD on all functional parameters investigated.

We therefore aimed to explore the possibility of whether CytoD could be applied in a concentration that would retain its positive effect on the conserved morphology but would not display functional impairments. For initial screening we analysed dose-dependent changes in the amplitude of electrically induced Ca^{2+} transients (◘ Fig. 20.2A). As depicted in ◘ Fig. 20.2A, we performed the experiments at various CytoD concentrations (up to 3.6 μM) up to DIV3. This period was recently identified as sufficient for expression of exogenous proteins by adenoviral gene transfer[11,20]. The intersection between the DIV3 and DIV0 relationship thus indicated the most appropriate CytoD concentration (0.5 μM) for preserving the DIV0 behavior. Panels 20.2Ab–d compare the amplitude of typical transients (left column) and the kinetic properties after normalisation (right column). In the following, we investigated whether additional important properties, including Ca^{2+} handling (◘ Fig. 20.2B and 20.3), AP (◘ Fig. 20.4), and contractility (◘ Fig. 20.5), were also well preserved at this CytoD concentration.

Global Ca^{2+} handling was analysed in greater detail with a post-rest behavior protocol stimulation as depicted in ◘ Fig. 20.2B. After a resting period, the first as well as the steady state amplitude of electrically-induced Ca^{2+} transients displayed a significant drop at DIV3 when compared to DIV0. In the presence of 0.5 μM CytoD in the culture medium, such properties were preserved at typical levels for DIV0 (compare black and open columns in ◘ Fig. 20.2Ca and b, DIV3). When we analysed the Ca^{2+} transient duration, we found that 0.5 μM CytoD caused a prolongation as early as DIV0, but the massive prolongation observed during 3 days of culture without CytoD was significantly reduced. In additional experiments, we further assessed characteristics of Ca^{2+} handling by

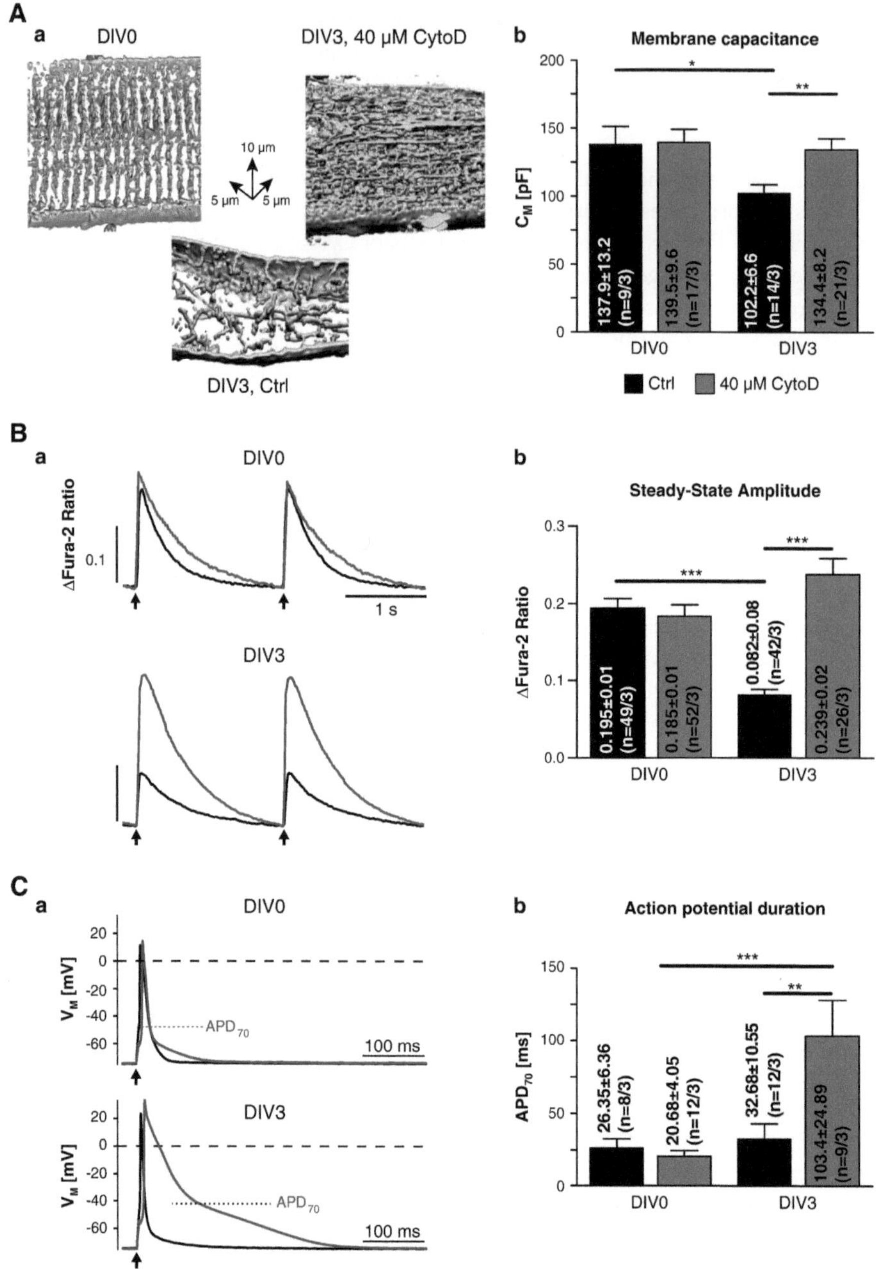

◘ **Fig. 20.1** Effects of 40 µM CytoD on adult cultured rat ventricular myocytes. Panel (A) depicts changes in membrane topology. Representative 3D reconstructions of confocal sections at DIV0 and DIV3 with and without 40 µM CytoD are plotted in (Aa), while (Ab) shows the statistical analysis of patch-clamp-based capacitance measurements under the conditions mentioned. Panel (B) reflects the Ca^{2+} handling under steady state conditions. (Ba) depicts examples of Fura-2 traces in cardiomyocytes at DIV0 and DIV3 with 40 µM CytoD (red) and without CytoD (black), while (Bb) summarises the statistics for the calcium transient amplitude. The black arrows in (Ba) indicate the time point of electrical field stimulation. Panel (C) compares the action potential at DIV0 and DIV3 with 40 µM CytoD (red) and without CytoD (black) in representative examples (Ca) and the statistics of APD70 (Cb). (For interpretation of the references to color in this figure legend, the reader is referred to the web version of the original article.)

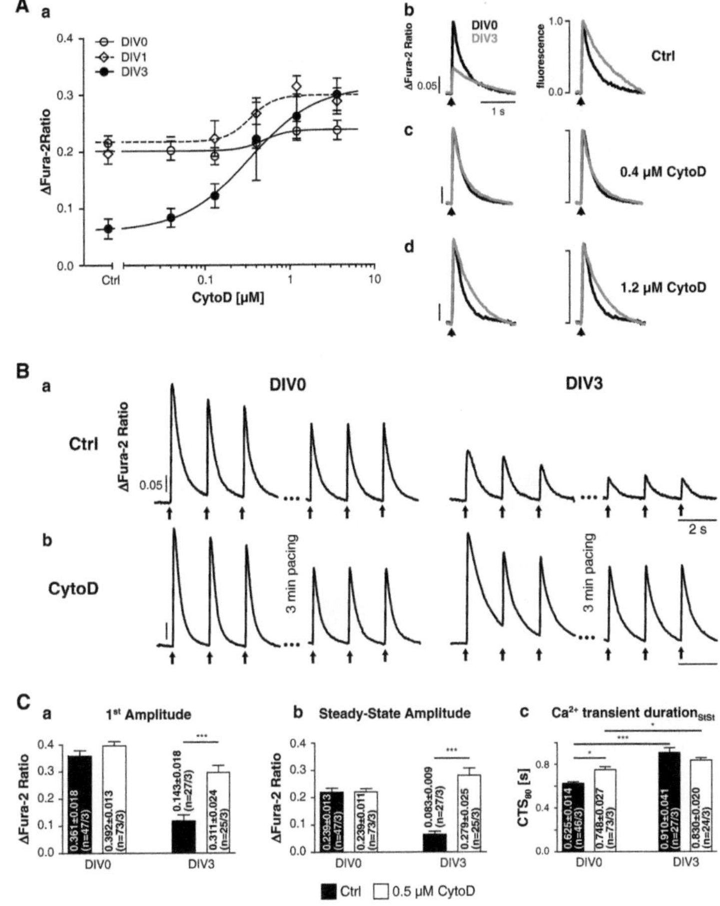

Fig. 20.2 Electrically evoked Ca²⁺ transients. Panel (Aa) depicts the dose–response curves for the given concentrations of CytoD at different days of culture (DIV0–DIV3). Panels (Ab–d) provide example traces for control conditions (Ab), for 0.4 µM CytoD (Ac) and 1.2 µM CytoD (Ad). The black arrows indicate the time point of electrical field stimulation. Panel (B) depicts a more detailed post-rest protocol to compare calcium transients in the presence and absence of 0.5 µM CytoD in the culture medium at DIV0 and DIV3. Panel (Ba) depicts representative traces under control conditions and (Bb) in the presence of 0.5 µM CytoD at both DIV0 and DIV3. The black arrows indicate the time point of electrical field stimulation. Panel (C) displays the statistical analysis of the Fura-2-based calcium amplitudes, namely the 1st amplitude after rest (Ca), the steady state amplitude (Cb) and the calcium transient duration (Cc).

discharging the sarcoplasmic reticulum (SR)-Ca²⁺ store under steady state conditions (■ Fig. 20.3). For this, we first electrically stimulated the cells and then in steady state briefly (5–10 s) applied 10 mM caffeine. For DIV3 cells treated with CytoD, we found that the amplitude (■ Fig. 20.3Bb) and the fractional Ca²⁺ release (ratio of the Ca²⁺ transient amplitudes evoked by electrical stimulation over the transient amplitude induced by caffeine, ■ Fig. 20.3Bc) resembled DIV0 values with CytoD, while the apparent NCX activity appeared to be slightly decreased (■ Fig. 20.3Bd).

■ Fig. 20.4 summarises the properties of APs in the absence (black traces and columns) and in the continuous presence (gray traces, open columns) of 0.5 µM CytoD in the culture medium. While without CytoD the AP amplitude displayed a significant drop at DIV3, the presence of 0.5 µM CytoD prevented that effect, indicating that this concentration resulted in good preservation of basic electrophysiological parameters (resting potential (■ Fig. 20.4Ba), AP amplitude (■ Fig. 20.4Bb) and APD (■ Fig. 20.4Bc–f)) over a culture time of 3 days.

Myocyte contractility was analysed as electrically-induced changes of the sarcomere length (■ Fig. 20.5). After a resting period, the amplitude of the first contraction (■ Fig. 20.5Ba) as well as the steady state amplitudes (■ Fig. 20.5Bb) were decreased at DIV3 under control conditions. In the presence of 0.5 µM CytoD, both parameters as well as the contraction duration (■ Fig. 20.5Bd and Be) were similar to those found at DIV0. Thus, CytoD in the culture medium preserved the contractile properties.

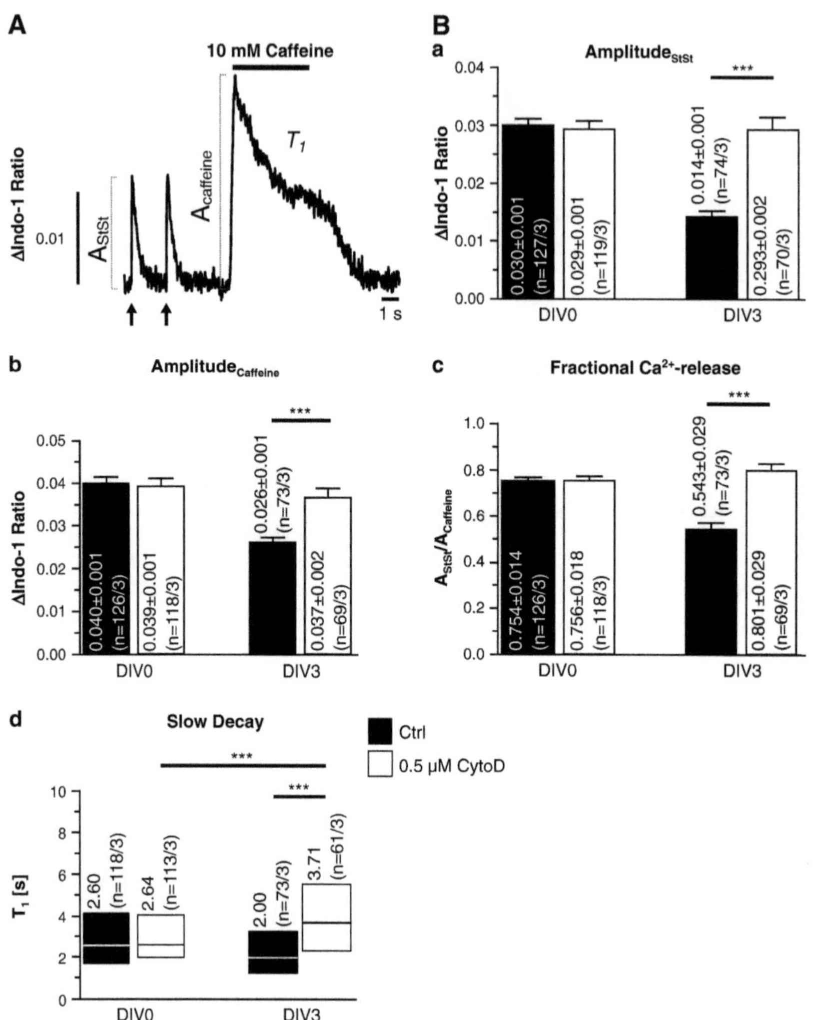

Fig. 20.3 Caffeine-induced Ca^{2+} release of rat ventricular myocytes with and without 0.5 µM CytoD in the culture medium. Panel (A) plots an example trace under control conditions with all parameters indicated tested for statistical analysis (B). The black arrows below the traces indicate the time points of electrical stimulation, while the black bar above the trace represents the period of caffeine application. The following read-outs were analysed: steady state amplitude (Ba), caffeine-induced amplitude (Bb), fractional Ca^{2+} release (Bc) calculated as the ratio of the former two parameters and the decay time constant (Bd) during caffeine application.

Because reports regarding the interaction of CytoD and actin have been controversial (see 20.1 - Introduction) and CytoD is often referred to as a cytoskeleton disrupter, we decided to directly analyse the actin cytoskeleton by probing actin filaments with phalloidin. We performed high resolution and super resolution imaging of the resulting staining as summarised in ▪ Fig. 20.6. ▪ Fig. 20.6Aa displays a single confocal slice through a typical, freshly isolated myocyte with a representative fluorescence profile along the longitudinal axis (yellow line, ▪ Fig. 20.6Ba). ▪ Fig. 20.6Ab–d depict standard high-resolution confocal sections through typical cardiac myocytes labeled with phalloidin (red color) and GPI-YFP (green color). The latter staining was employed to indicate plasma membrane

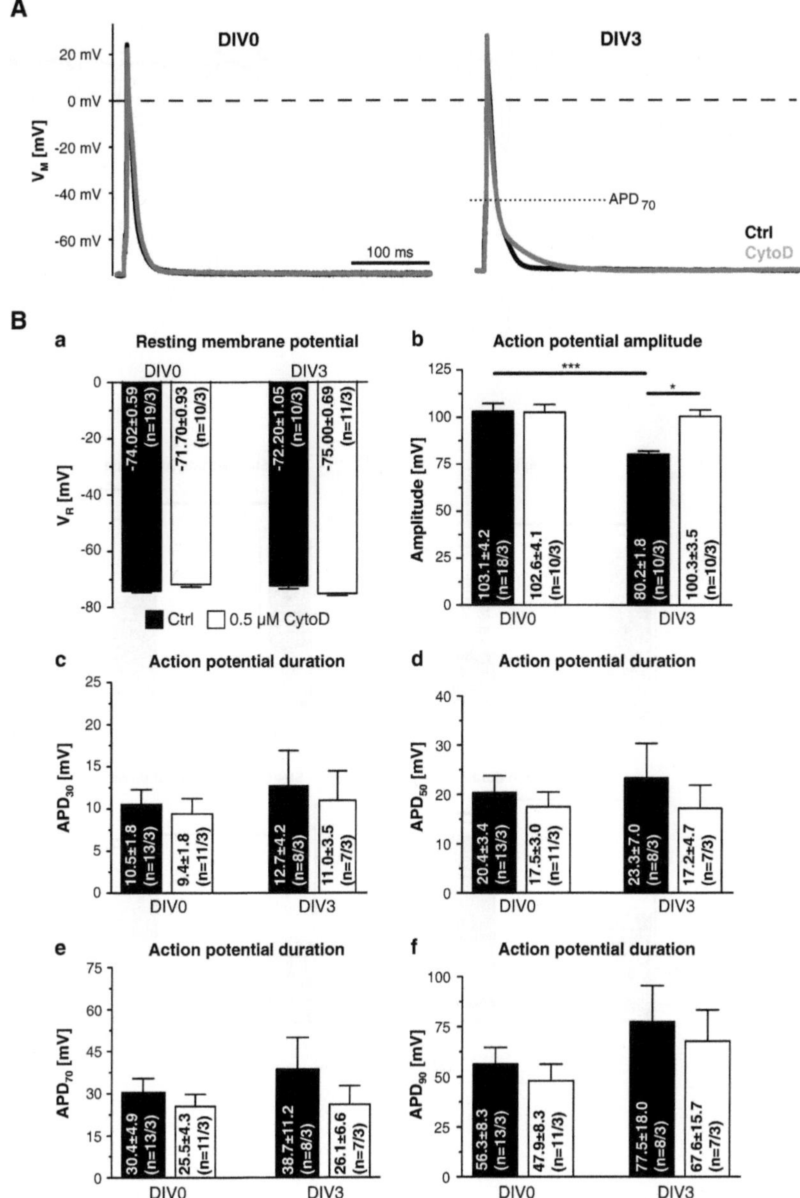

◘ Fig. 20.4 Action potentials of rat ventricular myocytes with and without 0.5 µM CytoD in the culture medium at DIV0 and DIV3. Panel (A) provides representative action potentials of the four conditions probed. Panel (B) shows the statistical analysis for the resting membrane potential (Ba), the action potential amplitude (Bb), the APD_{30} (Bc), APD_{50} (Bd), APD_{70} (Be) and the APD_{90} (Bf).

topology[11] in the same cell and to allow identification of structurally intact cells. Viral transduction was mandatory because standard membrane probes, such as those used for ◘ Fig. 20.1A, did not withstand the permeabilisation necessary for the application of the actin probe. Quantification of the actin staining was performed by analysing the pattern of fluorescence along the longitudinal

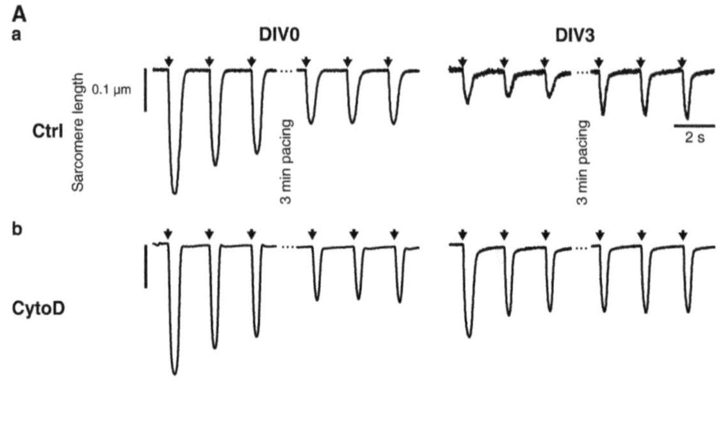

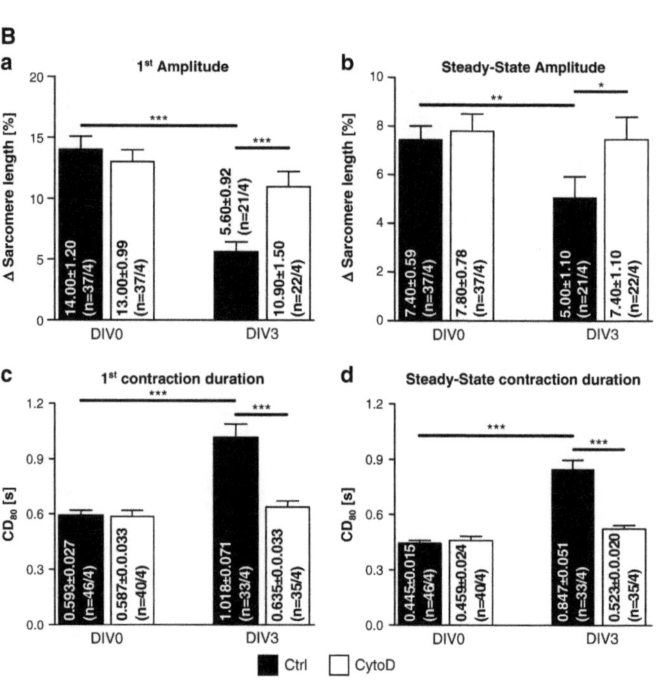

☐ Fig. 20.5 Contraction of rat ventricular myocytes with and without 0.5 µM CytoD in the culture medium at DIV0 and DIV3. Panel (A) depicts representative sarcomere length measurements under control conditions (Aa) and in the presence of 0.5 µM CytoD (Ab) at both DIV0 and DIV3. The black arrows indicate the time points of electrical field stimulation. Panel (B) displays the statistical analysis for the first amplitude after rest (Ba), the steady state amplitude (Bb), the contraction duration D_{80} of the first contraction after rest (Bc) and in the steady state (Bd).

axis of the cells (yellow lines in ☐ Fig. 20.6A). The patterns displayed in ☐ Fig. 20.6Ba–d were quantified by power spectral analysis (☐ Fig. 20.6Ca with color coding similar to ☐ Fig. 20.6B). Such analysis clearly indicated that without CytoD, actin arrangement was already modulated as early as DIV1. For freshly isolated cells (DIV0, green line in ☐ Fig. 20.6Ca) and cells in 0.5 µM and 40 µM CytoD at DIV1 (☐ Fig. 20.6Ca, grey and red line, respectively), the power spectrum shows similar peaks at the sarcomeric frequency (0.58 µm^{-1}).

The diminished peak in the power spectrum for untreated myocytes at DIV1 (☐ Fig. 20.6Ca, black line) indicated a loss of structure as a result of early remodeling processes. This observation was documented by results of a population analysis, as presented in ☐ Fig. 20.6Cb. This revealed a significant difference between untreated cells at DIV1 and all other conditions tested. We rescanned the regions marked by white boxes (☐ Fig. 20.6Ab–d) at a higher magnification using STED imaging (☐ Fig. 20.6Da–c) to enable easier identification of typical banding

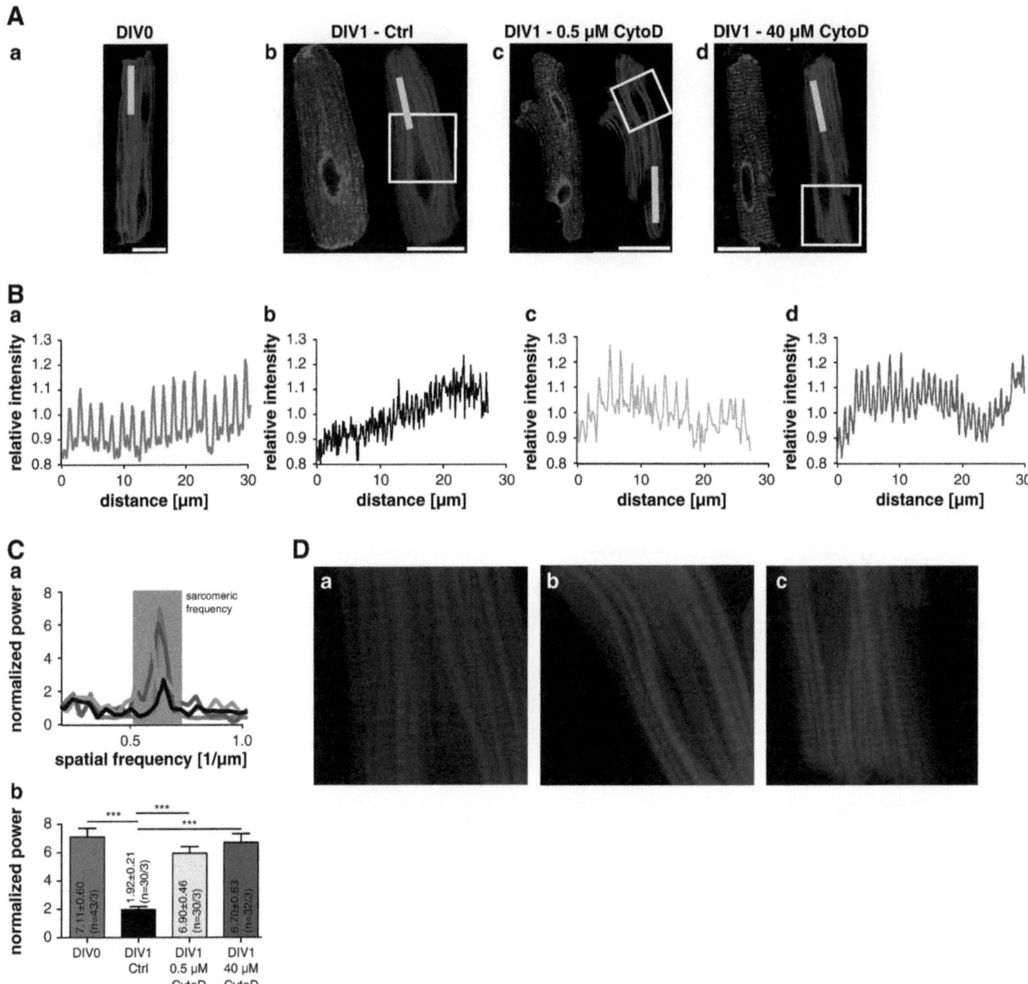

◘ **Fig. 20.6** Actin filaments in cultured adult rat ventricular myocytes. Representative images of rat cardiomyocytes expressing GPI-YFP as a membrane marker and stained by phalloidin-ATTO-647N to visualise the actin filaments. Cells are fixed at DIV0 (Aa) or DIV1 for all other conditions at different CytoD concentrations, as indicated above the image columns. The white bar in the confocal images (Aa)–(Ad) represents 20 µm. The yellow bars indicate 2 µm-wide stripes from which intensity profiles were taken and plotted below (Ba)–(Bd). Panel (Ca) depicts a power spectrum analysis of such profiles. The color code corresponds to the profiles in (Ba), (Bb) and (Bc). Panel (Cb) represents a statistical analysis of the normalised intensity of the power spectrum at the sarcomeric frequency. The subsections indicated by the white rectangle in (Ab), (Ac) and (Ad) were rescanned in the STED imaging mode and displayed in images (Da), (Db) and (Dc), respectively. (For interpretation of the references to color in this figure legend, the reader is referred to the web version of the original article.)

patterns. Thus, alterations in the cytoskeleton have already been identified as early as DIV1, at which time all cells displayed an unaltered T-tubular network (see green staining in ◘ Fig. 20.6A). In the presence of 0.5 µM CytoD the analysed properties of actin at DIV1 resembled those at DIV0.

In a further set of experiments, we addressed the question to what degree our optimised CytoD concentration modulated the remodeling of T-tubules during culture, e.g., DIV3[11]. To that end, we probed the morphology of the plasma membrane with di-8-ANEPPS. The results are summarised in

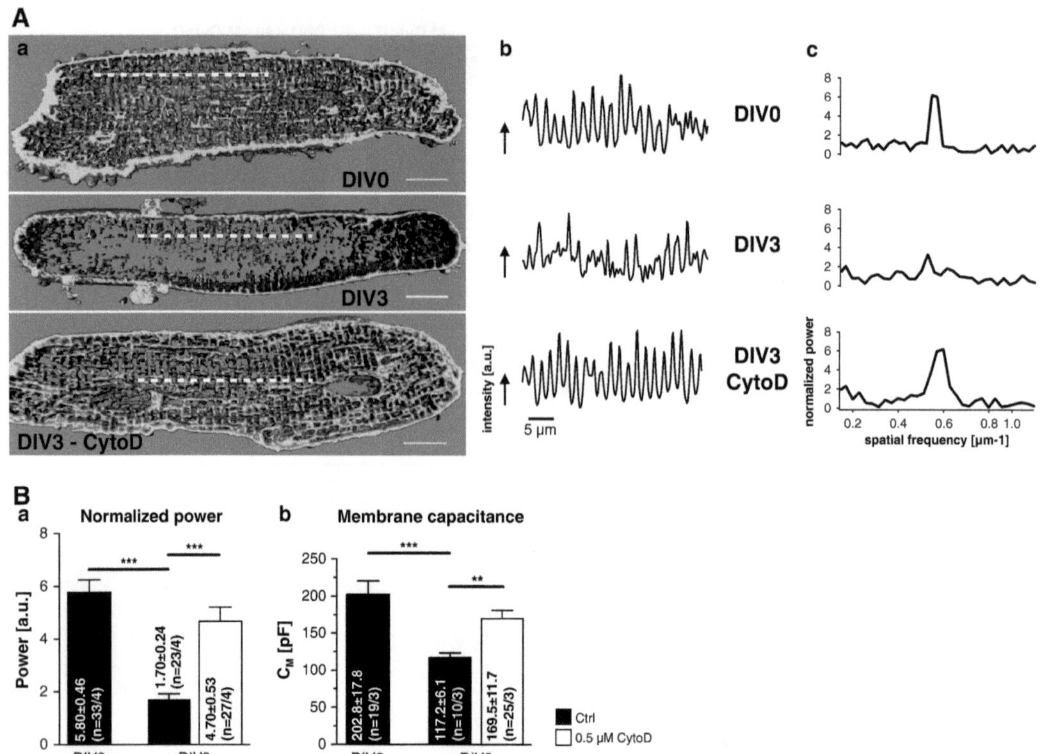

◘ Fig. 20.7 T-tubular structure of rat ventricular myocytes with and without 0.5 μM CytoD in the culture medium at DIV0 and DIV3. Panel (A) displays representative confocal sections of di-8-ANEPPS stained cells. (Aa) depicts 3D rendered cells that were cut open to reveal the T-tubular structure (DIV0 and DIV3 in the presence of 0.5 μM CytoD) or the residues of the T-tubules (DIV3 in the absence of CytoD). (Ab) shows the corresponding intensity profile along the dashed white line in the images (Aa). From these, intensity profile power spectra were calculated and plotted in (Ac). Panel (Ba) summarises the statistical analysis of such power spectra under the conditions given, while panel (Bb) depicts the corresponding patch-clamp-based capacitance measurements.

◘ Fig. 20.7. The analysis of 3D surface renderings of typical cells suggested a conservation of T-tubular structures by 0.5 μM CytoD (◘ Fig. 20.7Aa). We quantified the images using power spectral analysis of the fluorescence distribution along the longitudinal axis of the myocytes (see dashed white lines in ◘ Fig. 20.7Aa) and found that the characteristic sarcomeric frequency peak at approximately 1.8 μm was much more prominent at DIV0 and DIV3 with 0.5 μM CytoD than at DIV3 without the supplement. We verified those results using a larger population of cells (◘ Fig. 20.7Ba). To further obtain additional independent support for our findings, we also employed electrophysiological characterisation of the plasma membrane. We investigated the membrane capacitance of a population of myocytes and found that our optimised CytoD concentration also conserved DIV0 conditions of the membrane capacitance (compare DIV3 with DIV0 values in ◘ Fig. 20.7Bb).

Finally, we were interested in the putative physiological consequences of preventing subcellular remodeling in the presence of CytoD and studied the subcellular properties of electrically evoked Ca^{2+} transients. For these experiments, we performed confocal line scans in the absence and the presence of 0.5 μM and 40 μM CytoD, as depicted in ◘ Fig. 20.8. In this study, DIV0 was compared to DIV3 and the statistical analysis also included the post-rest and the steady state behavior. Similar to

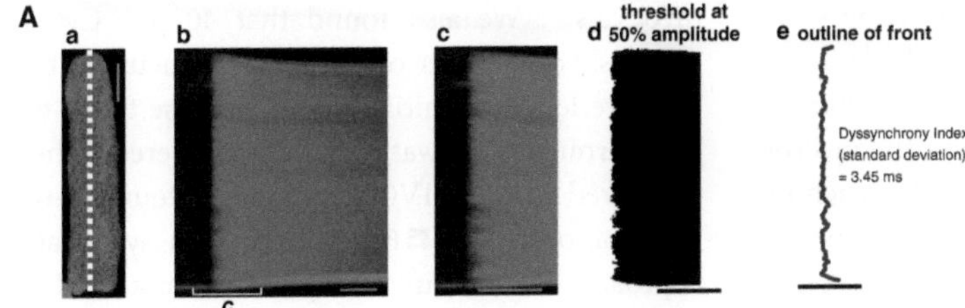

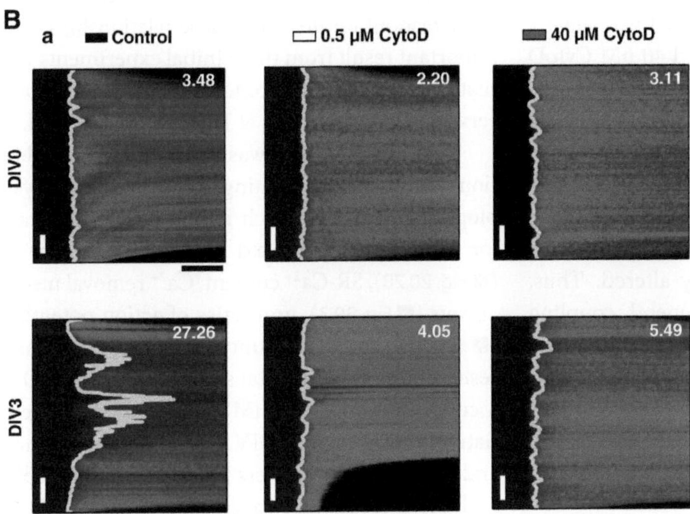

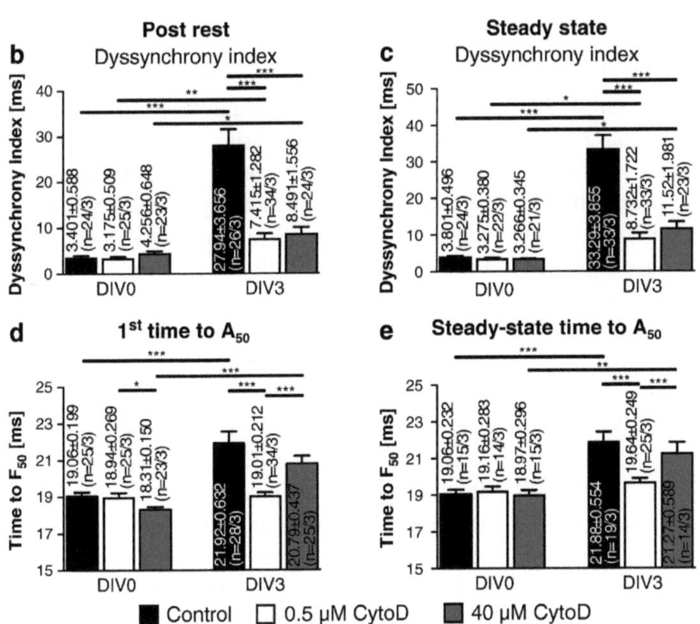

Fig. 20.8 CytoD effects on subcellular Ca^{2+} release. Panel (A) illustrates the procedure of data acquisition and analysis. In a 2D image of a cardiomyocyte, a line was chosen (Aa). The scale bar in (Ba) depicts 10 µm. Along this line, a line scan was performed over time (Ab). Image (Ac) depicts a temporal enlargement to point out spatial differences in the calcium release. A binary image with the threshold of 50% amplitude is illustrated in (Ad). The standard deviation of the 50% line is defined as the dyssynchrony index (Ae). Typical line scans in the absence of CytoD, with 0.5 µM CytoD and 40 µM CytoD in the culture medium at DIV0 and DIV3 are plotted in (Ba). The yellow lines indicate the outline of the F_{50} profiles. The number in the top right corner denotes the dyssynchrony index of this particular line scan. The statistical analysis of the dyssynchrony index (Bb) and (Bc) as well as the time to 50% of the calcium amplitude (Bd) and (Be) under post-rest and steady state conditions, respectively, are shown. Throughout the line scans vertical scale bars depict 15 µm and horizontal scale bars 50 ms. (For interpretation of the references to color in this figure legend, the reader is referred to the web version of the original article.)

a recent study[17], we also characterised the spatial properties of the Ca^{2+} transients by calculating the dyssynchrony index, that is, a quantitative measure of the transient's homogeneity during the upstroke period (◘ Fig. 20.8Ab–e). While the dyssynchrony index is a parameter that reflects the concerted action of the molecular players in CICR, the time to half amplitude is more indicative of the general coupling between calcium entry and calcium release. For both post-rest and steady state conditions, the dyssynchrony index at DIV3 was slightly increased in the presence of 0.5 and 40 µM CytoD compared to DIV0 measurements but much less pronounced than in the absence of CytoD at DIV3.

When analysing the time to half-amplitude (◘ Fig. 20.8Bd–e), we found that while in the presence of 0.5 µM CytoD, the values were unchanged between DIV0 and DIV3. Under all other conditions, this value was significantly altered. Thus, 0.5 µM CytoD preserved the general coupling between Ca^{2+} influx and Ca^{2+} release and consequently maintained a robust EC coupling.

20.4   Discussion

The current report represents a significant improvement in the culture conditions of adult ventricular myocytes to enable the conservation of both morphology and physiology. Moreover, it also provides initial evidence for possible mechanisms of T-tubular remodeling in general. Such alterations of the T-tubular characteristics have been described in various cardiac diseases, including the transition from hypertrophy to heart failure[13] and following myocardial infarction[17], but until now, no mechanistic insights have been provided concerning how this plasma membrane remodeling might occur.

20.4.1   Improved culture of adult cardiac myocytes

Our starting point was based on reports in which high CytoD concentrations not only disrupted excitation–contraction coupling but were also beneficial for long-term culture of adult cardiac myocytes[7,8]. We also found that 40 µM CytoD diminished the loss of T-tubular structures, but we were left with initial doubts because the T-tubular arrangement was significantly altered. When compared to the DIV0 situation, we found T-tubular »crowding« (◘ Fig. 20.1Aa). In a systematic approach, we set out to explore the potential of CytoD as a culture supplement by employing an important physiological parameter, i.e., the amplitude of electrically evoked Ca^{2+} transients, as an assay readout for dose–response relationships. The important result from these initial experiments was that CytoD at sub-micromolar concentrations preserved DIV0 conditions at DIV3 extremely well.

This initial notion was substantiated in additional studies investigating a plethora of physiological parameters, such as the post-rest behavior of electrically evoked global Ca^{2+} transients (◘ Fig. 20.2B), SR-Ca^{2+} content, Ca^{2+} removal mechanisms (◘ Fig. 20.3), properties of action potentials (◘ Fig. 20.4) and contractility (◘ Fig. 20.5). The basic result of this experimental series was that CytoD at a concentration of 0.5 µM prevented de-differentiation effects seen at DIV3, preserving structural and functional parameters found at DIV0. Despite a slight prolongation in calcium transient duration and an increase in NCX activity at DIV3 with 0.5 µM CytoD, this treatment, to our knowledge, caused the best functional preservation reported after such a period *in vitro*. Interestingly, such a prolongation did not translate into extended contraction transients, as seen in ◘ Fig. 20.5Bc and Bd.

Reports on Ca^{2+} measurements in the presence of CytoD are rather sparse[7,21,22]. Although using 2-photon excitation recording of intracellular Ca^{2+} transients was successfully reported for Langendorff mouse hearts with 50 µM CytoD, such transients could not be compared to control conditions because the mechanical uncoupling was a necessary requirement for these technically demanding experiments[21]. Leach *et al.* recorded line scans in the presence and absence of 40 µM CytoD during a culture period of 4 days and described severe alterations in Ca^{2+} handling, but did not include a statistical/quantitative analysis[7]. In contrast to our current study, very often, acute effects of CytoD had been investigated, e.g., Howarth *et al.* reported such immediate effects using 40 µM CytoD[22].

Under physiological Na^+ concentrations, heterologously expressed NCX displayed 40% reduced activity with 1 μM CytoD[23]. In ventricular myocytes, we did not find changes in the apparent NCX activity after 0.5 μM CytoD treatment at DIV0, but a slight increase of the time constant τ (decreased activity) at DIV3 was found.

Following acute application of 20–40 μM and 10 μM CytoD, previous investigations reported a 25% reduction in the Na^+ current[4] and an abolishment of the Ca^{2+} mediated inward rectification of K^+ channels[24], respectively. For alterations in both membrane currents, we would have expected a significant impact on the amplitude and shape (APD) of action potentials. Under our conditions, we could not find any changes in the shape of APs induced by the chronic treatment with 0.5 μM CytoD. Nevertheless, our results are difficult to compare with previous reports because they all focused on acute CytoD effects. In our experiments, CytoD was only present in the culture medium but was washed out for the acute experiments. If there had been altered AP properties in the presence of CytoD, these effects would have been reversible immediately after washout.

In addition, the obvious differences in the functional parameters of the cardiac myocytes with CytoD treatment between our study and previous reports can most likely be explained by different concentrations of CytoD. While previous reports mainly used between 10 and 80 μM CytoD[2,4,7,21,24], our culturing was performed at 0.5 μM CytoD (◘ Fig. 20.2B–◘ Fig. 20.5 and ◘ Fig. 20.7).

The action of CytoD on ventricular myocytes was clearly dose-dependent (◘ Fig. 20.2A). We were particularly surprised by our finding that 0.5 μM CytoD conserved the contractility of myocytes because CytoD was used as a mechanical uncoupler[2,3,5,6]. However, similar to above, we believe that this is a question of the concentration because a concentration of 4 μM CytoD had no effect on the contraction in cat cardiomyocytes[24]. In addition, in Langendorff mouse hearts, 5 μM CytoD was determined to be the minimum concentration that reliably reduced the developed pressure by more than 90%[5].

Based on this, we conclude that 0.5 μM CytoD is the optimal concentration to maximally conserve cellular function over a culture period of at least 3 days. This is a sufficient time to enable the expression of fusion proteins and genetically encoded biosensors[9,11,20] and therefore can serve as an improved cellular model for genetic as well as pharmacological manipulations *in vitro*.

20.4.2 The interplay of CytoD and actin

CytoD primarily acts on actin[26] and leads to mechanical uncoupling at a concentration of 40 μM[4,6,21] but does not show mechanical uncoupling at concentrations of 0.5 μM[5,25] (◘ Fig. 20.5). To further investigate changes in the actin microarchitecture, we visualised the actin cytoskeleton in ventricular myocytes (◘ Fig. 20.6). In all conditions studied, we surprisingly did not find a disruption of the actin arrangement. Initial reports on the interaction of CytoD and actin described binding of CytoD to the barbed end of actin filaments, thus preventing actin polymerisation[27,28]. We believe that disruption and prevention of polymerisation are not necessarily the same and that these interpretation differences contribute to the controversy found in the literature. Reports that focus on the molecular mechanism of CytoD action asserted that CytoD binding to actin prevented further polymerisation/filament elongation by inducing net depolymerisation[29-31]. Most likely, the binding of CytoD to actin together with its effect on mechanical uncoupling might have led to the concept of CytoD as an actin disruptor. Instead, Calaghan *et al.* suggested that mechanical uncoupling might be due to decreased myofilament sensitivity to Ca^{2+} through interaction with sarcomeric actin[6,32]. Therefore, we conclude that at the concentrations used in our report, CytoD leads to actin conservation rather than disruption.

We detected actin remodeling (more diffuse staining in ◘ Fig. 20.6Ab, Bb, Da) as early as after 1 day in culture and quantified that by power spectral analysis (◘ Fig. 20.6C, black line and column). This cell culture-induced modulation was prevented with the optimised CytoD concentration identified in our current report. The addition of 0.5 μM CytoD resulted in a conservation of actin filament banding (◘ Fig. 20.6Ac, Bc, Db) very similar to the one found at DIV0 (◘ Fig. 20.6Aa and Ba). Our power spectral analysis substantiated this notion (◘ Fig. 20.6Cb, compare grey and green columns). Due to the »freezing« of the actin filaments, the culture in the presence of CytoD excludes the study of actin filament dynamics in cardiomyocyte culture.

20.4.3 CytoD as a pharmacological tool to study cardiomyocyte remodeling

As depicted in ◘ Fig. 20.2, ◘ Fig. 20.3, ◘ Fig. 20.4 and ◘ Fig. 20.5, the presence of 0.5 μM CytoD in the culture medium conserved the functional properties of ventricular myocytes in culture. Although it was reported before that T-tubular remodeling could be prevented by high CytoD concentrations, our results strongly discourage the application of such high concentrations because in the current study, we found that 40 μM CytoD, despite preventing T-tubular loss, clearly induced highly artificial T-tubular »crowding« (◘ Fig. 20.1Aa). From this, we conclude that the application of 40 μM CytoD substituted one kind of remodeling, i.e., loss of T-tubular structures, by another one, i.e., T-tubular »crowding«. Here, we present evidence that after »titration« of the CytoD concentration, one can find an optimal balance that resulted in a conservation of the regular cross-striational arrangement of T-tubules (◘ Fig. 20.7A).

As a culture supplement, 0.5 μM CytoD prevented culture-dependent remodeling in all morphological and functional parameters that were analysed. Culture-dependent alterations of functional properties can be primarily related to the loss of T-tubules and the resulting loss of L-type Ca^{2+} channel–RyR interactions. To provide further evidence for this assumption, we performed confocal line scans (◘ Fig. 20.8). Here, we compared three different conditions: control medium and medium supplemented with either 0.5 or 40 μM CytoD. To characterise the effects of the culture conditions on subcellular aspects of EC coupling, we analysed rapid line scans during electrical stimulations by means of the following two different parameters: the dyssynchrony index (DI) and the time to half

amplitude (◘ Fig. 20.8). Notably, DI changed for all conditions when comparing DIV0 and DIV3. Nevertheless, the relative change was vastly different. While in the presence of CytoD, DI doubled; this value increased almost an order of magnitude without CytoD. When we analysed EC coupling on a more global basis (time to half amplitude), both control as well as 40 μM CytoD conditions showed alterations of this parameter, while our optimised concentration maintained DIV0 properties.

In summary, these findings suggest a possible link between remodeling of the cytoskeletal components and the observed T-tubular remodeling resulting in the appearance of orphaned RyR as in many cardiac diseases. It will thus be interesting to extend the findings and concepts described in the current report using cultured cardiac cells to mechanisms of plasma membrane remodeling observed, e.g., following myocardial infarction[17], in diabetic cardiomyopathy[33,34], during heart failure in general[12,14,35] or particularly during the transition from hypertrophy to heart failure[13]. We propose the short-term culture of adult ventricular myocytes (3 days to additionally allow virus-mediated genetic manipulation) as an appropriate model system to obtain additional insights into cellular structural remodeling during cardiac diseases.

20.5 References

[1] Medical Subject Headings, US National Library of Medicine, U.S. National Institutes of Health, Bethesda (2009).

[2] M. Biermann, M. Rubart, A. Moreno, J. Wu, A. Josiah-Durant and D.P. Zipes, Differential effects of cytochalasin D and 2,3 butanedione monoxime on isometric twitch force and transmembrane action potential in isolated ventricular muscle: implications for optical measurements of cardiac repolarization. J Cardiovasc Electrophysiol, 9 12 (Dec 1998), pp. 1348–1357.

[3] J. Wu, M. Biermann, M. Rubart and D.P. Zipes, Cytochalasin D as excitation–contraction uncoupler for optically mapping action potentials in wedges of ventricular myocardium. J Cardiovasc Electrophysiol, 9 12 (Dec 1998), pp. 1336–1347.

[4] A.I. Undrovinas, G.S. Shander and J.C. Makielski, Cytoskeleton modulates gating of voltage-dependent sodium channel in heart. Am J Physiol, 269 1 Pt 2 (Jul 1995), pp. H203–H214.

[5]   L.C. Baker, R. Wolk, B.R. Choi, S. Watkins, P. Plan and A. Shah, et al. Effects of mechanical uncouplers, diacetyl monoxime, and cytochalasin-D on the electrophysiology of perfused mouse hearts. Am J Physiol Heart Circ Physiol, 287 4 (Oct 2004), pp. H1771–H1779.

[6]   S.C. Calaghan, E. White, S. Bedut and J.Y. Le Guennec, Cytochalasin D reduces Ca^{2+} sensitivity and maximum tension via interactions with myofilaments in skinned rat cardiac myocytes. J Physiol, 529 Pt 2 (Dec 1 2000), pp. 405–411.

[7]   R.N. Leach, J.C. Desai and C.H. Orchard, Effect of cytoskeleton disruptors on L-type Ca channel distribution in rat ventricular myocytes. Cell Calcium, 38 5 (Nov 2005), pp. 515–526.

[8]   K.Y. Chung, M. Kang and J.W. Walker, Contractile regulation by overexpressed ETA requires intact T tubules in adult rat ventricular myocytes. Am J Physiol Heart Circ Physiol, 294 5 (May 2008), pp. H2391–H2399.

[9]   C. Viero, U. Kraushaar, S. Ruppenthal, L. Kaestner and P. Lipp, A primary culture system for sustained expression of a calcium sensor in preserved adult rat ventricular myocytes. Cell Calcium, 43 1 (Jan 2008), pp. 59–71.

[10] L. Kaestner, A. Scholz, K. Hammer, A. Vecerdea, S. Ruppenthal and P. Lipp, Isolation and genetic manipulation of adult cardiac myocytes for confocal imaging. J Vis Exp, 31 (2009).

[11] K. Hammer, S. Ruppenthal, C. Viero, A. Scholz, L. Edelmann and L. Kaestner, et al. Remodelling of Ca(2+) handling organelles in adult rat ventricular myocytes during long term culture. J Mol Cell Cardiol, 49 3 (Sep 2010), pp. 427–437.

[12] M.B. Cannell, D.J. Crossman and C. Soeller, Effect of changes in action potential spike configuration, junctional sarcoplasmic reticulum micro-architecture and altered t-tubule structure in human heart failure. J Muscle Res Cell Motil, 27 5–7 (2006), pp. 297–306.

[13] Wei S, Guo A, Chen B, Kutschke W, Xie YP, Zimmerman K, et al. T-tubule remodeling during transition from hypertrophy to heart failure. Circ Res. Aug 20; 107(4): 520–31.

[14] A.R. Lyon, K.T. MacLeod, Y. Zhang, E. Garcia, G.K. Kanda and M.J. Lab, et al. Loss of T-tubules and other changes to surface topography in ventricular myocytes from failing human and rat heart. Proc Natl Acad Sci U S A, 106 16 (Apr 21 2009), pp. 6854–6859.

[15] Q. Tian, M. Oberhofer, S. Ruppenthal, A. Scholz, V. Buschmann and H. Tsutsui, et al. Optical QT-screens based on adult ventricular myocytes. Cell Physiol Biochem, 27 (Feb 21 2011), pp. 281–290.

[16] L. Kaestner and P. Lipp, Non-linear and ultra high-speed imaging for explorations of the murine and human heart, J. Popp, G. von Bally, Editors , Optics in life science, SPIE, Munich (2007) 66330K-1–K-10.

[17] W.E. Louch, H.K. Mork, J. Sexton, T.A. Stromme, P. Laake and I. Sjaastad, et al. T-tubule disorganization and reduced synchrony of Ca^{2+} release in murine cardiomyocytes following myocardial infarction. J Physiol, 574 Pt 2 (Jul 15 2006), pp. 519–533.

[18] J.C. Reil, M. Hohl, M. Oberhofer, A. Kazakov, L. Kaestner and P. Mueller, et al. Cardiac Rac1 overexpression in mice creates a substrate for atrial arrhythmias characterized by structural remodelling. Cardiovasc Res, 87 3 (Aug 1 2010), pp. 485–493.

[19] O. Muller, Q. Tian, R. Zantl, V. Kahl, P. Lipp and L. Kaestner, A system for optical high resolution screening of electrical excitable cells. Cell Calcium, 47 3 (Mar 2010), pp. 224–233.

[20] L. Kaestner, S. Ruppenthal, S. Schwarz, A. Scholz and P. Lipp, Concepts for optical high content screens of excitable primary isolated cells for molecular imaging, K. Licha, C.P. Lin, Editors , SPIE Biomedical Optics; 2009, SPIE, Munich (2009), pp. 737–745.

[21] M. Rubart, E. Wang, K.W. Dunn and L.J. Field, Two-photon molecular excitation imaging of Ca^{2+} transients in Langendorff-perfused mouse hearts. Am J Physiol Cell Physiol, 284 6 (Jun 2003), pp. C1654–C1668.

[22] F.C. Howarth, M.R. Boyett and E. White, Rapid effects of cytochalasin-D on contraction and intracellular calcium in single rat ventricular myocytes. Pflugers Arch, 436 5 (Oct 1998), pp. 804–806.

[23] J.P. Reeves, M. Condrescu, G. Chernaya and J.P. Gardner, Na^+/Ca^{2+} antiport in the mammalian heart. J Exp Biol, 196 (Nov 1994), pp. 375–388.

[24] M. Mazzanti, R. Assandri, A. Ferroni and D. DiFrancesco, Cytoskeletal control of rectification and expression of four substates in cardiac inward rectifier K^+ channels. FASEB J, 10 2 (Feb 1996), pp. 357–361.

[25] H. Tsutsui, K. Ishihara and Gt. Cooper, Cytoskeletal role in the contractile dysfunction of hypertrophied myocardium. Science, 260 5108 (Apr 30 1993), pp. 682–687.

[26] H. Ohmori and S. Toyama, Direct proof that the primary site of action of cytochalasin on cell motility processes is actin. J Cell Biol, 116 4 (Feb 1992), pp. 933–941.

[27] S.L. Brenner and E.D. Korn, Substoichiometric concentrations of cytochalasin D inhibit actin polymerization. Additional evidence for an F-actin treadmill. J Biol Chem, 254 20 (Oct 25 1979), pp. 9982–9985.

[28] S.S. Brown and J.A. Spudich, Cytochalasin inhibits the rate of elongation of actin filament fragments. J Cell Biol, 83 3 (Dec 1979), pp. 657–662.

[29] S.S. Brown and J.A. Spudich, Mechanism of action of cytochalasin: evidence that it binds to actin filament ends. J Cell Biol, 88 3 (Mar 1981), pp. 487–491.

[30] M.D. Flanagan and S. Lin, Cytochalasins block actin filament elongation by binding to high affinity sites associated with F-actin. J Biol Chem, 255 3 (Feb 10 1980), pp. 835–838.

[31] A. Morris and J. Tannenbaum, Cytochalasin D does not produce net depolymerization of actin filaments in HEp-2 cells. Nature, 287 5783 (Oct 16 1980), pp. 637–639.

[32] S.C. Calaghan, J.Y. Le Guennec and E. White, Cytoskeletal modulation of electrical and mechanical activity in cardiac myocytes. Prog Biophys Mol Biol, 84 1 (Jan 2004), pp. 29–59.

[33] T.O. Stolen, M.A. Hoydal, O.J. Kemi, D. Catalucci, M. Ceci and E. Aasum, et al. Interval training normalizes cardiomyocyte function, diastolic Ca^{2+} control, and SR Ca^{2+} release synchronicity in a mouse model of diabetic cardiomyopathy. Circ Res, 105 6 (Sep 11 2009), pp. 527–536.

[34] K.F. McGrath, A. Yuki, Y. Manaka, H. Tamaki, K. Saito and H. Takekura, Morphological characteristics of cardiac calcium release units in animals with metabolic and circulatory disorders. J Muscle Res Cell Motil, 30 5–6 (2009), pp. 225–231.

[35] L.S. Song, E.A. Sobie, S. McCulle, W.J. Lederer, C.W. Balke and H. Cheng, Orphaned ryanodine receptors in the failing heart. Proc Natl Acad Sci U S A, 103 11 (Mar 14 2006), pp. 4305–4310.

IV

Calcium Signalling in Red Blood Cells

The non-selective voltage-activated cation channel in the human red blood cell membrane: reconciliation between two conflicting reports and further characterisation

Lars Kaestner, Palle Christophersen, Ingolf Bernhardt, Poul Bennekou

Reprint from J. Bioelectrochem. (2000) **52**: 117-125.

■ Abstract

Using the patch-clamp technique, the non-selective, voltage-activated cation channel in the human red blood cell (RBC) membrane was further characterised. Activity of the cation channel could be demonstrated at a range of salt concentrations with the current–voltage characteristics for monovalent cations going from linear to superlinear functions, depending on the cation concentration in the range of 100–500 mM. The non-selective voltage-activated cation channel was demonstrated to be permeable to the divalent cations Ca^{2+} and Ba^{2+}, and even Mg^{2+}. The current–voltage relations for the divalent cations were superlinear even at 75 mM salt concentration, but indicated outward rectification in contrast to the I–V curve for monovalent cations. The degree of activation at a given membrane potential depended strongly on the prehistory of the channel. The gating exhibited hysteretic-like behaviour, since the quasi steady-state deactivation and activation curves were displaced by ~25 mV. This result fully explains apparent discrepancies between $V_{0.5}$-values previously obtained by slightly different experimental protocols. The possible physiological/pathophysiological role of the channel is discussed in the context of the demonstrated permeability for divalent cations.

21.1   Introduction

A voltage-activated cation channel in the human red blood cell (RBC) membrane was originally proposed by Halperin *et al.*[1] based on flux experiments in solutions of low ionic strength, causing depolarisation of the RBC membranes. Using the patch-clamp technique, Christophersen and Bennekou[2] found first evidence for a non-selective, voltage-activated cation channel in the RBC membrane. The channel was further described by Bennekou[3] and Kaestner *et al.*[4] as well as related to the findings of Halperin *et al.*[1].

The channel is coupled to an acetylcholine receptor of nicotinic type[3]. However, the investigations of Kaestner[4] are made in the nominal absence of the agonist. This can explain, why the occurrence of the channel in this paper was measured to be much smaller compared to the study of Bennekou[3], where the number of channels per cell was estimated to be in the range of some 300.

The conductance of the channel for Na^+ as well as for K^+ was shown to be linear with a single channel conductance of 21 pS at physiological salt concentrations[4] and superlinear with a zero current single channel conductance of 35 pS in 500 mM salt solutions[2]. In addition, the channel was shown to exhibit a set of conductance substates[2-4]. In order to resolve the ambiguities with regard to the identity of the channel as reported in these studies, we focus here on the apparent discrepancies between Christophersen and Bennekou[2] and those of Kaestner *et al.*[4]. So, one purpose of this study is to reconcile the previously reported contradictions in the conductance and the open probability of the non-selective voltage-activated cation channel in the human red blood cell. We demonstrate, that the open channel properties (single channel conductance and I–V characteristics) as well as the voltage-dependent gating are extremely dependent on the salt concentrations and the direction of voltage-changes and that previous observations are in accordance with the extended description presented here.

The physiological relevance of the channel is not known, but it has been proposed that this channel could be involved in the maintenance of the red cell steady state membrane potential[3], and about a possible involvement in pathological increased leak cation fluxes has been speculated[4,5]. Apart from the non-selective voltage-activated cation channel, a Ca^{2+}-activated K^+ channel, the so-called Gardos channel (e.g., [6-8]), is present in the human erythrocyte membrane. A possible functional interaction between these channels is discussed.

21.2 Experimental

21.2.1 Blood and solutions

Freshly drawn or bank blood (3–6 days old) from healthy human donors was used for the experiments. The blood was washed three times by centrifugation (1500×g, 8 min) at room temperature in the physiological NaCl solution containing (mM): 145 NaCl, 10 glucose, 10 morpholinoethane sulfonic acid/tris(hydroxymethyl)aminomethane (Mes/TRIS), pH 7.4. Plasma and buffy coat were removed by aspiration. The cells were then washed once in the appropriate medium used for the experiment.

For the patch-clamp experiments, a small aliquot of 5–10 µl of the cell suspension was added to the measuring chamber giving a final haematocrit in the range of 2–5×10^{-4}%. Since a variety of different solutions have been used, the concrete composition of solutions is given in each figure separately. Basically, all solutions contained between 75 and 500 mM of XCl (X being K, Ca, Mg, Ba) and 5 to 10 mM of a buffer (3-(N-morpholino)propanesulfonic acid (MOPS) or Mes/TRIS). The pH was adjusted to 7.4 by addition of the appropriate hydroxides or N-methyl-d-glucamine (NMDG). All solutions were filtered using 0.2 µm syringe filters (Nalgene, USA). If present, small amounts of additional substances (carbachol (2 µM), ethylene glycol bis(β-aminoethyl ether)-N,N,N',N'-tetraacetic acid (EGTA, 1 mM)) were added to the solutions before the cells.

For flux experiments, cells were suspended (5% haematocrit) either in physiological solutions corresponding to a membrane potential of about −10 mV or in sucrose-substituted solutions of low ionic strength (LIS) corresponding to a membrane potential of about +45 mV[9,10]. During some measurements, bumetanide and ouabain were present in the solutions in order to inhibit the $Na^+/K^+/2Cl^-$ cotransporter and the Na^+/K^+ pump, respectively.

21.2.2 Patch-clamp measurements

Pipettes were pulled from soft glass as well as from borosilicate glass (different brands). The pipettes had a resistance in the range of 10–20 MΩ (in 150 mM KCl solution). After a gigaseal (5–50 GΩ) was reached, the inside-out configuration was formed spontaneously or by moving the pipette tip through the air–water interface. A variety of voltage steps and ramps in the range of ±120 mV were applied. The currents were low-pass filtered between 0.5 and 3 kHz (Bessel filter) and digitised at sample rates between 5 and 44 samples/ms. For data analysis, all-point amplitude histograms were used.

21.2.3 Flux measurements

For measurements of unidirectional K^+ efflux, the cells were loaded for at least 2 h at a 50% haematocrit in the physiological NaCl solution in the presence of ^{86}Rb. After incubation at 37°C, the cells were washed (10 s, 12,000 × g) five times at 4°C immediately before the experiment. K^+ fluxes were started by adding the cells to the prewarmed flux medium (5% haematocrit, 37°C). Aliquots were centrifuged through 0.5 ml $MgCl_2$ solution (106 mM) layered on 0.25 ml dibutylphalate. ^{86}Rb in the supernatant was determined by liquid scintillation counting (TRI-CARB 1600 TR, Packard). At the end of the experiment, an aliquot of cells was removed for determination of total radioactivity. The rate constant was calculated as the negative slope of the linear regression line obtained by semi-logarithmic plot of counts/unit time against time. Effluxes were calculated from the rate constants using an intracellular K^+ concentration of 90 $mmol/l_{cells}$[11].

21.2.4 Reagents

Inorganic salts, sucrose and glucose were of analytical grade. Ouabain, bumetanide, EGTA, NMDG and carbachol were from Sigma (St. Louis, USA), TRIS from Fluka Chemie (Buchs, Switzerland) and MOPS from SERVA (Heidelberg, Germany). ^{86}Rb (in RbCl) was obtained from Amersham International (Amersham, UK).

21.3 Results

21.3.1 Single channel conductance and the dependence on ion concentration

The results of the current–voltage experiments, which covered a wide range (100–500 mM) of symmetric K^+ concentrations, are shown in ◘ Fig. 21.1. In the potential range of 0–120 mV, the curves were superlinear for KCl concentrations above 200 mM only, whereas linear functions were obtained at lower concentrations (◘ Fig. 21.1A) (in the voltage range −120–0 mV, the non-selective voltage-activated channel stays closed, also cp. Ref. [4] and ◘ Fig. 21.4). To avoid the ambiguity due to this effect of the membrane potential, the conductance vs. concentration curve (◘ Fig. 21.1B) is based on the estimated zero current conductances (g_0). The zero current conductance vs. concentration is a saturating function with a maximal conductance of ~70 pS and a $K_{0.5}$-value of ~150 mM.

21.3.2 The voltage-dependent gating and the open state probability

Since the channel under investigation showed several substates[2-4], the channel is regarded as open

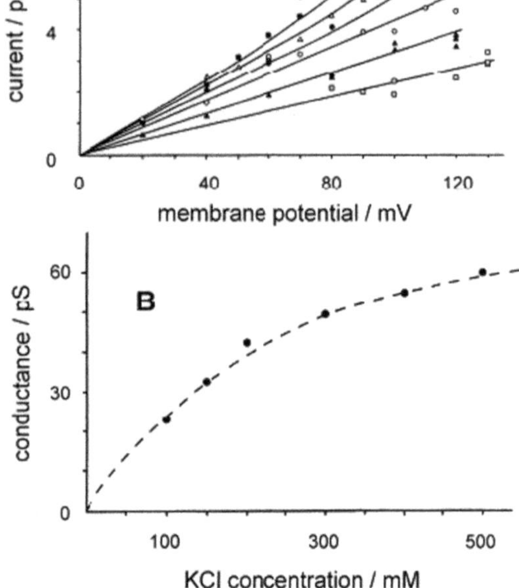

◘ **Fig. 21.1** Current–voltage curves and zero current conductances for the non-selective cation channel at varied concentrations of symmetrically distributed K^+. (A) Current–voltage curves obtained at 100 (□), 150 (▲), 200 (○), 300 (•), 400 (△) and 500 (■) mM K^+. The current corresponding to each datapoint was determined as the peak-position of the fitted Gaussian distribution of all samples from a 1- to 5-min run at the indicated potential. (B) Zero current conductances of the main state plotted vs. the concentration of K^+. Pipette and bath solutions (mM): 100–500 KCl, 5 MOPS, ~4 NMDG, 1 EGTA (pH=7.4), 0.02 Ca^{2+}.

with regard to calculation of open state probability when not in the well defined closed state. To test the channel for possible non-stationary gating, single channel activity was observed for hours. As an example, data from one patch clamped at a membrane potential of +60 mV has been analysed and plotted in ◘ Fig. 21.2. The mean open state probability was calculated in segments of 30 s duration and plotted vs. time. During the first 45 min, the open state probability varies around a constant value of 0.85. The jacked appearance of the curve demonstrate that 30 s is actually too short a period for calculation of a stable open state probability. However, at 45 min, the channel suddenly exhibits an unprovoked shift to a very low

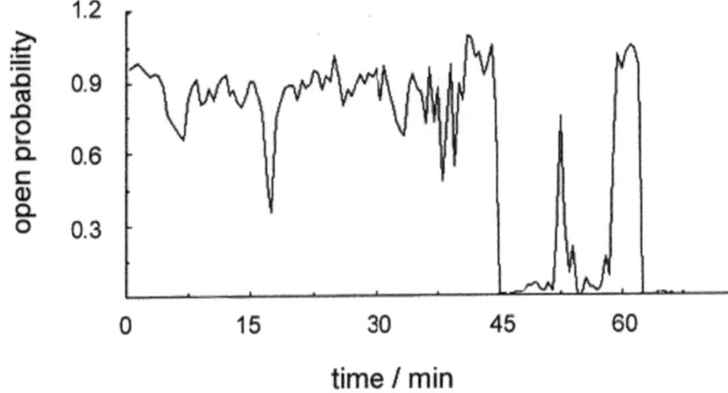

Fig. 21.2 Open state probability vs. time of the voltage-activated non-selective cation channel. For convenience, the open probability was calculated as mean current in 30 s periods of baseline corrected current traces, divided by the main-state current. Pipette and bath solutions (mM): 500 KCl, 5 MOPS, ~4 NMDG, 1 EGTA (pH=7.4), 0.02 Ca^{2+}. V_m=+60 mV. Filter frequency: 500 Hz.

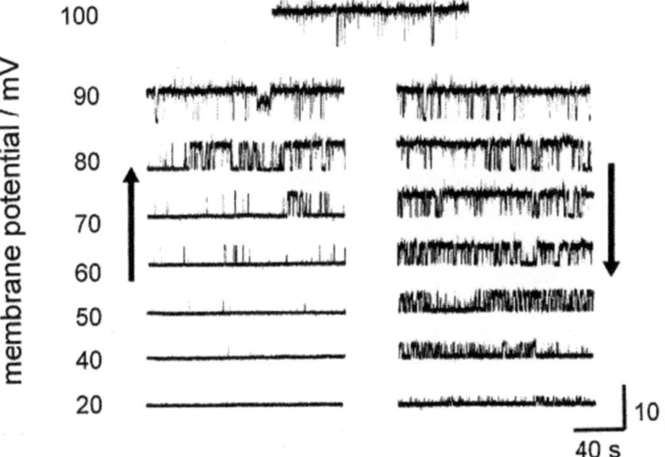

Fig. 21.3 Anomalous voltage dependence of the non-selective cation channel. Current traces from an experiment where the membrane potential was increased stepwise to +100 mV and back to 0 mV again. The arrows indicate the order of applied membrane potentials. Filter frequency: 500 Hz. Salt solutions as in Fig. 21.2.

open state probability. It is not known at present whether these long shutdown periods are part of the normal kinetic scheme and correspond to an extra exponential in the open state distribution, since very few are observed during the lifetime of a patch. However, the conclusions in the following paragraphs are based on experiments in which neither prolonged inactive periods nor other indications of instability occurred.

Kaestner *et al.*[4] reported that the channel starts to open at a membrane potential of about +30 mV and reaches an open state probability of close to 1 at about +70 mV at physiological salt concentrations, while Christophersen and Bennekou[2] detected a similar characteristic but observed the first channel openings at low negative potentials, using 500 mM salt concentrations. More detailed measurements, as presented in ◘ Fig. 21.3 and ◘ Fig. 21.4, show that the voltage dependent open state probability depends critically on the experimental protocol. ◘ Fig. 21.3 shows current traces from an experiment in which the cation channel was activated by increasing the membrane potential from 0 to +100 mV in steps of 10 or 20 mV and then deactivated again by the reverse procedure. During hyperpolarisation of the membrane only a few channel openings were observed at potentials below +60 mV, whereas the activity increased strongly at higher potentials. During depolarisations, however, channel activity could be observed even at +20 mV and generally, at each potential, the activity was higher than the corresponding

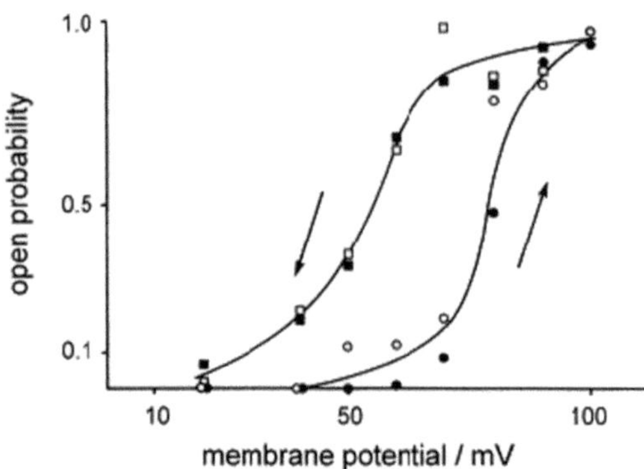

Fig. 21.4 The open state probability as function of the membrane potential. Filled symbols represent data from the experiment shown in Fig. 21.3. Open symbols represent data from an identical experiment performed on the same patch after 15 min at a potential of −15 mV. In both series, the open probability was calculated from 3 min of continuous recording at each potential. The curves were drawn by eye.

activity observed during activation. In **Fig. 21.4**, the open state probabilities (filled symbols) has been plotted vs. the membrane potential. The sigmoid curves show a marked hysteresis with the half maximal activity potentials being +80 mV for activation and +55 mV for deactivation. To investigate if the displacement of the activation curve was due to an irreversible process, the same patch was thereafter clamped at −15 mV for 15 min to deactivate the channel. The experiment was then repeated exactly as before. The results are plotted in **Fig. 21.4** by the open symbols, clearly showing that the phenomenon is repeatable.

21.3.3 The ion selectivity for divalent cations

The channel is permeable even to divalent cations. **Fig. 21.5A** shows representative current traces from experiments where the bathing solutions contained 75 mM $CaCl_2$, $BaCl_2$, or $MgCl_2$ and the membrane potential was clamped at +50 mV. The pipette solution contained 75 mM $CaCl_2$. **Fig. 21.5B** shows the corresponding all-point amplitude histograms. In **Fig. 21.6**, a current–voltage diagram for symmetrical 75 mM $CaCl_2$ solutions is shown. The plotted curve at positive

membrane potentials is an exponential regression; the conductance of the main state varies between 12 and 18 pS. Thus, the channel conductance for divalent cations is similar to what is found for monovalent cations at physiological salt concentrations (cf. **Fig. 21.1**). In contrast to the current traces for monovalent cations, the traces for divalent cations show some very short (about 10 ms) channel opening events at negative membrane potentials between −80 and −100 mV. However, the open probability was less than 0.0001. The regression line for negative membrane potentials (**Fig. 21.6**, dashed line, linear plot) gives a conductance of 5.4 pS. It is not obvious whether this conductance represents a substate or the mainstate, consequently, the fitted current–voltage traces in **Fig. 21.6** have not been connected. The current–voltage relationships for Mg^{2+} and Ba^{2+} at positive membrane potentials are identical to what is found for Ca^{2+} (not shown). The characteristics of the voltage-dependent gating observed for divalent cations were similar to the behaviour for monovalent cations, but generally with lower open state probabilities.

21.3.4 Flux experiments

It has been reported that nicotinic acetycholin receptor agonists increase the open state probability of the channel in inside-out patches[3]. It was, therefore, attempted to activate the channel in suspensions of intact red cells, at normal (−10 mV) as well as at depolarised membrane potential (LIS solution, +45 mV) with carbachol (**Table 21.1**). In LIS solution a direct effect of carbachol on the K^+ efflux could be expected. However, a significant stimulation of the K^+ efflux by carbachol was neither seen in the presence of ouabain plus bumetanide nor in their absence (**Table 21.1**). The experiments in the absence of ouabain plus bumetanide were performed to exclude a possible channel inhibition by ouabain and/or bumetanide. Furthermore, at a membrane potential of −10 mV in the presence of Ca^{2+}, a small Ca^{2+} influx realised by the non-selective cation channel (cf. Appendix 21.A) could result in an activation of the Gardos channel leading to an enhanced K^+ efflux. However, such an effect has not been observed (**Tab. 21.1**).

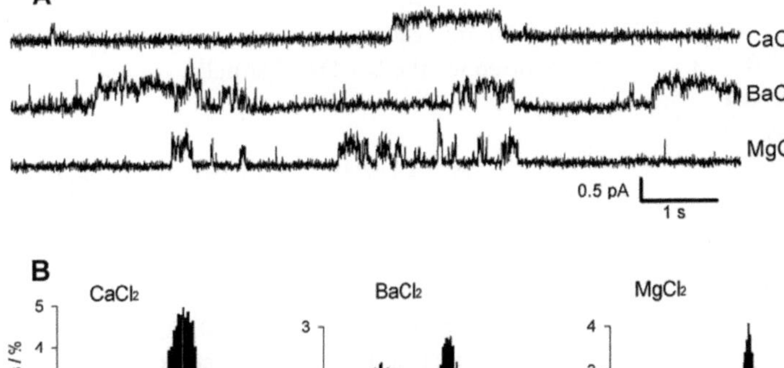

◘ Fig. 21.5 Current traces of the non-selective cation channel for different divalent cations. In (A), 10 s of a representative recording at a membrane potential of +50 mV are presented. The pipette solution contained 75 mM $CaCl_2$, 10 mM TRIS and 2 µM carbachol (pH=7.4) in all cases and the bath solution contained 75 mM of the indicated divalent cation and 10 mM TRIS (pH=7.4). (B) shows the corresponding all-point amplitude histograms for a time period of 1 min.

◘ Tab. 21.1 K (^{86}Rb)-effluxes of human red blood cells incubated in physiological NaCl solution (HIS) and low ionic strength (LIS) solution. HIS solution contained (in mM): NaCl 145, glucose 10, Mes/TRIS 10, pH 7.4. LIS solution contained (in mM): sucrose 250, glucose 10, Mes/TRIS 10, pH 7.4. If present, the concentrations of bumetanide, ouabain, calcium and carbachol were 100 µM, 100 µM, 2 mM and 2 µM, respectively. The efflux values represent mean±S.D. (three independent experiments).

solution	HIS		LIS			
Bumetanide+ouabain	+	+	+	+	-	-
Calcium	+	+	-	-	-	-
Carbachol	-	+	-	+	-	+
Efflux/mMol $(l_{cells}\,h)^{-1}$	1.7±0.4	1.8±0.6	80±4	82±4	82±7	91±11

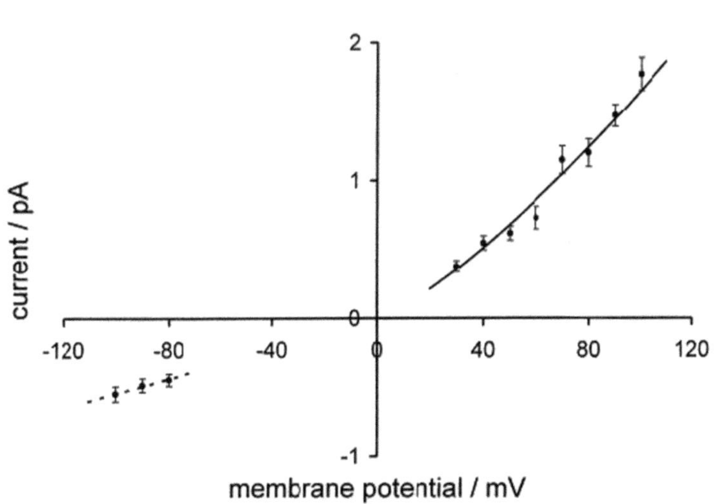

◘ Fig. 21.6 Current–voltage curve of the non-selective cation channel for calcium ions at symmetrical concentration of 75 mM $CaCl_2$ and 10 mM TRIS (pH=7.4). In addition, the pipette solution contained 2 µM carbachol. The dots at a positive membrane potential represent the main state of channel openings as shown in Fig. 21.5. The regression is an exponential fit and the conductance varies between 12 and 18 pS. It is not known if the dots at negative membrane potentials represent a substate or the mainstate of the channel, since the openings have been very short (about 10 ms) and occurred with a very small open probability (<0.0001). The values are taken from several patches (n=6).

21.4   Discussion

21.4.1   The dependence of the channel conductance on the ion concentration

The series of I–V curves obtained at different salt concentrations (❏ Fig. 21.1) demonstrate conventional saturation of the single channel conductance, as has been described for a multitude of ion channels[12]. Taking into account the change in the current–voltage characteristics (linear to superlinear, ❏ Fig. 21.1) as a function of the salt concentrations, we find that the open channel properties, as described by Christophersen and Bennekou[2,3] and by Kaestner *et al.*[4], respectively, are due to the varied experimental solutions used in the two studies.

21.4.2   Voltage dependence of the open state probability

The voltage-activated cation channel shows a very complex gating behaviour and the results reported here are far from giving an exhaustive picture. Instead, we have focused on a few characteristic properties, which seems different from other ion channels. The difference between the ascending and descending leg of the curves in ❏ Fig. 21.4 represents a left shift of the open state probability of about 25 mV after activation. Since the experiment could be reproduced on the same patch after a period at a deactivating (negative) potential, the shift is an inherent property of the gating mechanism, rather than the result of irreversible changes, for example, »wash out« of some regulatory component in the membrane. This means that the gating depends strongly upon the channels' prehistory. With the finding of the hysteresis-like open probability curve (❏ Fig. 21.4), the conflicting reports mentioned in the 21.1 – Introduction and references[2,4] can be reconciled, since the open probabilities of the channel represented the sigmoid descending[2] and ascending[4] leg of the hysteresis-like activity curve, respectively. The mechanism, which underlies the hysteresis-like behaviour of the voltage-dependent gating, is not known at present. A possible explanation could be a very slow equilibration of a voltage dependent step in the gating mechanism, which has been reported for the NMDA channel[13].

21.4.3   Selectivity for divalent cations

The channel is permeable to the divalent cations Ca^{2+}, Mg^{2+} and Ba^{2+}. As measured by both zero current permeability ratios (bath Mg^{2+} or Ba^{2+} *vs.* pipette Ca^{2+}) as well as conductance ratios, it is apparent that this channel does not discriminate between divalent cations. Further, the zero current conductances for the divalent ions are comparable to the conductance observed at physiological salt concentrations of monovalent cations. However, it is noteworthy that the I–V superlinearity is preserved at 75 mM divalent salt solutions, whereas that property is observable for monovalent ions only at much higher salt concentrations. Another peculiarity of divalent cation conduction is the effect on the gating properties of the channel. Although the basic voltage-dependent gating is clearly preserved for divalent cations, the open state probability observed in the limit of high positive voltages does not tend to one, as is the case with monovalent ions. Further, in the case of monovalent cations, clear inward currents are only observable as non-stationary tail-currents in response to deactivating voltage steps[2]. In contrast to that, rare gating events can be observed at large negative voltages in experiments with divalent cations. Overall, this may indicate that divalent cation conduction yields a less steep open probability *vs.* voltage curve than the monovalent ions do. We presently have no information regarding changes between open states or gating hysteresis for the divalent ions.

21.4.4   Relation to macroscopic flux data

A number of studies reporting results performed on cell suspensions have shown that depolarisation of the human red cell membrane elicit an unselective pathway for cations[1,14-16]. We have previously hypothesised[4,17] that the present channel might account for this property of the red cell membrane and the present observations further strengthens this hypothesis (this statement is not in contradiction with the existence of the $K^+(Na^+)/H^+$ exchanger in the human red blood cell membrane, which is stimulated in LIS solution but not affected by the transmembrane potential[18]). It has been reported that the unselective cation pathway—once activated—persists even when the normal membrane potential (−10 mV) is established by resuspending the cells in media with normal Cl^- concentrations[19]. Qualitatively, this behaviour may reflect the hysteresis-like gating behaviour reported in this study. Quantitatively, however, in the patch-clamp experiments, a complete deactivation upon the return to negative potentials is observed. A possible explanation could be temperature dependent gating kinetics (~21°C in patch-clamp studies vs. 37°C in flux studies). In support of this, LaCelle and Rothstein[14] found a left shift of about 60 mV for activation of cation fluxes going from 23°C to 37°C. The demonstrated permeability to Ca^{2+} is also consistent with observations from macroscopic studies in that it is possible to activate the Gardos channel due to Ca^{2+} influx following depolarisation[1] and own experiments (data not shown). In contrast, it has not been possible in macroscopic flux studies to show an effect of carbachol at any membrane potential of the red cells (◘ Tab. 21.1). This could be due to a blocking influence of some intracellular component, which is removed in inside-out patch-clamp experiments. Whatever the mechanism behind the effect found in single channel experiments[3], the behaviour of the non-specific channel is very different from the cloned nicotinic acetylcholin receptors, especially with regard to the voltage-dependent gating and the complete lack of selectivity towards cations. It must therefore be concluded that the non-specific channel does not belong to the known family of nicotinic receptors.

With regard to the (patho-)physiological importance, the permeability of the channel for Ca^{2+} is an interesting property, which might relate this channel to unsolved ion transport problems in the RBC membrane. The residual Ca^{2+} flux in normal erythrocytes is in the range of 1–10 µmol $(l_{cells}\,h)^{-1}$ ref. [20], which has often been considered as the »ground permeability« of the lipid bilayer of the red cell membrane. However, knowing that this channel is in fact permeable to both mono- and divalent cations and having a very low (but probably not zero) open state probability at the RBC resting potential (−10 mV), it may conduct a significant portion of the basic Ca^{2+} influx, as well as the leak fluxes of monovalent cations, which determines the RBC resting membrane potential[21]. Considering the characteristics of the channel, it is possible to calculate the open state probability necessary to explain the reported Ca^{2+} flux. This hypothetical open state probability is 9.3×10^{-8} (see Appendix 21.A). To record an event with such a low probability, 1 year of continuous recording would be necessary to observe a channel opening of 2 ms.

Furthermore, a number of pathological situations occur where the red cell membrane permeability for Ca^{2+} increases. Among these are sickle cell anaemia, in which a key event is deoxygenation-induced erythrocyte dehydration at least partly due to Ca^{2+} influx by an unknown route and subsequent activation of the Gardos channel[22] and the increased Ca^{2+} permeability found in malaria infected erythrocytes[23,24]. Desai *et al.*[24] have reported channel-opening events in excised patches of Plasmodium falciparum-infected RBC with characteristics in good agreement with the properties of the channel described here.

21.5 Conclusion

Considering the comprehensive description of the red cell voltage-activated non-specific cation channel given in this paper, it is safe to conclude that only one channel has been identified so far. A puzzling aspect of this channel is the hysteresis-like voltage dependent gating. Although the basis of this effect is not understood at present, it reconciles contrary reports mentioned in the Introduction (21.1). Another important feature of this work is the unequivocal identification, at the single channel level, of a conductive pathway for the physiologically important divalent cations Mg^{2+} and Ca^{2+}. Considering the voltage gating characteristics of this channel, it is evident that the characterisation with regard to the function in the intact cell under physiological conditions will be difficult, but could provide firm answers to remaining questions about both the normal and pathological red cell ion homeostasis, which tentatively have been presented here.

21.6 References

[1] J.A. Halperin, C. Brugnara, M.T. Tosteson, T.V. Ha and D.C. Tosteson, Voltage-activated cation transport in human erythrocytes. Am. J. Physiol., 257 (1989), pp. C987–C996.

[2] P. Christophersen and P. Bennekou, Evidence for a voltage-gated, non-selective cation channel in human red cell membrane. Biochim. Biophys. Acta, 1065 (1991), pp. 103–106.

[3] P. Bennekou, The voltage-gated non-selective cation channel from human red cells is sensitive to acetylcholine. Biochim. Biophys. Acta, 1147 (1993), pp. 165–167.

[4] L. Kaestner, C. Bollensdorff and I. Bernhardt, Non-selective voltage-activated cation channel in the human red blood cell membrane. Biochim. Biophys. Acta, 1417 (1999), pp. 9–15.

[5] H. Staines, Cation Transport in Plasmodium falciparum-infected Human Erythrocytes, PhD Thesis, Oxford University, 1998.

[6] G. Gardos, The function of calcium in the potassium permeability of human erythrocytes. Biochim. Biophys. Acta, 30 (1958), pp. 653–654.

[7] O.P. Hamill, Potassium channel currents in human red blood cells. J. Physiol., 319 (1981), pp. 97P–98P.

[8] R. Grygorczyk, W. Schwarz and H. Passow, Ca^{2+}-activated K^+ channels in human red cells. Comparison of single-channel currents with ion fluxes. Biophys. J., 45 (1984), pp. 693–698.

[9] I. Bernhardt, A.C. Hall and J.C. Ellory, Transport pathways for monovalent cations through erythrocyte membranes. Stud. Biophys., 126 (1988), pp. 5–21.

[10] I. Bernhardt, Alteration of cellular features after exposure to low ionic strength medium, J. Bauer, Editor, Cell Electrophoresis, CRC Press, Boca Raton (1994).

[11] F. Marongiu, H.J. Holtmeier and A. von Klein-Wiesenberg, Zur Bestimmung des Mineralgehaltes in Erythrozythen. Klin. Wochenschr., 44 (1966), pp. 1405–1412.

[12] R. LaTorre and C.J. Miller, Conduction and selectivity in potassium channels. J. Membrae Biol., 71 (1983), pp. 11–30.

[13] L.M. Nowak and J.M. Wright, Slow voltage-dependent changes in channel open-state probability underlie hysteresis of NMDA responses in Mg(2+)-free solutions. Neuron, 8 (1992), pp. 181–187.

[14] P.A. LaCelle and A. Rothstein, The passive permeability of red blood cells to cations. J. Gen. Physiol., 50 (1966), pp. 171–188.

[15] J.A. Donlon and A. Rothstein, The cation permeability of erythrocytes in low ionic strength media of various tonicities. J. Membrae Biol., 1 (1969), pp. 37–52.

[16] G.S. Jones and P.A. Knauf, Mechanism of the increase in cation permeability of human erythrocytes in low-chloride media. Involvement of the anion transport protein capnophorin. J. Gen. Physiol., 86 (1985), pp. 721–738.

[17] P. Bennekou and P. Christophersen, A human red cell cation channel showing hysteresis like voltage activation/inactivation. Acta Physiol. Scand., 146 608 (1992), p. 56.

[18] S. Richter, J. Hamann, D. Kummerow and I. Bernhardt, The monovalent cation »leak« transport in human erythrocytes: an electroneutral exchange process. Biophys. J., 73 (1997), pp. 733–745.

[19] G.R. Kracke and P.B. Dunham, Effect of membrane potential on furosemide-inhibitable sodium influxes in human red blood cells. J. Membrae Biol., 98 (1987), pp. 117–124.

[20] V.L. Lew, I.M. Glynn and J.C. Ellory, Net synthesis of ATP by reversal of the sodium pump. Nature, 225 (1970), pp. 865–866.

[21] D.C. Tosteson and J.F. Hoffman, Regulation of cell volume by active cation transport in high and low potassium sheep red cells. J. Gen. Physiol., 44 (1960), pp. 169–194.

[22] V.L. Lew, O.E. Ortiz and R.M. Bookchin, Stochastic nature and red cell population distribution of the sickling-induced Ca^{2+} permeability. J. Clin. Invest., 99 (1997), pp. 2727–2735.

[23] R. Kramer and H. Ginsburg, Calcium transport and compartment analysis of free and exchangeable calcium in Plasmodium falciparum-infected red blood cells. J. Protozool., 38 (1991), pp. 594–601.

[24] S.A. Desai, E.W. McCleskey, P.H. Schlesinger and D.J. Krogstad, A novel pathway for Ca^{2+} entry into Plasmodium falciparum-infected blood cells. Am. J. Trop. Med. Hyg., 54 5 (1996), pp. 464–470.

21.7 Appendix: A Calculation of the channel open probability to explain the residual Ca^{2+} influx

The residual Ca^{2+} influx (J) into RBC is about 10 µmol $(l_{cells}$ h$)^{-1}$ at the resting membrane potential (U) of about -10 mV[20]. Dividing this flux by the number of red blood cells present in 1 l (n=8×10^{12}) gives the residual flux of a single cell (J_{sc}):

$$J_{SC} = \frac{J}{n} \tag{21.1}$$

On the other hand, the flux of a single cell (J_{sc}) can be derived from microscopic data and is equal to the product of the number of channels per cell (x), the open probability of the channels (p) at the applied membrane potential, and the number of cations (N) that pass the channel in a time (t) divided by Avogadro's number (N_a):

$$J_{SC} = \frac{xpN}{N_a t}$$ (21.2)

Since:

$$N = \frac{Q}{q_{el}}$$ (21.3)

where Q and q_{el} are the total charge passing the channel and the elementary charge, respectively, and:

$$Q = It = gUt$$ (21.4)

where I, g, U, and t are the current, conductance, applied membrane potential, and time, respectively, it follows:

$$J_{SC} = \frac{xpgU}{q_{el} N_a}$$ (21.5)

From (21.1) and (21.5), it results for the open probability p of a single channel:

$$p = \frac{N_a q_{el}}{ngUx} J$$ (21.6)

One has to take into account that the fluxes presented in (21.1) and (21.5) are given in mmol/h and mmol/s, respectively. Assuming a number of channels per cell[3] of 300 and a single-channel conductance for Ca^{2+} of 12 pS at a membrane potential of −10 mV (estimated from ◘ Fig. 21.6) and a maximal residual Ca^{2+} flux of 10 μmol/h for 1 l of red blood cells (see above), the open probability for a single channel would be 9.3×10^{-8}. One has to consider that this rough calculation is a slight underestimation, since sodium and potassium to some extent compete with calcium under physiological conditions.

Ion channels in the human red blood cell membrane: Their further investigation and physiological relevance

Lars Kaestner and Ingolf Bernhardt

Reprint from J. Bioelectrochem. (2002) 55: 71-74.

▪ Abstract

Using the patch-clamp technique, two different ion channels have been characterised further in the human red blood cell (RBC) membrane. We demonstrate that the non-selective cation channel (NSC) is permeable to Ca^{2+} and can be activated by prostaglandin E_2 (PGE_2). Therefore, the physiological role of this channel could be, together with the Ca^{2+}-activated K^+ channel, the participation in the process of blood clot formation. We give also evidence that another channel in the RBC membrane, so far assumed to be a small conductance anion channel, is more likely to be a proton or a hydroxyl ion channel.

22.1   Introduction

In the human red blood cell (RBC) membrane, three different types of ion channels have been described so far. Besides the well-known Ca^{2+}-activated K^+ channel (Gardos channel)[1], conflicting reports[2,3] about a non-selective cation channel (NSC) have been reconciled recently[4]. In addition, there are two reports about a small conductance anion channel in the RBC membrane[5,6]. The classification of this channel as an anion channel was however only based on different measured conductances in chloride and nitrate media.

The aim of the present paper was therefore to further characterise the NSC and the small conductance anion channel. Using the patch-clamp technique, we demonstrate that NSC is permeable to Ca^{2+} and can be activated by prostaglandin E_2 (PGE_2). Based on these findings, we present an idea on the possible physiological role of both the NSC and the Gardos channel in the process of blood clot formation. We take into account a report demonstrating in flux experiments that the Gardos channel is activated by PGE_2[7]. However, no idea of the mechanism of activation has been presented. In addition, we give evidence that the putative anion channel is more likely to be a proton or a hydroxyl ion channel.

22.2   Experimental

For experiments, freshly drawn blood from healthy human donors was used. The RBCs were washed three times by centrifugation ($1500 \times g$, 8 min) in physiological NaCl solution containing (in mM): 145 NaCl, 10 glucose, 10 morpholinoethane sulfonic acid/tris(hydroxymethyl)aminomethane (MES/Tris), pH 7.4. Plasma and buffy coat were removed by aspiration. The solutions used in the patch-clamp experiments contained (in mM): 150 or 75 KCl (or 75 $CaCl_2$), 10 MES/Tris, 2.5 $BaCl_2$ and 10^{-7} or 10^{-4} PGE_2 (as indicated). The pH of all solutions was 7.4. Inorganic salts were of analytical grade. EGTA and PGE_2 were obtained from Sigma (St. Louis, USA). MES/Tris was purchased from Fluka Chemie (Buchs, Switzerland). Pipettes were pulled from borosilicate glass (GC150F-10, Harvard Apparatus, UK) and had a resistance of about 10 MΩ. The measurements were carried out at room temperature (23 °C). Inside–out patches were performed by the following procedure. After a gigaseal (5–20 GΩ) was reached, the inside–

out configuration was formed spontaneously or by moving the pipette tip through the air–water interface. The measured currents were low-pass filtered at 1 kHz (Bessel filter) and digitised at 3–5 kHz.

22.3   Results and discussion

Patch-clamp experiments on human RBCs were performed in the presence of PGE_2 in the extracellular solution (pipette solution). First, experiments were done with a PGE_2 concentration of 10^{-10} M. The physiological PGE_2 concentration is about one order of magnitude lower, the concentration that occurs when the activated platelet release of PGE_2 is about one order of magnitude higher[8]. Although an effect on channel openings could be seen at a PGE_2 concentration of 10^{-10} M, we were not able to evaluate the obtained current traces quantitatively. Therefore, the PGE_2 concentration was increased to 10^{-7} M. In contrast to almost all other inside–out measurements of the NSC in the human RBC membrane[2-4] in the presence of PGE_2, channel openings could also be detected at negative membrane potentials (for current traces at different membrane potentials, see insert of ◘ Fig. 22.1). The data were analysed using all point amplitude histograms. The current–voltage (I–V) diagram is shown in ◘ Fig. 22.1. The channel is slightly outward rectifying, i.e. while at the positive membrane potentials, the conductance is almost constant (35 pS), it is decreasing and reaches a value of about 24 pS as the membrane potential becomes more negative. Although channel openings are detected at negative membrane potentials, it can be assumed that the recorded channel is identical with the previously described NSC in the human RBC membrane. This assumption is based on a comparison of the conductances and characteristics of current traces.

Besides NSC, a Ca^{2+}-activated K^+ channel exists in the human RBC membrane. Its conductance is in the same range as that for NSC[1]. However, since 2.5 mM Ba^{2+} is present in the bath solution (Ba^{2+} inhibits the Ca^{2+}-activated K^+ channel), an activation of the Ca^{2+}-activated K^+ channel can be excluded. An activation of the anion channel described in the RBC membrane (see below) can

be also excluded since its conductance in physiological chloride solutions is less than 10 pS[5].

Interestingly, an activation of the Ca^{2+}-activated K^+ channel in the human RBC membrane by PGE_2 was demonstrated in flux experiments[7]. However, this report is lacking for an idea of the mechanism of activation. Furthermore, it has been shown that NSC is permeable for divalent cations[4]. The I–V diagram for Ca^{2+} is shown in ◘ Fig. 22.1. In symmetrical 75 mM $CaCl_2$ solutions, NSC shows a conductance between 15 and 18 pS. The argument that the conductance is indeed caused by NSC is similar as that for the conductance in the presence of PGE_2 (see also [4]). Based on the data shown above, it seems reasonable to assume a possible mechanism of ion channel activation that could be of importance in the process of blood clot formation. PGE_2 activates the non-selective cation channel. The reversal potential of NSC for intact RBCs in their own plasma is approximately 0 mV. Since the resting potential of human RBCs is about −10 mV, an inward current is expected and cations present in the blood plasma including Ca^{2+} can enter the cells. Although this report shows that the open probability of NSC at a PGE_2 concentration of 10^{-10} M is very low, it was shown in Ref. [4] that even a small open probability of NSC leads to remarkable Ca^{2+} influxes, and hence, to the increase of the intracellular Ca^{2+} concentration resulting in the activation of the Ca^{2+}-activated K^+ channel. Since for the Ca^{2+}-activated K^+ channel, the transport ratio of K^+ compared to Na^+ is 16:1[1], the reversal potential for this channel under physiological conditions in the blood is about +70 mV (calculated from the Nernst equation). Therefore, at a resting potential of the human RBC of about −10 mV, an outward current of K^+ is provoked. The maximal resulting efflux can be calculated to be 270 mmol $(l_{cells} \, h)^{-1}$. Although this enormous value is an upper limit of the possible K^+ efflux, the K^+ loss of the cells is so high that it leads to significant shrinkage of the RBCs in a very short time (minutes). This effect seems to play an important role in the thrombus formation. Although it is believed that RBCs themselves are primarily not involved in the clot formation, it is evident that they make up an enormous part of it. However, such proposed mechanism of interaction could

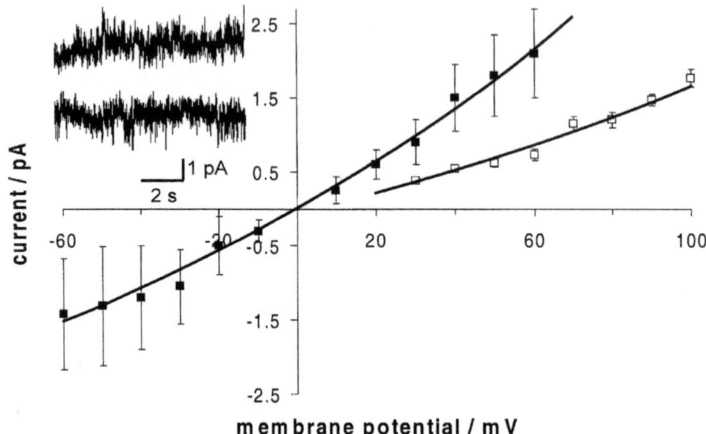

◘ Fig. 22.1 Current–voltage diagrams of NSC channels from inside–out patches of red blood cells. ■, Symmetrical solution containing (in mM): 150 KCl and 10 MES/Tris. In addition, the pipette and bath solution contained 10^{-7} M PGE_2 and 2.5 mM Ba^{2+}, respectively. The data are mean±S.D. values from three patches. □, Symmetrical solution containing (in mM): 75 $CaCl_2$ and 10 MES/Tris. In addition, the pipette solution contained 2 µM carbachol to stimulate the channel openings (the channel is coupled to an acetylcholine receptor of nicotinic type[9]). The data are mean±S.D. values from six patches. The insert shows representative examples of current traces of symmetrical KCl solutions in the presence of PGE_2 in the pipette at membrane potentials of +30 mV (upper trace) and −30 mV (lower trace).

◘ Fig. 22.2 Current–voltage diagrams of small conductance channels from inside–out patches of red blood cells. ○, 150 mM KCl bath and pipette solution. ●, 75 mM KCl bath solution and 150 mM KCl pipette solution. In addition, all solutions in the figure contained (in mM): 10 MES/Tris and 2.5 Ba^{2+}. The data are mean±S.D. values from eight patches. The insert shows representative examples of current traces of the small conductance channel at the indicated membrane potentials.

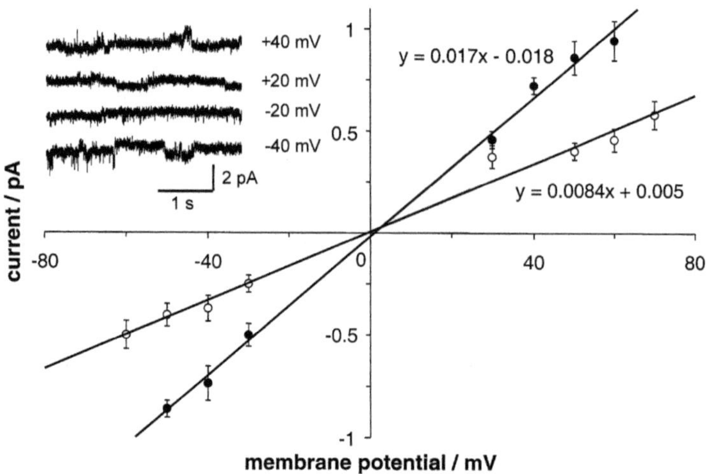

explain the physiological importance of both NSC as well as the Ca^{2+}-activated K^+ channel. Their physiological functions were so far unknown.

Besides the two above-mentioned channels, a small anion channel[5,6] was reported to be present in the RBC membrane. This channel showed a conductance of 6 and 15 pS in chloride and nitrate media, respectively. The channel could be inhibited by persantin[6], and the number of channels per cell was estimated to be 100[5]. Representative examples of current traces of the small conductance channel at different membrane potentials in the chloride media are shown in the insert of ◘ Fig. 22.2. This figure also shows the *I–V* diagrams of the measurements of the channel in symmetrical 150 mM KCl solutions and after the replacement of the bath solution (same patches) by a 75 mM KCl solution. These $I\text{-}V$ curves allow the conclusion that the channel cannot be either a conventional $K^+(Na^+)$ channel or a chloride channel since a division of the KCl concentration in the bath solution by two does not lead to a significant shift in the reversal potential. Based on the Nernst equation, the reversal potential should change from 0 to approximately +17 and −17 mV for a cation and anion channel, respectively. Even more surprising, the reduction of the KCl concentration in the bath leads to an enhancement of the channel conductance from 8.4 to 17 pS. Thus, the $I\text{-}V$ curves can only be interpreted assuming that a proton or a hydroxyl ion channel is involved. The classification of this channel as an anion channel[5,6] is only based on the different above-mentioned conductances in chloride and nitrate media. However, this finding already excludes a conventional $K^+(Na^+)$ channel. This statement is supported by the fact that under conditions where the pipette and the bath solution contained 20 and 70 mM Na–tartrate, respectively (in addition, both solutions contained 2.5 mM $BaCl_2$, 10 mM MES/Tris (pH 7.4); sucrose was added to adjust the osmolarity of the solutions to 300 mOsM), a linear $I\text{-}V$ curve was obtained giving a reversal potential close to 0 mV and a conductance of 9.7 pS (not shown).

Summarising the results, it seems more likely that the low conductance channel described in the present paper represents a proton or a hydroxyl ion channel in the human RBC membrane. In order to support this finding, it would be helpful to measure the channel conductance at different pHs in the bath and pipette solution. To have a pH gradient present at least for several minutes, one has to inhibit the anion exchanger (band 3) in the RBC membrane. In the first approaches, we used DIDS or dipyridamol in the pipette solution (for inside-out patches), but under these conditions, we were not able reach a gigaseal (it seems likely that these substances prevent a seal formation).

A possible physiological relevance of the channel could be its participation in the oxygen and carbon dioxide transport. CO_2 produced in the tissue diffuses into the erythrocytes. Using the enzyme carbonic anhydrase, CO_2 will be hydrated and H^+ as well as HCO_3^- is produced. Normally, the protons are buffered by haemoglobin[10] and HCO_3^- leaves the cells in exchange for Cl^- via band 3. However, since the process is connected with an acidification of the red blood cells, a physiological relevance could be that protons leave the cell via the proton channel.

22.4 References

[1] R. Grygorczyk and W. Schwarz, Properties of the Ca^{2+}-activated K^+ conductance of human red cells as revealed by the patch-clamp technique. Cell Calcium, 4 (1983), pp. 499–510.

[2] P. Christophersen and P. Bennekou, Evidence for a voltage-gated, non-selective cation channel in human red cell membrane. Biochim. Biophys. Acta, 1065 (1991), pp. 103–106.

[3] L. Kaestner, C. Bollensdorff and I. Bernhardt, Non-selective voltage-activated cation channel in the human red blood cell membrane. Biochim. Biophys. Acta, 1417 (1999), pp. 9–15.

[4] L. Kaestner, P. Christophersen, I. Bernhardt and P. Bennekou, The non-selective voltage-activated cation channel in the human red blood cell membrane: reconciliation between two conflicting reports and further characterization. J. Bioelectrochem., 52 (2000), pp. 117–125.

[5] W. Schwarz, R. Grygorczyk and D. Hof, Recording single-channel currents from human red cells. Methods Enzymol., 173 (1989), pp. 112–121.

[6] H. Passow, W. Schwarz, M. Glibowicka, N. Aranibar and M. Raida, Studies of erythroid band 3 protein-mediated anion transport in red blood cells and in xenopus oocytes, F. Palmieri, E. Quagliariello, Editors , Molecular Basis of Biomembrane Transport, Elsevier, Amsterdam (1988), pp. 121–139.

[7] Q. Li, V. Jungmann, A. Kiyatkin and P.S. Low, Prostaglandin E_2 stimulates a Ca^{2+}-dependent K^+ channel in human erythrocytes and alters cell volume filterability. J. Biol. Chem., 271 (1996), pp. 18,651–18,656.

[8] J.B. Smith, C. Ingerman, J.J. Kocsis and M.J. Silver, Formation of prostaglandins during the aggregation of human blood platelets. J. Clin. Invest., 52 (1973), pp. 965–969.

[9] P. Bennekou, The voltage-gated non-selective cation channel from human red cells is sensitive to acetylcholine. Biochim. Biophys. Acta, 1147 (1993), pp. 165–167.

[10] A.M. Saleh and D.C. Battle, Basic mechanisms of intracellular pH homeostasis in lymphocytes. Semin. Nephrol., 11 (1991), pp. 3–15.

Prostaglandin E$_2$ activates channel-mediated calcium entry in human erythrocytes: An indication for a blood clot formation supporting process

Lars Kaestner, Wiebke Tabellion, Peter Lipp, Ingolf Bernhardt

Reprint from Thrombosis and Haemostasis (2004) **92**: 1269-1272.

■ **Abstract**

Prostaglandin E$_2$ (PGE$_2$) is released from platelets when they are activated. Using fluorescence imaging and the patch-clamp technique, we provide evidence that PGE$_2$ at physiological concentrations (10^{-10} M) activates calcium rises mediated by calcium influx through a non-selective cation-channel in human red blood cells. The extent of calcium increase varied between cells with a total of 45% of the cells responding. It is well known that calcium increases elicited the calcium-activated potassium channel (Gardos channel) in the red cell membrane. Previously, it was shown that the Gardos channel activation results in potassium efflux and shrinkage of the cells. Therefore, we conclude that the PGE$_2$ responses of red blood cells described here reveal a direct and active participation of erythrocytes in blood clot formation.

23.1 Introduction

Prostaglandin E$_2$ (PGE$_2$) is known to be released by activated platelets[1] as well as by erythrocytes themselves under mechanical stress[2]. Therefore PGE$_2$ can act on the erythrocyte membrane. It has been hypothesised that erythrocytes could also play an active role in blood coagulation. However, the effect of PGE$_2$ on erythrocytes is discussed controversially.

Li and colleagues[3] could show that PGE$_2$ stimulation activated the calcium-dependent potassium channel (Gardos channel) in human erythrocytes. This work failed to demonstrate an increased intracellular calcium concentration. Kaestner and Bernhardt[4] hypothesised a mechanism of calcium entry via the non-selective cation channel[5,6] that was shown to be permeable to calcium[4,7]. This lead to the suggestion that the potassium efflux found by Li *et al.* was due to an activation of the Gardos channel elicited by an increased intracellular calcium concentration.

So far, two independent research groups tried to prove this hypothesis, but could not detect any PGE$_2$-induced calcium influx in erythrocyte cell suspensions using fura-2 as the calcium indicator[8,9]. Therefore it was concluded that the proposed mechanism of a PGE$_2$ induced calcium increase does not exist in erythrocytes[9].

It has been shown that the haemoglobin of erythrocytes causes considerable problems with fura-2 measurements in human erythrocytes[10]. In contrast, fluo-3 a precursor of fluo-4 has proven to act as a sensitive calcium indicator in human erythrocytes[11]. Thus, in the present study we used fluo-4 as the calcium indicator in the video-imaging experiments.

23.2 Material and methods

For the experiments, freshly drawn blood from healthy human donors was used and prepared as described elsewhere[4,7]. Inorganic salts were obtained from Merck (Darmstadt, Germany), PGE$_2$ and A23187 from Sigma-Aldrich (Taufkirchen,

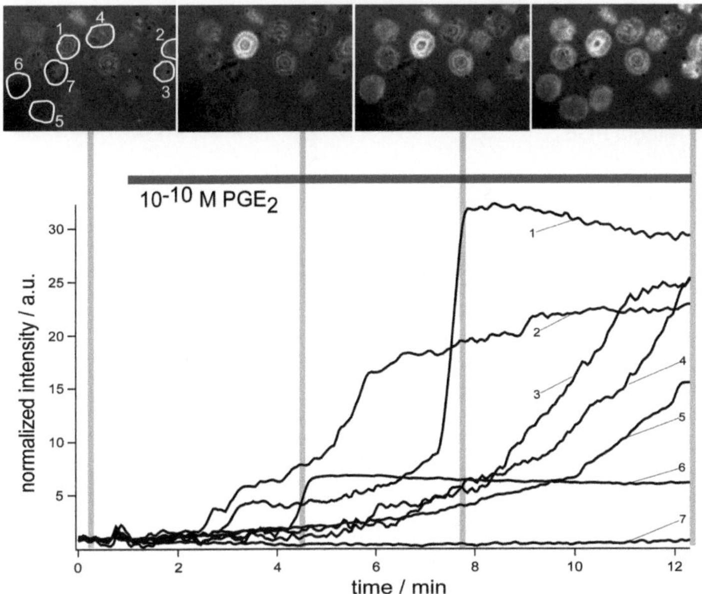

Fig. 23.1 Fluorescence-imaging of human erythrocytes treated with PGE$_2$ using the calcium flourophor fluo-4. Erythrocytes of healthy donors were loaded with the fluo-4 AM and placed on a cover slip. Fluorescence imaging was performed with an excitation wavelength of 490 nm while the resulting fluorescence was imaged through a 100x objective lens. In the first fluorescence image seven erythrocytes are marked with regions of interest. The fluorescence intensity of these regions of interest over time is plotted in the graph, which is background corrected and normalised to the resting fluorescence. The addition of 10^{-10} M PGE$_2$ is indicated by the horizontal bar. For selected times, individual fluorescence images are provided. From these data it became evident that six out of seven erythrocytes investigated responded to the PGE$_2$ stimulation with an increase in intracellular calcium.

Germany) and fluo-4 AM from Molecular Probes (Leiden, Netherlands).

Prior to the experiments, the blood cells were loaded with the indicator (fluo-4 AM, 5 μM) for 60 min followed by a period of 15 min for de-esterification of the internalised dye. The blood cells were allowed to settle on glass cover slips and were finally transferred onto the stage of an inverted microscope. Fluorescence imaging was performed by an integrated system supplied by Till Photonics (Gräfelfing, Germany) consisting of a monochromator (Polychrome IV), a Nikon inverted microscope (Eclipse TE2000-U), and a CCD-camera (Imago).

The patch-clamp experiments were performed as detailed elsewhere[4,7]. All measurements were carried out at room temperature (21°C).

23.3 Results and discussion

Fig. 23.1 shows data from an experiment, where fluo-4 loaded erythrocytes were exposed to PGE$_2$ at a concentration of 10^{-10} M. In this particular experiment six out of seven marked erythrocytes displayed a clear response to PGE$_2$ indicated the increased fluo-4 fluorescence and hence an increased intracellular calcium concentration. It can be read from the traces that individual cells displayed quite a variable response, both in their kinetics and amplitude. However, this variability was representative for the 577 erythrocytes tested for their response to PGE$_2$, 258 out of the total number of cells responded (44.7%) In control experiments calcium entry was verified using the calcium ionophore A23187 and the calcium chelator EDTA (data not shown).

So far, there is just one transporter in the human erythrocyte membrane that is known to

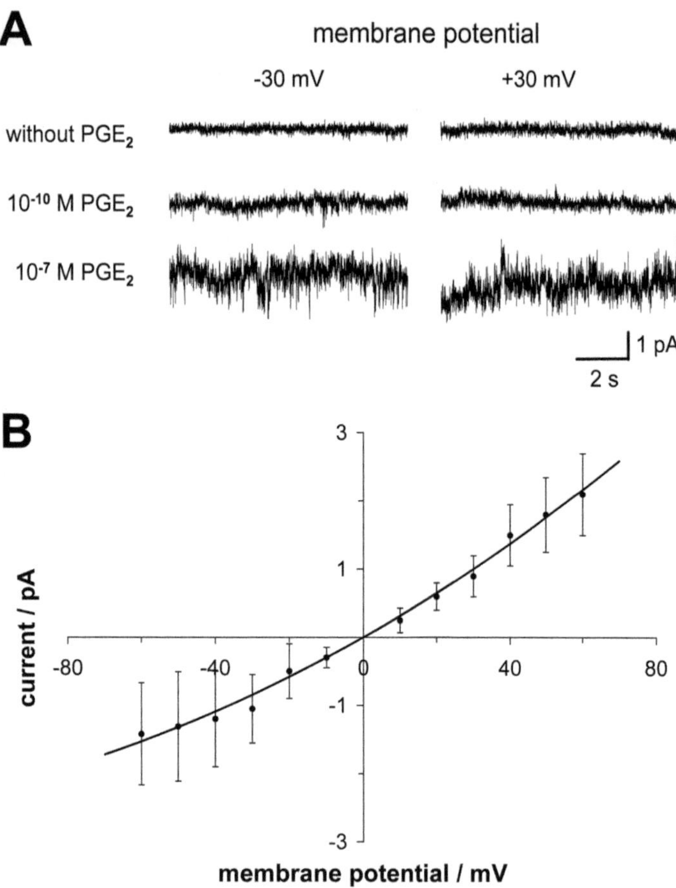

■ **Fig. 23.2** Excised patch-clamp measurements of a PGE$_2$ activated ion-channel. Panel A shows typical membrane current traces from inside-out patches in the absence (upper traces) and presence (lower traces) of two different PGE$_2$ concentrations at -30 mV (left column) and +30 mV (right column). Panel B shows the current voltage relationship of the PGE$_2$-activated channel after stimulation with 10^{-7} M PGE$_2$. The conductance of the channel at positive membrane potentials is nearly linear with a value of 35 pS. As the membrane potential becomes more negative the conductance decreases slightly to about 24 pS (at -60 mV). The values are derived from amplitude histograms of three patches. All measurements were performed in symmetrical 150 mM KCl solutions buffered to a pH value of 7.4. Additionally, the pipette solution contained the indicated concentration of PGE$_2$ and the bath solution contained 2.5 mM BaCl$_2$ to ensure the inhibition of the Gardos channel.

mediate calcium influx, the non-selective cation channel[4,7]. However, due to experimental evidence the existance of an additional calcium carrier has been suggested[12]. In order to determine the nature of the PGE$_2$-induced calcium influx, patch-clamp experiments were performed. In ■ Fig. 23.2A the response of erythrocytes to the addition of PGE$_2$ is exemplified. In the presence of 10^{-10} M PGE$_2$ in the pipette, inside-out patches show channel like current fluctuations, which were absent without PGE$_2$ (■ Fig. 23.2A middle row of current traces). However, a quantitative evaluation of these traces was impossible and therefore the PGE$_2$ concentration was raised to 10^{-7} M (■ Fig. 23.2A, lower traces). Under these conditions the recorded current traces were analysed using amplitude histograms enabling us to construct current voltage diagrams

as shown in ■ Fig. 23.2B. The channel was slightly outward rectifying, had a zero voltage conductance of 35 pS, and a reversal potential of 0 mV.

Although the involvement of a so far unknown ion-channel can not be completely ruled out, it seems evident that the PGE$_2$-activated channel is the same as the so far described non-selective cation channel[4-7]. Thus notion is supported when comparing the PGE$_2$-activated channel with the entirety of all known ion-channels in human erythrocytes[13,14]: (i) it is the only known channel that has the ability to permeate calcium, (ii) the appearance of the current traces is similar and it has (iii) the same single channel conductance. Furthermore, the Gardos channel can be ruled out to be the calcium mediating channel, because the presence of barium in the bath solution inhibits the Gardos channel[15].

Although we were surprised to find such a pronounced variability between the erythrocytes, we believe that this is most likely due to different expression ratios of PGE_2 receptors, ion channels mediating calcium influx, and calcium pumps, which are responsible to extrude calcium from the cell. Nevertheless, the exact reason for such variations is still unclear.

Calcium entry and the resulting rise in intracellular calcium activates the Gardos channel and the lipid-scramblase in the plasma membrane[16,17]. The ensuing breakdown of the lipid asymmetry as well as the potassium efflux through the Gardos channel, which provokes cell shrinkage, are well known symptoms for apoptosis. Therefore calcium influx through the non-selective cation channel can trigger apoptotic-like behaviour in erythrocytes[18]. Subsequently, by activation of the non-selective cation-channel, PGE_2 induces a 'suicidal' mechanism in erythrocytes. We are still unclear why only a subpopulation of erythrocytes responded to PGE_2 stimulation, neither is the identity of this subpopulation known.

The PGE_2-activation of a suicidal mechanism might shed light on a long standing quarrel about the possible active participation of erythrocytes in blood coagulation[19-21]. The argumentation so far culminated in the work of Andrews and Low[11,22], describing possible mechanisms whereby erythrocytes might participate in haemostasis. One of them is strongly supported by the results of this paper. The intracellular calcium increase leads, according to the above described mechanism, to cellular shrinkage. Calcium rises also trigger a remodelling of the plasma membrane lipid composition, which in itself might be the actual mechanism resulting in an augmented aggregation of erythrocytes while the shrinkage actively supports the formation of a solid blood clot.

As mentioned in the introduction, PGE_2 can even be produced by the erythrocytes themselves[2] without the activation of platelets and the subsequent coagulation cascade. This fact may lead to the speculation that the mechanism explained above plays a role during the thrombus formation in coronary and cerebral arteries. Therefore it could be a cellular mechanism participating in such life threatening diseases like cardiac infarction and stroke.

23.4 References

[1] Smith JB, Ingerman C, Kocsis JJ, et al. Formation of prostaglandins during the aggregation of human blood platelets. J Clin Invest 1973; 52: 965-9.

[2] Oonishi T, Sakashita K, Ishioka N, et al. Production of Prostaglandins E_1 and E_2 by adult human red blood cells. Prostaglandins Other Lipid Mediat 1998; 56: 89-101.

[3] Li Q, Jungmann V, Kiyatkin A, et al. Prostaglandin E_2 stimulates a Ca^{2+}-dependent K^+ channel in human erythrocytes and alters cell volume and filterability. J Biol Chem 1996; 271: 18651-6.

[4] Kaestner L, Bernhardt I. Ion channels in the human red blood cell membrane: their further investigation and physiological relevance. J Bioelectrochem 2002; 55: 71-4.

[5] Christophersen P, Bennekou P. Evidence for a voltage-gated, non-selective cation channel in the human red cell membrane. Biochim Biophys Acta 1991; 1065: 103-6.

[6] Kaestner L, Bollensdorff C, Bernhardt I. Non-selective voltage-activated cation channel in the human red blood cell membrane. Biochim Biophys Acta 1999; 1417: 9-15.

[7] Kaestner L, Christophersen P, Bernhardt I, et al. The non-selective voltage-activated cation channel in the human red blood cell membrane: reconciliation between two conflicting reports and further characterisation. J Bioelectrochem 2000; 52: 117-25.

[8] Soldati L, Lombardi C, Adamo D, et al. Arachidonic acid increases intracellular calcium in erythrocytes. Biochem Biophys Res Commun 2002; 293: 974-8.

[9] Kucherenko YV, Weiss E, Bernhardt I. Effect of the ionic strength and prostaglandin E_2 on the free Ca^{2+} concentration and the Ca^{2+} influx in human red blood cells. J Bioelectrochem 2004; 62: 127-33.

[10] Blackwood AM, Sagnella GA, Markandu ND, et al. Problems associated with using Fura-2 to measure free intracellular calcium concentrations in human red blood cells. J Hum Hypertens 1997; 11: 601-4.

[11] Yang L, Andrews DA, Low PS. Lysophosphatidic acid opens a Ca(++) channel in human erythrocytes. Blood 2000; 95: 2420-5.

[12] Tiffert T, Bookchin RM, Lew VL. Calcium Homeostasis in Normal and Abnormal Human Red Cells. In: Red Cell Membrane Transport in Health and Disease. Springer Verlag 2003; 373-405.

[13] Bennekou P, Christophersen P. Ion Channels. In: Red Cell Membrane Transport in Health and Disease. Springer Verlag 2003; 139-52.

[14] Kaestner L. Passive elektrogene Transportprozesse durch die Membran von Humanerythrozyten und Erythrozyten von gesunden und Malaria-infizierten Hühnern. Berlin, Logos Verlag; 2001.

[15] Sarkadi B, Gardos G. Drug actions on Potassium fluxes in red cells. In: The red cell membrane. A model for solute transport. Humana press 1989; 369-96.

[16] Haest CWM. Distribution and Movement of Membrane Lipids. In: Red Cell Membrane Transport in Health and Disease. Springer Verlag 2003; 1-26.

[17] Woon LA, Holland JW, Kable EP, et al. Ca^{2+} sensitivity of phospholipid scrambling in human red cell ghosts. Cell Calcium 1999; 25: 313-20.

[18] Lang KS, Duranton C, Poehlmann H, et al. Cation channels trigger apoptotic death of erythrocytes. Cell Death Differ 2003; 10: 249-56.

[19] Hellem AJ. The adhesiveness of human blood platelets in vitro. Scand J Clin Lab Invest 1960; 12(Suppl): 1-117.

[20] Hellem AJ, Borchgrevink CF, Ames SB. The role of red cells in haemostasis: the relation between haematocrit, bleeding time and platelet adhesiveness. Br J Haematol 1961; 7: 42-50.

[21] Saniabadi AR, Lowe GD, Barbenel JC, et al. Further studies on the role of red blood cells in spontaneous platelet aggregation. Thromb Res 1985; 38: 225-32.

[22] Andrews DA, Yang L, Low PS. Phorbol ester stimulates a protein kinase C-mediated agatoxin-TK-sensitive calcium permeability pathway in human red blood cells. Blood 2002; 100: 3392-9.

Functional NMDA receptors in rat erythrocytes

Asya Makhro, Jue Wang, Johannes Vogel, Alexander A. Boldyrev, Max Gassmann, Lars Kaestner, Anna Bogdanova

Reprint from the American Journal of Physiology - Cell Biology (2010) **298**: C1315-C1325.

■ **Abstract**

N-methyl-D-aspartate (NMDA) receptors are ligand-gated non-selective cation channels mediating fast neuronal transmission and long-term potentiation in the central nervous system. These channels have a 10-fold higher permeability for Ca^{2+} compared with Na^+ or K^+ and binding of the agonists (glutamate, homocysteine, homocysteic acid, NMDA) triggers Ca^{2+} uptake. The present study demonstrates the presence of NMDA receptors in rat erythrocytes. The receptors are most abundant in both erythroid precursor cells and immature red blood cells, reticulocytes. Treatment of erythrocytes with NMDA receptor agonists leads to a rapid increase in intracellular Ca^{2+} resulting in a transient shrinkage via Gardos channel activation. Additionally, the exposure of erythrocytes to NMDA receptor agonists causes activation of the nitric oxide (NO) synthase facilitating either NO production in L-arginine-containing medium or superoxide anion ($O_2^{\cdot-}$) generation in the absence of L-arginine. Conversely, treatment with an NMDA receptor antagonist MK-80, or the removal of Ca^{2+} from the incubation medium causes suppression of Ca^{2+} accumulation and prevents attendant changes in cell volume and $NO/O_2^{\cdot-}$ production. These results suggest that the NMDA receptor activity in circulating erythrocytes is regulated by the plasma concentrations of homocysteine and homocysteic acid. Moreover, receptor hyperactivation may contribute to an increased incidence of thrombosis during hyperhomocysteinemia.

24.1 Introduction

NMDA receptors (NRs) are ligand-gated non-selective cation channels that mediate fast neuronal transmission and long-term potentiation in the central nervous system (CNS)[4]. Glutamate is a primary regulator of activity of the NRs in the brain. This specific class of glutamate receptors plays an essential role in regard to memory, cognition, sensation, and motor control. Itako contributes to the glutamate excitotoxicity that is involved with numerous neurological disorders including stroke, schizophrenia, Alzheimer's disease, and Parkinson's disease[39]. Recently, NR expression has been reported to exist also outside the CNS in a number of peripheral tissues such as the lungs, heart, kidney, liver, spleen, bone, vascular endothelium, and lymphoid cells[7,11,18,19,29,32]. The physiological role of NRs in non-neuronal tissues remains largely unclear.

The nature of NR agonist(s) in control of receptor function in non-neuronal tissues is a matter of debate. Glutamate concentration in the synaptic cleft of the brain reaches millimole levels, whereas the circulating plasma concentrations of glutamate in healthy human subjects ranges from 14 to 70 μM[13,47]. These concentrations are below the 100 μM threshold recognised as the IC_{50} for the receptor activation in neurons[26]. Glutamate, thus, is most likely not a major contributor to the regulation of the NRs in blood cells in healthy humans. In addition to glutamate, homocysteine (HC), and homocysteic acid (HCA) have also been shown to function as NR agonists in mammals. Total plasma HC concentration in healthy subjects is ~7–10 μM, which can increase to 22.3 ± 12.6 μM in response to normal aging[42]. Pathological increases of HC have also been observed as a result of either a hereditary predisposition affecting methionine metabolism or conditions of folate or vitamin B12 deficiency[16]. For instance, in hyperhomocysteinemic patients HC has been observed to reach as high as 476 μM[43]. Both HC and HCA possess a high affinity to bind to NRs such that even a modest increase in plasma concentrations may result in activation of the receptors[6].

As previously mentioned, NRs are non-selective ligand-gated cation channels; however, the selectivity of the NR channels for Ca^{2+} exceeds that for Na^+ 10-fold and thus ligand-mediated activation of the receptors typically results in an induction of transient inward Ca^{2+} current. Consequently, both activation and inactivation of NRs, irrespective of host tissue, are linked to the local changes in intracellular Ca^{2+} levels. Such changes in intracellular Ca^{2+} have been reported to assist in facilitating the proliferation of the vascular smooth muscle cells and aortic endothelia cells in response to HC- or HCA-mediated NR activation[11,46]. Megakaryoblasts initiate differentiation and reduction in pro-platelet formation in response to NMDA exposure, whereas proliferation rate and survival remain unaltered[25].

Lymphocytes and thrombocytes are among the blood cells expressing NRs[7,17]. To the best of our knowledge no studies have been performed to assess the presence of functional NRs in erythrocytes, although interaction of noncompetitive and competitive NR antagonists [^{3}H]MK-801 and [^{3}H]CGS-19755, respectively, with red blood cells (RBCs) has previously been reported[36]. Furthermore, dehydration of erythrocytes was reported in patents with hyperhomocysteinemia[34]. In this study, with the use of rat erythrocytes, we have demonstrated the presence of functional NRs in RBCs and investigated their distribution pattern in addition to their role in regulation of cell volume, osmotic resistance, endothelial nitric oxide synthase (eNOS, NOS3) activity, and cellular redox state.

24.2   Materials and Methods

24.2.1 Cell isolation and handling

For all experiments we used RBCs from adult male Wistar rats (200–300 g). The animals were purchased from Elevage Janvier (Le-Genest-Saint-Ile, France) and kept on commercial rodent chow in the sterile laboratory animal facilities at the Institute of Veterinary Physiology. Animal handling and experimentation were reviewed, approved, and carried out in accordance with the Swiss animal protection laws and institutional guidelines. Blood samples were withdrawn from the caudal vena cava into heparinised syringes, and erythrocytes were isolated via centrifugation at 1,000 g for 5 min. The buffy coat and plasma were discarded while the remaining RBCs were washed three times with an incubation medium of the following composition (in mM): 145 NaCl, 5 KCl, 1 $CaCl_2$, 0.15 $MgCl_2$, 15 glucose, and 10 Tris·HCl, pH 7.4. Packed cells were then resuspended in the incubation medium supplemented with 0.1% BSA to a hematocrit of 40%.

UT-7/Epo cells were cultured in α-MEM supplemented with 20% fetal calf serum and 3 U/ml human recombinant erythropoietin (EPREX, Janssen-Cilag, Neuss, Germany). Bone marrow cells were harvested from the femur bone. Cells were suspended, filtered through a 200-nm mesh, concentrated by centrifugation at 2,000 g for 5 min, and used for either [^{3}H]MK-801 binding experiments or immunoblotting.

Cerebellar tissue was isolated from rat pups at postnatal day 10 (n = 5) and used for immunoblotting. Cerebellar granule cells were isolated from rat pups at postnatal day 10 as described elsewhere[40] and used for [^{3}H]MK-801 binding experiments.

24.2.2 Erythrocyte cytoskeleton-free membrane preparation

Packed RBCs (about 500 μl) were hemolysed in 2 ml of ice-cold lysis buffer containing (in mM) 10 Tris·HCl (pH 7.4), 1 EDTA, 0.1 PMSF, 10 $Na_4P_2O_7$, 10 NaF, plus 10 μg/ml pepstatin A, 10 μg/ml leupeptin, and 5 μg/ml aprotinin. Membranes were pelleted at 47 000 g for 20 min at 4°C. To enrich the membrane fraction for the NR protein, the obtained ghosts were deprived of most of the cytoskeletal proteins as described elsewhere[23]. Protein concentration in the obtained smooth membrane samples was determined by using BCA Protein Assay (Pierce; Rockford, Ill) with BSA as standard.

24.2.3 **Radiolabeling with MK-801**

A radioactive labeling assay was used to detect the number of putative NRs in erythrocytes, UT-7/Epo cells, and cerebellar granule cells. The respective cells were washed in PBS and incubated with 5 nM [^{3}H]MK-801 (20 Ci/mmol American Radiolabeled Chemicals) per 10^6 cells. The unspecific binding was assessed by treatment of cells with the radiolabeled antagonist in the presence of 100 μM of unlabeled dl-5-methyl-10,11-dihydro-5H-dibenzo[a,b]-cyclohepten-5,10-imine (MK-801) for 1 h. The amount of bound [^{3}H]MK-801 was assessed with a Packard 1600 TR liquid scintillation counter, and the number of receptors were calculated thereafter using the following equation:

$$N_{NR} = \frac{A_{cells}}{A_{sp}(MK-801) * N_A * N_{cells}} \qquad (24.1)$$

where N_{NR} is a number of NRs, A_{cells} is activity of the MK-801 bound to the cells in [Bq], A_{sp} is a specific activity of the [^{3}H]MK-801 in [Bq/mmol], N_A is the Avogadro number (6.022×10^{23} mol^{-1}), and N_{cells} are the number of cells in the sample.

24.2.4 **Western blot analysis**

For Western blot analysis cells were washed two times in PBS (2,000 g, 5 min) and lysed in Laemmli buffer. Membrane proteins were first separated on the 7.5% SDS-PAGE (500 μg protein/lane) and then transferred to Protan BA83 nitrocellulose membranes. Even protein transfer was controlled by Ponceau red staining. Membranes were then blocked for 1 h at room temperature followed by an overnight incubation at 4°C with primary antibodies against the NR1 subunit of NRs (dilution 1:1,000; 1–564 fragment, Novus Biologicals) dissolved in TBS containing 5% milk. The membranes were then washed three times in TBST and incubated for 1 h at room temperature with suitable horseradish peroxidase-conjugated secondary antibodies (1:1,000 dilution; antimouse, Jackson ImmunoResearch Laboratories, West Grove, PA). The enhanced chemiluminescent detection Western blotting system (Fujifilm LAS-3000 System, FUJIFILM Life Science) was used for detection

and quantification of protein. Actin served as a loading control.

24.2.5 **Immunohistochemistry and flow cytometry**

To assess the NR levels in the membranes of reticulocytes and mature erythrocytes, staining for both the NR1 and NR2 subunits of the NR and for the transferrin receptor (TrR) as a marker of reticulocytes was performed. For fluorescent imaging the cells were suspended in the incubation medium at hematocrit of 0.4% and incubated for 60 min on ice with a cocktail of the following primary antibodies: mouse monoclonal anti-NMDA NR1 Pan antibody (1:100 dilution, Novus Biologicals), rabbit polyclonal anti-NMDA NR2 antibody (1:100 dilution; Affinity BioReagents), and rat monoclonal anti-transferrin receptor antibodies (1:100 dilution; Abcam). Thereafter, the cells were washed from the primary antibodies (centrifugation for 5 min at 500 g at 10°C) and incubated for an additional 60 min on ice with the corresponding secondary antibodies: anti-rabbit Cy3-conjugated, and anti-mouse Cy5-conjugated, and anti-rat Cy2-conjugated, all in dilution 1:50. After one wash of the secondary antibodies was completed, the specific staining was assessed microscopically (Zeiss Axiovert 200M). Flow cytometry was used to assess the distribution of the NR in RBC population. Erythrocytes in suspension were treated with the mouse monoclonal anti-NR1 antibody (dilution 1:50; Novus Biologicals) for 60 min on ice and thereafter with the secondary anti-mouse Cy-5 conjugated antibody (1:50 dilution) and fluo-4 (final concentration 10 μM) for further 30 min at room temperature. Cells treated with only secondary antibodies were used as a negative control.

24.2.6 **Unidirectional K$^+$(^{86}Rb) influx measurements**

Unidirectional influx rates for K$^+$ were assessed by using ^{86}Rb as a tracer. Unidirectional residual (ouabain-resistant Cl$^-$-independent) K$^+$ influx was detected in erythrocytes incubated at hematocrit of 5–8% in the chloride-free medium containing (mM) 145 $NaCH_3SO_4$, 5 KCH_3SO_4, 0.15 $MgSO_4$, 1 Ca-gluconate (when not Ca^{2+} free), 10 glucose, 10 sucrose, and 10 HEPES-Tris, pH 7.4 at room temperature. Incubation medium was always supplemented with 1 mM ouabain to block active K^+ influx and with 100 µM L-arginine. Various amounts of NMDA, HCA or glutamate, 1 µM clotrimazole, or 50 µmol MK-801 were added to some of the samples. Cells were preincubated for 15 min to achieve full inhibition of the Na-K-ATPase and allow agonists and inhibitor of the NRs and the blocker of Gardos channels, clotrimazole, to bind to their targets. Influx was then initiated and measured by addition of ^{86}Rb. Aliquots of the RBC suspension (400 µl) were collected 10, 20, 30, 45, and 60 min after the onset of incubation with the tracer and washed from the extracellular ^{86}Rb with ice-cold washing medium containing 100 mM $Mg(NO_3)_2$ and 10 mM imidazol buffered with HNO_3 to the pH 7.4 when on ice. After washing was completed, the cell pellet was lysed in 5% trichloracetic acid (TCA), and the amount of ^{86}Rb accumulated in erythrocytes was assessed in deproteinised supernatant and normalised to the amount of the radioactive tracer in the incubation medium. Unidirectional K^+ influx was then calculated from the slope of the linear part of the radioactive tracer uptake plot.

24.2.7 Intracellular calcium content monitoring

Live cell imaging was performed to monitor intracellular Ca^{2+} kinetics in individual cells treated with NMDA. Cells were loaded with fluo-4 AM (Molecular Probes) at a concentration of 5 µM for 1 h at 37°C. Then cells were washed in Tyrode solution containing (mM) 135 NaCl, 5.4 KCl, 10 glucose, 1 $MgCl_2$, 1.5 $CaCl_2$, and 10 HEPES. The pH was adjusted to 7.35 by using NaOH. Cells were plated on poly-L-lysin (Sigma)-coated coverslips in Tyrode solution, and then the cell sedimentation and dye de-esterification were allowed to occur. Fluorescence was finally measured on an inverted microscope (TE2000, Nikon) equipped with a ×100 Plan Apo 1.4 objective. Attached to the microscope was a video imaging setup (TILL Photonics) consisting of a monochromator (Polychrome IV), a camera (Imago), and the control hard- and software (TILL-Vision). Cells were observed at an excitation of 480 nm and images were collected every 5 s. A 505-nm long-pass dichroic mirror separated the emission light from the excitation light, and a 535/40 bandpass filter was used to further improve the image quality. A local perfusion system was utilised to quickly exchange solutions in the field of view and to apply the agonist NMDA at a concentration of 100 µM. Images were processed in ImageJ (Wayne Rasband, National Institute of Mental Health) and traces handled by Igor Pro software (WaveMetrics).

24.2.8 Osmotic resistance and hemolysis

The effect of agonists and antagonists of the NR on osmotic resistance was assessed acutely by preincubated cells with 1 mM of NMDA or HCA for 1 min at room temperature in the incubation medium. Thereafter, 3 µl of packed cells were added to the cuvette equipped with a magnetic stirrer containing 2 ml of 0.9% NaCl solution (final cell number $\sim 10^7$ cells/2 ml). A bolus of distilled water was added to reduce the osmolarity from 300 to 130 mosM. Optical density was monitored at 630 nm every 0.25 s after the initiation of hypoosmotic stress to assess hemolysis. A detailed description of this method may be found elsewhere[48].

Cellular fragility was determined by monitoring shear stress-induced hemolysis. The cells were suspended in either Ca^{2+}-containing or nominally Ca^{2+}-free medium at hematocrit of 40% and incubated for 1 h at 37°C in an Eppendorff Thermomixer (900 rpm shaking speed) in the presence and absence of 100 µM NMDA and/or 100 µM MK-801. Hemolysis was assessed at the end of the incubation as the amount of haemoglobin released into the medium. After centrifugation (5,000 g for 10 min at 4°C), supernatant was collected and haemoglobin concentration was determined spectrophotometrically using Drabkin reagent.

24.2.9 Morphological changes of RBCs exposed to the NR agonist

Erythrocytes were suspended in the incubation medium at a ratio 1:200, allowed to settle on the imaging chamber bottom, and pretreated with 10 µM clotrimazole or 5 mM Na_3VO_4 for 15 min at room temperature. Thereafter, HCA at a final concentration of 1 mM was added just before the measurement, and time course of the morphological alterations was followed.

24.2.10 Reduced and oxidized glutathione measurement

Cellular reduced (GSH) and oxidised (GSSG) glutathione content was determined by using Ellmann's reagent as described elsewhere[44]. After incubation in the presence or absence of 100 µM NMDA, 100 µM HCA and/or 50 µM MK-801 in L-arginine-free or L-arginine-containing medium for 60 min, 100-µl aliquots of erythrocytes suspension (40% hematocrit) were mixed with 900 µl of 5% TCA, and the obtained lysates were centrifuged for 10 min at 16,000 g to pellet denatured proteins. Formation of the coloured complex of GSH with the Ellmanns reagent was assessed spectrophotometrically at 412 nm.

24.2.11 Nitrite and nitrate assessment

Erythrocytes in suspension (40% hematocrit) or whole blood were incubated with 100 µM of NMDA, or HCA, or 50 µM MK-801 for 60 min at 25°C in presence or in absence of 200 µM L-arginine or N^G-nitro-L-arginine methyl ester (L-NAME). After the incubation nitrite levels were assessed in the incubation medium or plasma by using a chemiluminescence detection technique. After incubation the erythrocytes were pelleted by centrifugation (16,000 g) and the supernatant (medium or plasma) was collected. Nitrate was reduced to nitrite using Cd coated with Cu and (NO_2^- +NO_3^-) assessed by a chemiluminiscence detector CLD 88 (Eco Medics AG, Switzerland) as described elsewhere[33].

24.2.12 Statistical analysis

All data are mean values of at least 5 independent experiments and are presented as means ± SE when not stated specially. The comparison between the experimental groups was performed by using a normality test followed by the two-tailed Student's t-test for paired or unpaired samples (GraphPad Instat.V3.05). The level of statistical significance was set at $p < 0.05$, $p < 0.01$, or $p < 0.001$. Fitting of the data obtained for the dose dependence of the HCA and glutamate effects on the residual K^+ influx was performed by using SigmaPlot 8.0 nonlinear regression module (100 iteration steps). The equation chosen for fitting was $y = y_0 + ax/(b+x)$, assuming pseudo-single binding site kinetics for the agonist-receptor interaction.

24.3 Results

24.3.1 Presence of NMDA receptor in mature RBCs and bone marrow

The presence of NRs in rat erythrocytes and in erythroid precursor cells has been addressed by using several independent techniques including: immunoblotting, immunohistochemistry, flow cytometry, and radiolabled antagonist binding assay. Double staining of the bone marrow-derived precursor cells with the antibodies against CD36, a marker of erythroid lineage, and against the NR1 subunit of the NRs revealed the presence of cells expressing both CD36 and the NR (◘ Fig. 24.1A). These observations were confirmed via immunoblotting (◘ Fig. 24.1B). NRs were expressed in rat bone marrow and UT-7/Epo human erythroleukemic cell line and could also be detected in the erythrocyte membrane fraction. The size of NR1 subunit (~120 kDa) in erythroid cells was similar to that in rat cerebellum (◘ Fig. 24.1B). The number of receptor copies per cell was assessed from the binding of the radiolabeled NR antagonist [3H]MK-801 covalently interacting with the receptor unit (a tetramer composed of two NR1 and two NR2 subunits)

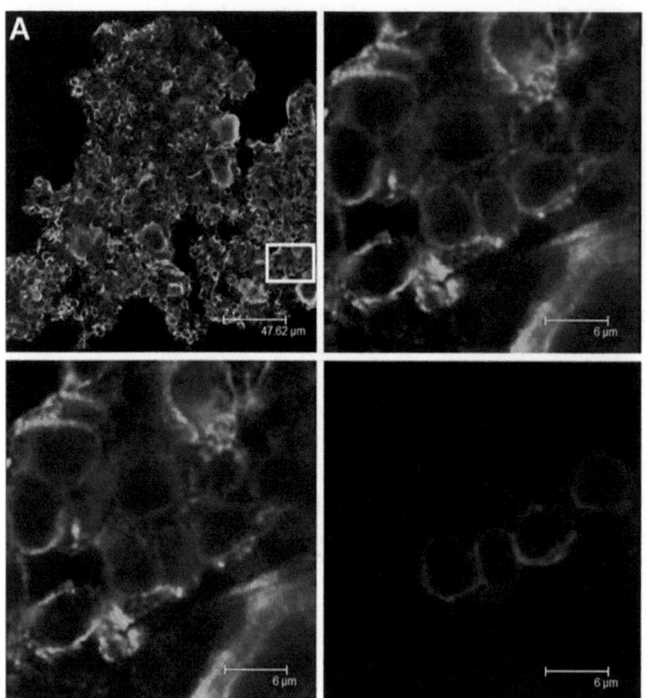

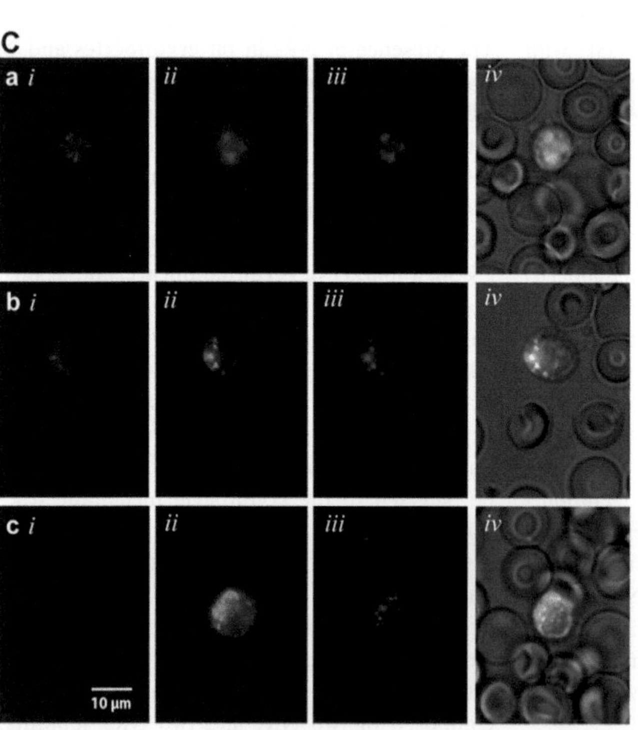

□ Fig. 24.1 Presence of the N-methyl-D-aspartate (NMDA) receptors (NR) in the membranes of erythroid precuror cells and circulating erythrocytes. A: confocal image of the bone marrow tissue stained using the antibodies against CD36 (red channel) and NR1 (cyan channel). The magnified area is framed, and the single channels and the overlay of the magnified area are shown. The staining was repeated four times. B: presence of the NR1 subunit at the protein level in rat erythrocyte cytoskeleton-free smooth membranes, UT-7/Epo cell line, rat bone marrow cells, and rat cerebellum. The data represent two independent gels where NR1 staining was performed. NR1 presence in the red blood cell (RBC) membrane lysate (RBC, left) and in the protein extracts of UT-7 cell line and bone marrow (BM, right) was compared with that of rat cerebellum (Cereb.) used as a positive control. Size of the protein of interest (~110 kDa) may be estimated from the markers lane shown in the right. The line in the right-hand panel indicates where irrelevant lanes were spliced from the gel. The experiment was repeated four times. C: abundance of NR in erythrocytes. Staining with the antibodies against TrR (blue channel, i), NR1 (green channel, ii), and NR2 (red channel, iii) was performed in nonfixed cells. Shown are the single channel recording and an overlay of all three fluorescent channels with the brightfield image. Cells in a and b subset of images contain both subunits of the NR and a TrR, whereas the cell in c subset contains NR1 and NR2 subunits but lacks TrR. Staining was repeated four times and at least 10 fields were analysed in a single experiment. (For interpretation of the references to color in this figure legend, the reader is referred to the web version of the original article.)

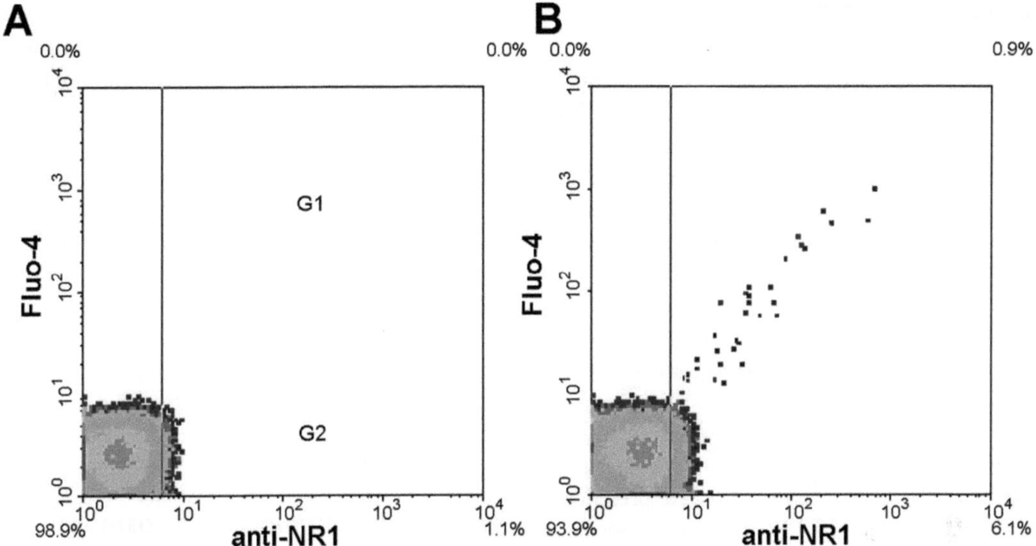

◘ Fig. 24.2 Correlation between the abundance of the NRs and the intracellular Ca^{2+} levels. A: negative control scatter dot plot showing red blood cell (RBC) population chosen according the size (forward scatter) and granularity (side scatter) after incubation with secondary Cy5-conjugated anti-mouse antibodies only. Gating conditions were chosen so that all the cells were located in the lower left quadrant when emission in green fluorescent channel (λ_{em} 488 nm) is shown in Y-axes and far red fluorescence intensity (λ_{em} 650 nm) plotted at X-axes. B: scatter dot plot of the RBCs loaded with fluo-4 and incubated with the anti-NR1 mouse monoclonal primary antibody followed by Cy5-conjugated anti-mouse secondary antibody. Two subpopulations seen in the dot plot represent subpopulation with the high intracellular Ca^{2+} and high number of NR1 copies shifting to the G1 quadrant and the subpopulation with low intracellular Ca^{2+} and low numbers of the NR1 copies in the G2 quadrant. Each plot represents readouts from 10,000 cells. Similar results were obtained in five independent experiments with RBCs obtained from different animals of various age and gender.

at a 1:1 stoichiometry. Binding studies revealed the presence of 3.5×105 copies/cell in UT-7/Epo cells and 8×10^6 copies/cell in dispersed cerebellar granule cells, whereas 8 ± 1.4 binding sites are retained in circulating erythrocytes when equal binding of the antagonist to all cells in the population was assumed. This assumption proved to be false when the distribution of NR (NR1 and NR2 subunits) was assessed in the RBC population with immunocytochemistry and flow cytometry. As depicted in ◘ Fig. 24.1C, very few cells contained a high number of NR1 and NR2 subunits. Most of these receptor-possessing cells showed positive staining for the reticulocytes marker TrR. Some cells retained high number of NRs, whereas TrRs was already lost.

◘ Fig. 24.2 depicts the scatter plots for RBCs stained with only secondary Cy5-conjugated anti-mouse antibody used as a negative control (◘ Fig. 24.2A), and the cells were loaded with fluo-4 (Y-axis) and exposed to the monoclonal mouse anti-NR1 antibodies (X-axis) followed by the above-mentioned secondary antibodies (◘ Fig. 24.2B). As can be seen from the ◘ Fig. 24.2B a low number ($2.37 \pm 0.32\%$, N = 5) of cells appearing in the G1 quadrant are characterised by high numbers of the NR copies and high intracellular Ca^{2+} levels. The amount of these receptor-enriched cells could be underestimated as they were found to be more susceptible to hemolysis under conditions of shear stress and could have been lost during centrifugation and the passage through the cytometer (see below). It varied from 0.5 and 4% in different animals. Many more cells in the population residing in G2 quadrant at the scatter plot contained significantly less copies of the receptor and low intracellular Ca^{2+} levels (◘ Fig. 24.2B).

24.3.2 **Function of the NRs**

NRs are non-selective cation channels that are more sensitive to transport Ca^{2+} into neuronal cells upon activation. High intracellular Ca^{2+} levels in RBCs containing the highest number of NR copies, as indicated in **◘** Fig. 24.2B, suggests that the NRs in erythrocyte membrane are functional and mediate basal Ca^{2+} influx into the cells even in the absence of the agonists in the incubation medium. Treatment of the cells with NMDA dramatically facilitated Ca^{2+} accumulation in some but not all cells. Kinetics of the NMDA-induced Ca^{2+} entry into single rat erythrocytes was assessed with live imaging. In agreement with flow cytometry data (**◘** Fig. 24.2B), live imaging revealed presence of two subpopulation of RBCs: (i) »responders,« in which the addition of 100 μM NMDA triggered a rapid Ca^{2+} accumulation; and (ii) NMDA-insensitive »nonresponders« (**◘** Fig. 24.3). The subpopulation of »responders« consituted 92 of 627 erythrocytes (14.7%). However, of the 92 responding cells only 60 could be used for the final analysis (**◘** Fig. 24.3B) since the other 32 lysed during the 15 min of observation. In the control group (524 cells recorded), the Ca^{2+} levels remained stable and low (**◘** Fig. 24.3B). Minor but statistically significant increase in the intracellular Ca^{2+} content was thus observed in the whole erythrocyte population treated with NMDA, although the separation of cells into »responders« and »nonresponders« subpopulations revealed that changes in Ca^{2+} were much more pronounced and restricted to the limited number of cells.

Treatment of the cells with HCA, glutamate, or NMDA caused a dose-dependent increase in a unidirectional, residual (ouabain-resistant, chloride-independent) K^+ influx, which was measured by using ^{86}Rb as a tracer for K^+ transport (**◘** Fig. 24.4). Doses of the agonists required to cause the half-activation of the K^+ influx (IC_{50}) was estimated by fitting the respective curves and were found to be 21.1 ± 0.78 μM for HCA and 88.2 ± 0.1 μM glutamate (**◘** Fig. 24.4A). Although the exact estimation of the IC_{50} value for NMDA could not have possibly been determined, maximal specific activation was already observed at 50 and 10 μM without an effect (**◘** Fig. 24.4B). This suggests an IC_{50} value

of about 30 μM. Higher supraphysiological concentrations (500–1,000 μM) of the synthetic NR agonist further resulted in a nonspecific increase in the residual K^+ influx.

The NMDA sensitivity of the K^+ influx was lost in the presence of MK-801, an antagonist of NRs (**◘** Fig. 24.5A). NMDA-mediated upregulation in K^+ influx was clearly secondary to the Ca^{2+} uptake as it was suppressed in Ca^{2+}-free medium (**◘** Fig. 24.5B). This finding along with sensitivity of NMDA-induced fluxes to clotrimazole, a Gardos channel blocker, suggests that entry of Ca^{2+} into the cells through the NR facilitates the opening of the Gardos channel.

24.3.3 **Cell volume control, osmotic resistance, and morphological alterations**

Activation of Gardos channels is known to have a profound effect on cell volume. The Ca^{2+} uptake kinetics (**◘** Fig. 24.3) suggests that NMDA-induced cellular shrinkage occurs within the first several minutes of treatment. The corresponding alterations in cellular morphology were observed in response to HCA treatment in some cells within the first 80 s of exposure to the NR agonist; however, this effect was not ubiquitous among all cells (**◘** Fig. 24.6A). The respective alterations in morphology were completely reversed within 180 s of cellular exposure to HCA alone but not during the simultaneous treatment of HCA and the Ca^{2+} pump inhibitor Na_3VO_4. Vanadate treatment did, however, increase the number of responding cells suggesting that the low Ca^{2+} levels in erythrocytes with few NRs (**◘** Fig. 24.2B) was maintained due to the efficiency of the Ca^{2+} pump. Activation of the Gardos channels was involved in the temporary morphological changes as treatment of the cells with clotrimazole almost entirely suppressed the HCA-induced cupping and echniocytosis (**◘** Fig. 24.6A). An acute osmotic resistance test also indicated that both HCA and NMDA result in RBC shrinkage thus making the cells more resistant to osmotic hemolysis (**◘** Fig. 24.6B). Hemolysis rate was assessed via the observation of haemoglobin liberation into the medium during the first 60 s of cells exposure to hypoosmotic shock.

A

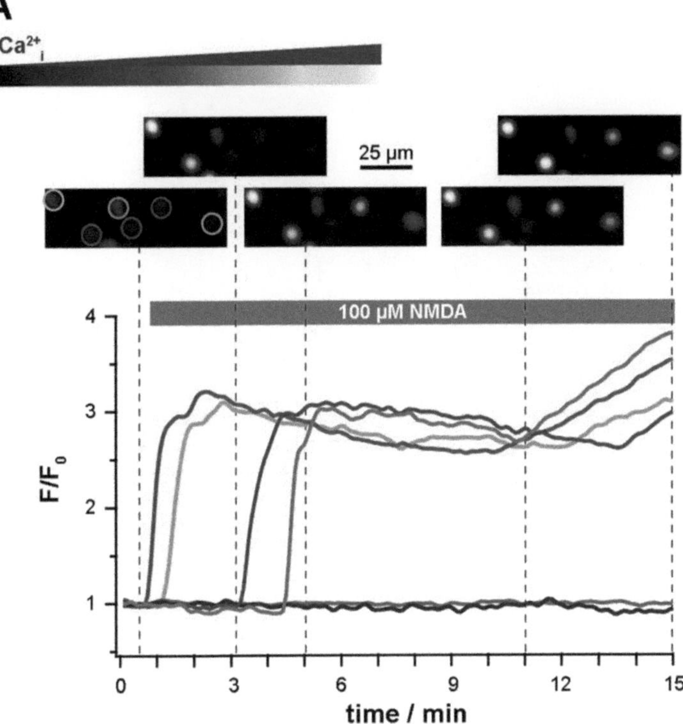

B

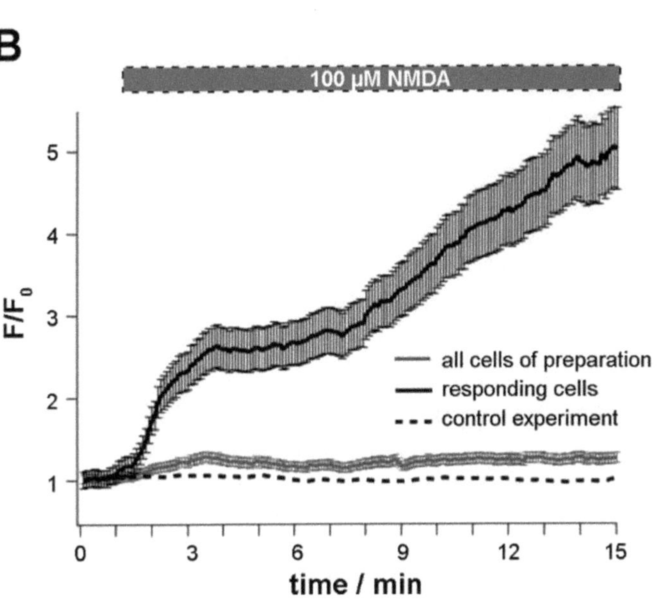

◻ Fig. 24.3 Kinetics of Ca^{2+} uptake by the NMDA-treated erythrocytes. A: original recordings of the changes in the intracellular Ca^{2+} levels in single cells visualised by monitoring fluorescence from the fluo-4 Ca^{2+} sensor in response to 100 µM NMDA treatment. Time of incubation with NMDA is shown with a grey bar. Images of single cells are shown and those that were chosen for analysis are marked as regions of interest with colored circles. The changes in the fluorescent intensity (F) normalised to the initial one (F_o) over time is plotted below in the corresponding colours. The colour coding scale for the relative changes in the intracellular Ca^{2+} levels is shown above. B: statistical analysis of cell populations exposed to 100 µM NMDA (grey and black plots) and control cells (dotted plot). The grey plot represents mean values ± SEM of 627 randomly chosen cells from three independent preparations of different animals exposed to NMDA for 15 min, whereas the black plot shows means ± SEM for the responding subpopulation of 60 cells (for details see text). The variety of the onset time of individual cells leads to a slower increase compared with single erythrocytes as shown in A. (For interpretation of the references to color in this figure legend, the reader is referred to the web version of the original article.)

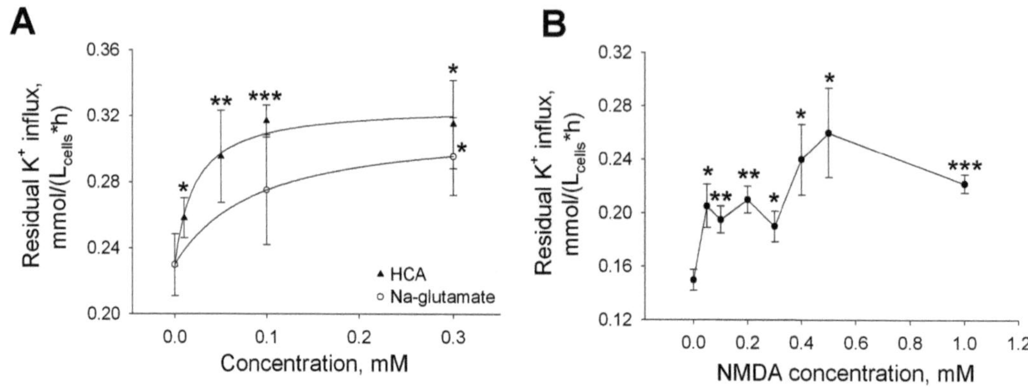

Fig. 24.4 Effect of NR agonists on the residual (ouabain-insensitive, chloride-independent) K^+ influx in rat erythrocytes. A: dose dependence of the effect of the physiological agonists, glutamate, and homocysteic acid (HCA) on the residual K^+ influx. Data are means of five independent experiments ± SEM. The curve was obtained by fitting using the equation $y = y_0 + ax/(b+x)$. The constants obtained for the HCA plot were $y_0 = 0.229$, $a = 0.0972$, $b = 0.0211$ and the corresponding coefficients for the glutamate curve were $y_0 = 0.229$, $a = 0.0854$, $b = 0.0882$. B: dose-response curve for the effect of the synthetic agonist NMDA on the residual K^+ influx. Data are means of 6–10 experiments ± SEM. *$p < 0.05$, **$p < 0.01$, and ***$p < 0.001$ in paired Student's t-test compared with the nontreated control.

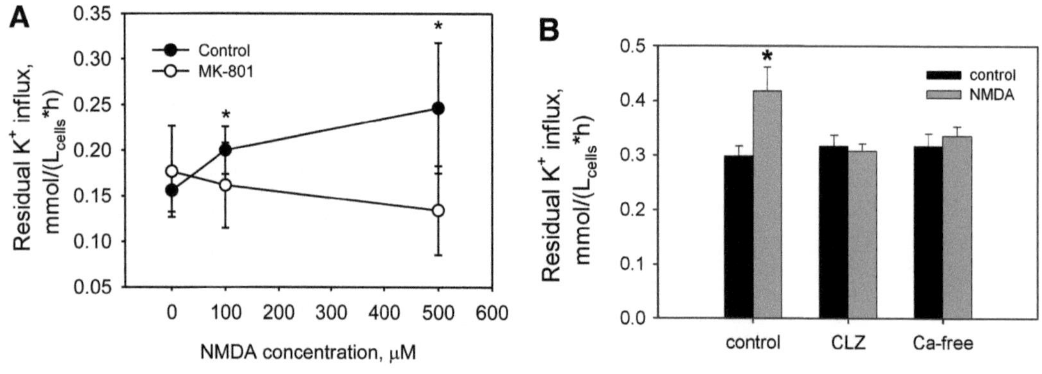

Fig. 24.5 Pharmacological inhibition of the NMDA-induced activation of the residual K^+ influx. A: inhibition of the NMDA-sensitive component of the K^+ influx by antagonist of the NR, MK-801 (10 µM). Data are means of six independent experiments ± SD. *$p < 0.05$ in paired Student's t-test when compared with the corresponding NMDA-free control. B: lack of the NMDA-induced stimulation of the residual K^+ influx in cell pretreated with the Gardos channel blocker clotrimazole (1 µM) as well as in Ca^{2+}-free incubation medium. Data are means of 5 independent experiments ± SEM. *$p < 0.05$ in paired Student's t-test when compared with the nontreated control.

Cellular shrinkage is known to trigger a regulatory volume increase response (RVI) mediated by the volume-sensitive Na/H exchanger or Na-K-2Cl cotransporter in rat erythrocytes[37]. We have also observed the late RVI response in rat erythrocytes treated with NMDA or HCA. Swelling of cells treated with NR agonists for 1 h was monitored and water content was seen to increase from 65.8 ± 1.0% in control to 66.7 ± 0.6% in cells exposed to 100 µM HCA and 67.2 ± 1.7% after treatment with 100 µM NMDA at 37°C. The fact that long-term treatment with the NR agonists made the cells more susceptible to the shear stress-induced hemolysis (**Fig. 24.6C**) could also be explained as a result of secondary swelling. This agonist-induced hemolysis could be completely

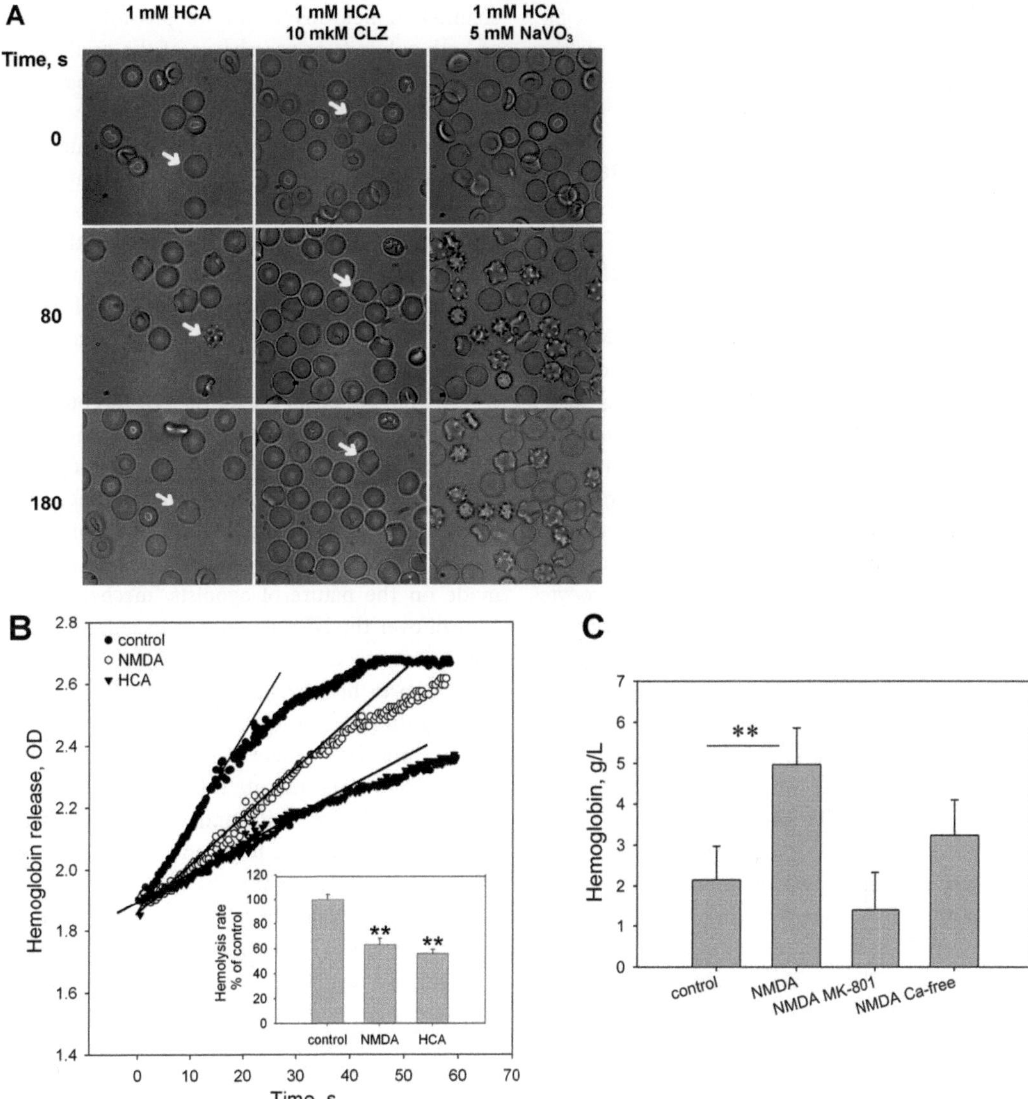

Fig. 24.6 Secondary changes in erythrocyte morphology, osmotic resistance, and resistance to shear stress caused by NR agonists. A: acute morphological alterations triggered by a high dose (1 mM) of NR agonist homocysteic acid (HCA) in the absence or in the presence of 10 μM clotrimazole or 5 mM Na_3VO_4. Images are taken from the cells before and 100 or 180 s after the addition of the HCA. Cells highlighted with arrows show characteristic reversible shape changes when exposed to HCA alone and or in the presence of clotrimazole. Inhibition of the Ca^{2+} pump with Na_3VO_4 facilitated the rate of echinocytic transformation and the number of responding cells compared with that in the presence of HCA alone as well as made these changes irreversible. B: osmotic resistance of rat erythrocytes treated with 1 mM HCA or NMDA. Osmotic resistance was assessed as a rate of hemolysis in response to acute hypoosmotic stress. Exposure of cells to agonists of the NR for 1 min caused a significant decrease in hemolytic rates calculated from the linear slopes of the plots of haemoglobin release. Shown are the representative haemoglobin liberation kinetics plots. Inset: statistical analysis of the rate of osmotic hemolysis for nine independent experiments; **$p < 0.01$ compared with the nontreated control. C: shear stress-induced hemolysis in cells after a long-term exposure to NMDA. Accumulation of haemoglobin in the incubation medium during 1 h at 37°C in an Eppendorff Thermomixer (900 rpm shaking speed) was assessed in control cells and in the cells exposed to 100 μM NMDA ± 100 μM MK-801 in the presence or absence of 1 mM Ca^{2+} in the incubation medium. Data are means of seven independent experiments ± SEM **$p < 0.01$.

abolished by treating erythrocytes with MK-801 and significantly reduced through the use of a nominally Ca^{2+}-free medium.

24.3.4 NO synthase activation and glutathione oxidation.

Function of the eNOS present in the red cell membrane of rodents and humans[27,33] is Ca^{2+} dependent[45]. Treatment of the whole blood with MK-801 resulted in a dose-dependent inhibition of NO_2^- + NO_3^- accumulation in plasma, whereas exposure to NMDA facilitated de novo NO production (◘ Fig. 24.7A). These observations were confirmed using erythrocytes suspension (◘ Fig. 24.7B). Conversely, exposure of erythrocytes to the NR agonists NMDA or HCA for 1 h resulted in an increased (NO_2^- + NO_3^-) generation in erythrocytes suspension (◘ Fig. 24.7B). Agonist-induced activation of NO production was initiated through the activation of the NR, which can be antagonised by MK-801. HCA-induced generation of NO required the presence of extracellular L-arginine and was blocked by L-NAME, which suggests that it was mediated by the eNOS (◘ Fig. 24.7B).

Oxidative burst in cells suspended in L-arginine-free medium was concomitant to the activation of the RBC-eNOS by either NMDA or HCA as observed by the increase in the intracellular GSSG levels (◘ Fig. 24.7C). In the L-arginine-supplemented medium, treatment of the cells with HCA (◘ Fig. 24.7C) or NMDA (data not shown) had negligible effects on the intracellular GSSG levels. HCA-induced oxidation was also inhibited by the NR antagonist MK-801 or by L-NAME (◘ Fig. 24.7C). In the absence of NR agonists L-NAME or L-arginine supplementation to the incubation medium did not cause any changes in the intracellular GSSG content (data not shown).

24.4 Discussion

In this study we present data that demonstrates the presence of functional NMDA receptors in mammalian RBCs. The combination of both subunits (NR1 and NR2) are required to form a functional channel in neurons[28], and we provide evidence of both subunits in membrane of the erythrocyte (◘ Fig. 24.1B and C). Pharmacological characteristics of these receptors resemble those reported for NRs in neurons. Agonists and antagonist both interact with these receptors at concentrations reported for neurons[14,26,30]. We have established evidence suggesting that the number and the activation state of NRs in circulating erythrocytes (the latter dependent on HC and/or HCA levels in plasma) are critical in control of the intracellular Ca^{2+} levels (◘ Fig. 24.2 and ◘ Fig. 24.3).

The physiological role of the NR receptor may only be fully understood once we know the abundance of the receptor present in erythrocytes as well as the mechanisms regulating the function of the receptor as well as the nature of agonists and antagonists. Whereas some speculations may be made on the nature of agonists, mechanisms in control over the NRs abundance in red cell membrane remains unknown. Plasma glutamate concentration in healthy subjects (14 to 47 μM[3,13,47]) are significantly lower than that measured in the synaptic cleft where it reaches several millimoles[14]. Our data indicate that glutamate binding to the NRs in rat brain and erythrocytes share similar IC_{50} of 96 ± 2 μM[26] and 88.2 ± 0.01 μM (◘ Fig. 24.4A), respectively. Plasma HC and HCA concentrations reported in healthy subjects range between 8 and 15 μM[24]. These values may rise substantially (up to 400 μM) as a result of folate or vitamin B12 deficiency, pregnancy, aging, or as a result of a hereditary methionine metabolism-related disorder[41]. Values reported for the NR's IC_{50} in the brain for HC and HCA are 14 ± 4 μM[51] and 14 μM[51], respectively, whereas it is 21.1 ± 0.78 μM for erythrocytes (◘ Fig. 24.4A). These data suggest that in healthy subjects these two agonists predominantly control NR activity in circulating RBCs, whereas glutamate will only be of importance in pathological cases such as stroke when plasma glutamate concentration may rise up to 200 μM[10]. Magnesium is one of the physiologically relevant inhibitors of NRs in the brain. In the present study we only tested responses of NRs present in erythrocytes to pharmacological inhibitors that are more selective and well-characterised. In erythrocytes MK-801 irreversibly inhibits the receptor-

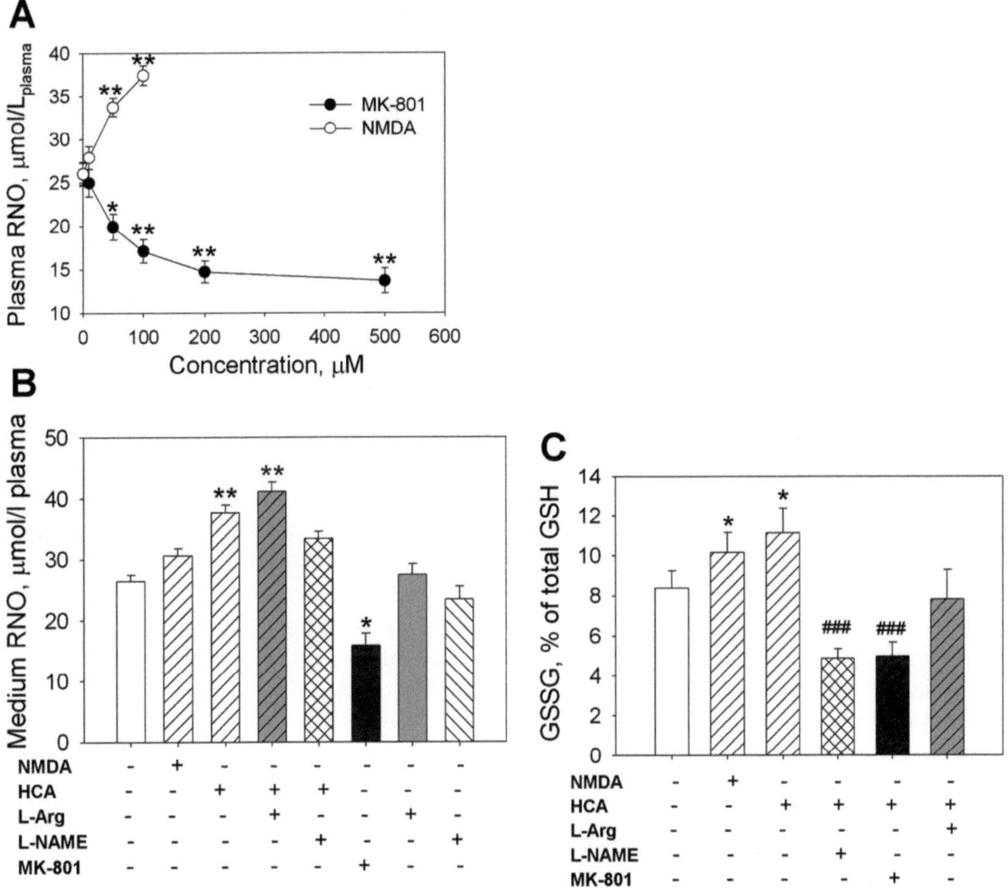

Abb. 24.7 Effect of the NR agonists and antagonist on the NO production and the intracellular redox state. A: total oxidised NO (RNO: NO_2^- + NO_3^-) levels in whole blood as a function of treatment with various concentrations of MK-801 or NMDA for 1 h. Data are means of five experiments ± SEM *p < 0.05 and **p < 0.01 compared with the basal levels of RNO in plasma. B: RNO levels in the incubation medium in which the RBC were exposed to the 100 μM of the NR agonists NMDA or HCA (right side-facing hatched bars) or 50 μM antagonist MK-801 (dark grey bar) in the presence or in the absence of either 100 μM L-arginine (grey bar) or nitro-L-arginine methyl ester (L-NAME) (left side-facing hatched bar) for 1 h. Combined treatment is represented as a combination of the corresponding bar filling. *p < 0.05 and **p < 0.01 compared with the untreated control (open bars). C: intracellular GSSG content in percentage of total GSH levels in erythrocytes treated similar as in B. Right side facing-hatched bars represent agonist-treated cells, left side-facing hatching stands for the L-NAME-treated ones, dark and light grey filling represents MK-801 and L-arginine exposure, respectively. The corresponding controls for the GSSG content in the cells treated with L-arginine, MK-801, or L-NAME alone are not shown and did not differ from the control shown as an open bar. Data are means of five independent experiments ± SE. *p < 0.05 compared with control (open bars) and ###p < 0.001 compared with the values in cells treated with HCA alone (right side-facing hatched bars).

mediated ion transport and downstream processes at micromolar concentrations used in electrophysiological studies on brain slices (e.g., [49]).

The NRs were not homogeneously distributed in the erythrocyte population and were most prevelant in reticulocytes (■ Fig. 24.1C). Although the number of receptor copies per reticulocytes were high compared with mature and senescent red cells, they were 3500-fold lower than that in UT7-(Epo) cell line suggesting that receptor levels are dramatically reduced during the late stages of differentiation of the erythroid precursors. Moreover, reduction in numbers has also been shown for other receptor types, some of which (e.g., TrR) are used as a marker of reticulocyte count. Although not all cells lacking TrR were also deprived of NR (■ Fig. 24.1C) suggesting that the NR is not present solely in reticulocytes and thus cannot be used as marker for them. Existence of erythrocytes with particularly either high or low levels of NRs may also be observed by using flow cytomery (■ Fig. 24.2). We have recently reported a similar distribution pattern for the erythropoietin receptors in mouse erythrocytes. These receptors are most abundant in reticulocytes and present at much lower levels in young and mature red cells while virtually absent in senescent cells[33].

Relatively high density of NRs in young cells may contribute to the process that selectively removes young erythrocytes when Epo receptors are dormant, which is known as neocytolysis[2,12]. In contrast to the common opinion that Ca^{2+} levels increase with red cell maturation and senescence[15], our data indicate that reticulocytes contain more Ca^{2+} than the either adult or senescent cells (■ Fig. 24.2). Thus Ca^{2+}-induced processes may control the rapid decrease in blood cell mass, whereas clearance of senescent cells is regulated by other factors[31]. Assessing the effects that Epo has on the activation state of NR and intracellular Ca^{2+} levels may shed light on the mechanism(s) involved in selective clearance of reticulocytes and young erythrocytes during conditions of Epo deficiency and the potential effects of Epo on the »non-selective cation channels« described my Myssina *et al.*[35].

Accumulation of Ca^{2+} in the cells following treatment with NR agonists is a common mechanism demonstrating NMDA/HCA cytotoxicity in erythrocytes and neurons[28] (■ Fig. 24.2 and ■ Fig. 24.3). Exposure of erythrocytes to NMDA or HCA gives rise to numerous Ca^{2+}-dependent responses. Downstream targets of the Ca^{2+}-sensitive signaling cascade have different Ca^{2+} sensitivity thresholds. Gardos channel activation requires 1–3 μM Ca^{2+} to reach half-activation[20]. These channels remain closed under most physiological conditions and are only activated by maneuvers that result in intracellular Ca^{2+} accumulation[5]. Activation of the Ca^{2+}-dependent, clotrimazole-

sensitive K^+ influx can be observed in the presence of 50–100 μM NMDA, whereas treatment of the cells with MK-801 or clotrimazole shows no effect (■ Fig. 24.5). Changes in the intracellular Ca^{2+} trigger characteristic alterations in erythrocyte morphology[21,22] resembling those we have observed in NMDA-treated cells (■ Fig. 24.6B). These reversible alterations in cell shape, which can be blocked by clotrimazole, reveal the result of Gardos channel activation.

De novo NO production by RBC-eNOS is Ca^{2+} dependent. Half-activation of eNOS requires 50–300 nM Ca^{2+} depending on the phosphorylation state of the enzyme[45]. Thus threshold levels of intracellular Ca^{2+} for eNOS activation are lower than those for Gardos channel activation (1–3 μM). Inhibition of Ca^{2+} uptake via NRs by MK-801 compromises eNOS function as Ca^{2+} levels drop below 50–100 nM due to its removal from the cytosol by the Ca^{2+} pump (■ Fig. 24.7A). On the other hand, treatment of the cells with NMDA further stimulates NO production in the presence of L-arginine, which suggests that steady-state Ca^{2+} levels are insufficient to support function of the eNOS at maximal rates. Changes in the NO production by erythrocytes in turn are reported to affect deformability, redox state, and oxygen-carrying capacity of RBCs[33,38].

In rat erythrocytes, alterations in activity of RBC-eNOS had an immediate effect on the intracellular redox balance. As shown before in mouse erythrocytes[33], activation of eNOS in rat RBCs by NR agonists was pro-oxidative under conditions of L-arginine deprivation as eNOS was forced into the $O_2^{\cdot-}$-generating mode (■ Fig. 24.6C). Similar effects were observed in mouse erythrocytes upon treatment with Epo[33]. Oxidative stress triggered by NR activation may possibly contribute to an increased susceptibility of erythrocytes to shear stress-induced hemolysis after a long-term exposure to NMDA or HCA (■ Fig. 24.5C).

Taken together our findings indicate that an increase in the number or activity of NRs may cause oxidative stress, abnormal volume regulation, and hemolysis. These events may contribute to the increased incidence of thrombosis and anemia that has been reported in patients with high concentration of HC, which can reach 200–300 μM[1,8].

Macrocytosis and abnormal RBC morphology are reported in patients with hyperhomocysteinemia[1] in which similar mechanisms as those described here may actually occur *in vivo*. So far the effect of HC on platelets has been suggested to be the only cause of thrombotic complications[34]. If our observations on rat erythrocytes reflect the conditions in human red cells, these cells would also contribute to HC-induced thrombus formation and hemolytic anemia in patients with vitamin B12 and folate deficiency[50]. The final impact of HC/HCA on the survival prognosis or mortality risk in hemodialysis patients was shown to increase by 3% per each additional 1 µM of HC in the plasma[9]. The next step therefore would be to assess the levels of NRs in erythrocytes of healthy subjects and patients with various forms of anemia. This work is currently in progress.

24.5 References

[1] Acharya U, Gau JT, Horvath W, Ventura P, Hsueh CT, Carlsen W. Hemolysis and hyperhomocysteinemia caused by cobalamin deficiency: three case reports and review of the literature. J Hematol Oncol 1: 26, 2008.

[2] Alfrey CP, Fishbane S. Implications of neocytolysis for optimal management of anaemia in chronic kidney disease. Nephron Clin Pract 106: c149–156, 2007.

[3] Andras IE, Deli MA, Veszelka S, Hayashi K, Hennig B, Toborek M. The NMDA and AMPA/KA receptors are involved in glutamate-induced alterations of occludin expression and phosphorylation in brain endothelial cells. J Cereb Blood Flow Metab 27: 1431–1443, 2007.

[4] Bashir ZI, Alford S, Davies SN, Randall AD, Collingridge GL. Long-term potentiation of NMDA receptor-mediated synaptic transmission in the hippocampus. Nature 349: 156–158, 1991.

[5] Bennekou P, Christophersen P. Ion channels. In: Red Cell Membrane Transport in Health and Disease, edited by Bernhardt I, Ellory JC. Berlin: Springer, 2003, p. 139–152.

[6] Boldyrev AA, Johnson P. Homocysteine and its derivatives as possible modulators of neuronal and non-neuronal cell glutamate receptors in Alzheimer's disease. J Alzheimers Dis 11: 219–228, 2007.

[7] Boldyrev AA, Kazey VI, Leinsoo TA, Mashkina AP, Tyulina OV, Johnson P, Tuneva JO, Chittur S, Carpenter DO. Rodent lymphocytes express functionally active glutamate receptors. Biochem Biophys Res Commun 324: 133–139, 2004.

[8] Brattstrom L, Wilcken DE. Homocysteine and cardiovascular disease: cause or effect? Am J Clin Nutr 72: 315–323, 2000.

[9] Buccianti G, Baragetti I, Bamonti F, Furiani S, Dorighet V, Patrosso C. Plasma homocysteine levels and cardiovascular mortality in patients with end-stage renal disease. J Nephrol 17: 405–410, 2004.

[10] Castellanos M, Sobrino T, Pedraza S, Moldes O, Pumar JM, Silva Y, Serena J, Garcia-Gil M, Castillo J, Davalos A. High plasma glutamate concentrations are associated with infarct growth in acute ischemic stroke. Neurology 71: 1862–1868, 2008.

[11] Chen H, Fitzgerald R, Brown AT, Qureshi I, Breckenridge J, Kazi R, Wang Y, Wu Y, Zhang X, Mukunyadzi P, Eidt J, Moursi MM. Identification of a homocysteine receptor in the peripheral endothelium and its role in proliferation. J Vasc Surg 41: 853–860, 2005.

[12] De Santo NG, Cirillo M, Kirsch KA, Correale G, Drummer C, Frassl W, Perna AF, Di Stazio E, Bellini L, Gunga HC. Anemia and erythropoietin in space flights. Sem Nephrol 25: 379–387, 2005.

[13] Divino Filho JC, Hazel SJ, Furst P, Bergstrom J, Hall K. Glutamate concentration in plasma, erythrocyte and muscle in relation to plasma levels of insulin-like growth factor (IGF)-I, IGF binding protein-1 and insulin in patients on haemodialysis. J Endocrinol 156: 519–527, 1998.

[14] Dzubay JA, Jahr CE. The concentration of synaptically released glutamate outside of the climbing fiber-Purkinje cell synaptic cleft. J Neurosci 19: 5265–5274, 1999.

[15] Foller M, Huber SM, Lang F. Erythrocyte programmed cell death. IUBMB life 60: 661–668, 2008.

[16] Fowler B. Homocysteine: overview of biochemistry, molecular biology, and role in disease processes. Seminars Vasc Med 5: 77–86, 2005.

[17] Genever PG, Wilkinson DJ, Patton AJ, Peet NM, Hong Y, Mathur A, Erusalimsky JD, Skerry TM. Expression of a functional N-methyl-D-aspartate-type glutamate receptor by bone marrow megakaryocytes. Blood 93: 2876–2883, 1999.

[18] Gill S, Veinot J, Kavanagh M, Pulido O. Human heart glutamate receptors–implications for toxicology, food safety, and drug discovery. Toxicol Pathol 35: 411–417, 2007.

[19] Gill SS, Pulido OM. Glutamate receptors in peripheral tissues: current knowledge, future research, and implications for toxicology. Toxicol Pathol 29: 208–223, 2001.

[20] Grygorczyk R, Schwarz W. Properties of the Ca^{2+}-activated K^+ conductance of human red cells as revealed by the patch-clamp technique. Cell Calcium 4: 499–510, 1983.

[21] Hagerstrand H, Danieluk M, Bobrowska-Hagerstrand M, Iglic A, Wrobel A, Isomaa B, Nikinmaa M. Influence of band 3 protein absence and skeletal structures on amphiphile- and Ca(2+)-induced shape alterations in erythrocytes: a study with lamprey (Lampetra fluviatilis), trout (Onchorhynchus mykiss) and human erythrocytes. Biochim Biophys Acta 1466: 125–138, 2000.

[22] Hagerstrand H, Mrowczynska L, Salzer U, Prohaska R, Michelsen KA, Kralj-Iglic V, Iglic A. Curvature-dependent lateral distribution of raft markers in the human erythrocyte membrane. Mol Membr Biol 23: 277–288, 2006.

[23] Hayashi H, Jarrett HW, Penniston JT. Peripheral proteins and smooth membrane from erythrocyte ghosts. Segregation of ATP-utilizing enzymes into smooth membrane. J Cell Biol 76: 105–115, 1978.

[24] Herrmann M, Taban-Shomal O, Hubner U, Bohm M, Herrmann W. A review of homocysteine and heart failure. Eur J Heart Fail 8: 571–576, 2006.

[25] Hitchcock IS, Skerry TM, Howard MR, Genever PG. NMDA receptor-mediated regulation of human megakaryocytopoiesis. Blood 102: 1254–1259, 2003.

[26] Jasek MC, Griffith WH. Pharmacological characterization of ionotropic excitatory amino acid receptors in young and aged rat basal forebrain. Neuroscience 82: 1179–1194, 1998.

[27] Kleinbongard P, Schulz R, Rassaf T, Lauer T, Dejam A, Jax T, Kumara I, Gharini P, Kabanova S, Ozuyaman B, Schnurch HG, Godecke A, Weber AA, Robenek M, Robenek H, Bloch W, Rosen P, Kelm M. Red blood cells express a functional endothelial nitric oxide synthase. Blood 107: 2943–2951, 2006.

[28] Kloda A, Martinac B, Adams DJ. Polymodal regulation of NMDA receptor channels. Channels 1: 334–343, 2007.

[29] Krizbai IA, Deli MA, Pestenacz A, Siklos L, Szabo CA, Andras I, Joo F. Expression of glutamate receptors on cultured cerebral endothelial cells. J Neurosci Res 54: 814–819, 1998.

[30] Lipton SA, Kim WK, Choi YB, Kumar S, D'Emilia DM, Rayudu PV, Arnelle DR, Stamler JS. Neurotoxicity associated with dual actions of homocysteine at the N-methyl-D-aspartate receptor. Proc Natl Acad Sci USA 94: 5923–5928, 1997.

[31] Lutz HU. Innate immune and non-immune mediators of erythrocyte clearance. Cell Mol Biol (Noisy-le-Grand, France) 50: 107–116, 2004.

[32] Merle B, Itzstein C, Delmas PD, Chenu C. NMDA glutamate receptors are expressed by osteoclast precursors and involved in the regulation of osteoclastogenesis. J Cell Biochem 90: 424–436, 2003.

[33] Mihov D, Vogel J, Gassmann M, Bogdanova A. Erythropoietin activates nitric oxide synthase in murine erythrocytes. Am J Physiol Cell Physiol 297: C378–C388, 2009.

[34] Mohan IV, Jagroop IA, Mikhailidis DP, Stansby GP. Homocysteine activates platelets in vitro. Clin Appl Thromb Hemost 14: 8–18, 2008.

[35] Myssina S, Huber SM, Birka C, Lang PA, Lang KS, Friedrich B, Risler T, Wieder T, Lang F. Inhibition of erythrocyte cation channels by erythropoietin. J Am Soc Nephrol 14: 2750–2757, 2003.

[36] Nasstrom J, Boo E, Stahlberg M, Berge OG. Tissue distribution of two NMDA receptor antagonists, [3H]CGS 19755 and [3H]MK-801, after intrathecal injection in mice. Pharmacol Biochem Behav 44: 9–15, 1993.

[37] Orlov SN, Kolosova IA, Cragoe EJ, Gurlo TG, Mongin AA, Aksentsev AL, SVK . Kinetics and pecularities of thermal inactivation of volume-induced Na/H exchange, Na,K,2Cl cotransport and K,Cl contrasport in rat erythrocytes. Biochim Biophys Acta 1151: 186–192, 1993.

[38] Ozuyaman B, Grau M, Kelm M, Merx MW, Kleinbongard P. RBC NOS: regulatory mechanisms and therapeutic aspects. Trends Mol Med 14: 314–322, 2008.

[39] Parsons CG, Danysz W, Quack G. Glutamate in CNS disorders as a target for drug development: an update. Drug News Perspect 11: 523–569, 1998.

[40] Petrushanko I, Bogdanov N, Bulygina E, Grenacher B, Leinsoo T, Boldyrev A, Gassmann M, Bogdanova A. Na-K-ATPase in rat cerebellar granule cells is redox sensitive. Am J Physiol Regul Integr Comp Physiol 290: R916–R925, 2006.

[41] Selhub J. The many facets of hyperhomocysteinemia: studies from the Framingham cohorts. J Nutr 136: 1726S–1730S, 2006.

[42] Seshadri S, Beiser A, Selhub J, Jacques PF, Rosenberg IH, D'Agostino RB, Wilson PW, Wolf PA. Plasma homocysteine as a risk factor for dementia and Alzheimer's disease. N Engl J Med 346: 476–483, 2002.

[43] Stabler SP, Marcell PD, Podell ER, Allen RH, Savage DG, Lindenbaum J. Elevation of total homocysteine in the serum of patients with cobalamin or folate deficiency detected by capillary gas chromatography-mass spectrometry. J Clin Invest 81: 466–474, 1988.

[44] Tietze F. Enzymic method for quantitative determination of nanogram amounts of total and oxidized glutathione: applications to mammalian blood and other tissues. Anal Biochem 27: 502–522, 1969.

[45] Tran QK, Leonard J, Black DJ, Nadeau OW, Boulatnikov IG, Persechini A. Effects of combined phosphorylation at Ser-617 and Ser-1179 in endothelial nitric-oxide synthase on $EC_{50}(Ca^{2+})$ values for calmodulin binding and enzyme activation. J Biol Chem 284: 11892–11899, 2009.

[46] Tsai JC, Wang H, Perrella MA, Yoshizumi M, Sibinga NE, Tan LC, Haber E, Chang TH, Schlegel R, Lee ME. Induction of cyclin A gene expression by homocysteine in vascular smooth muscle cells. J Clin Invest 97: 146–153, 1996.

[47] Tsai PJ, Huang PC. Circadian variations in plasma and erythrocyte glutamate concentrations in adult men consuming a diet with and without added monosodium glutamate. J Nutr 130: 1002S–1004S, 2000.

[48] Tyulina OV, Prokopieva VD, Dodd RD, Hawkins JR, Clay SW, Wilson DO, Boldyrev AA, Johnson P. In vitro effects of ethanol, acetaldehyde and fatty acid ethyl esters on human erythrocytes. Alcohol Alcohol 37: 179–186, 2002.

[49] Vander Jagt TA, Connor JA, Shuttleworth CW. Localized loss of Ca^{2+} homeostasis in neuronal dendrites is a downstream consequence of metabolic compromise during extended NMDA exposures. J Neurosci 28: 5029–5039, 2008.

[50] Ventura P, Panini R, Emiliani S, Salvioli G. Plasma homocysteine after insulin infusion in type II diabetic patients with and without methionine intolerance. Exp Clin Endocrinol Diabetes 112: 44–51, 2004.

[51] Yuzaki M, Connor JA. Characterization of L-homocysteate-induced currents in Purkinje cells from wild-type and NMDA receptor knockout mice. J Neurophys 82: 2820–2826, 1999.

Stimulation of human red blood cells leads to Ca^{2+}-mediated intercellular adhesion

Patrick Steffen, Achim Jung, Duc Bach Nguyen, Torsten Müller, Ingolf Bernhardt, Lars Kaestner, Christian Wagner

Reprint from Cell Calcium (2011) **50**, 54-61.

▪ Abstract

Red blood cells (RBCs) are a major component of blood clots, which form physiologically as a response to injury or pathologically in thrombosis. The active participation of RBCs in thrombus solidification has been previously proposed but not yet experimentally proven. Holographic optical tweezers and single-cell force spectroscopy were used to study potential cell–cell adhesion between RBCs. Irreversible intercellular adhesion of RBCs could be induced by stimulation with lysophosphatidic acid (LPA), a compound known to be released by activated platelets. We identified Ca^{2+} as an essential player in the signaling cascade by directly inducing Ca^{2+} influx using A23187. Elevation of the internal Ca^{2+} concentration leads to an intercellular adhesion of RBCs similar to that induced by LPA stimulation. Using single-cell force spectroscopy, the adhesion of the RBCs was identified to be approximately 100 pN, a value large enough to be of significance inside a blood clot or in pathological situations like the vasco-occlusive crisis in sickle cell disease patients.

25.1 Introduction

Previous studies of the interactions between red blood cells (RBCs) have relied on either hydrodynamic interactions[1] or interactions mediated by plasma macromolecules. Co-adhesion of RBCs under physiological conditions is known as rouleaux formation. These structures might be generated either by plasma polymers bridging between the cells[2,3] or, more likely, by depletion forces. The adhesion energies are small enough to allow for breakage in the shear flow, and the effect is thought to be reversible, even if some controversial data exist[4]. This reversibility is crucial for the ability of RBCs to cross capillary sections that are smaller than the cell diameter. However, in several pathologies, such as sickle cell disease or thalassemia, an increased propensity for red blood cell–cell adhesion has been observed[5]. Clot formation is a collective phenomenon based on the interplay of many components. In current models, the contribution of RBCs to the clotting process is thought to be purely passive, i.e., they are simply caged in the fibrin network due to their prevalence in the blood. Based on experimental and clinical studies that have shown a correlation between decreased hematocrit values and longer bleeding times[6-8] and their own experiments, Andrews and Low summarised the evidence for active participation of RBCs in thrombus formation. Kaestner *et al.*[9] proposed a signaling mechanism based on Ca^{2+} entry via a non-selective voltage-dependent cation (NSVDC) channel that is permeable to mono- and bivalent cations like Na$^+$, K$^+$ and Ca^{2+} ref[10,11]. This channel has been shown to be activated by prostaglandin E$_2$ (PGE$_2$), lysophosphatidic acid (LPA)[9,12] and by the mechanical deformations that occur when RBCs pass through capillaries smaller than their resting diameter[13]. PGE$_2$ and LPA are extracellular, local mediators that are released both by platelets after their activation within the coagulation cascade and by RBCs themselves under mechanical stress[14]. The aim of this study was to test the hypothesis that intercellular RBC adhesion can occur directly (without participation of platelets). To perform this at the level of individual cells, we used the complementary methods of non-invasive holographic optical tweezers (HOT)[15], using the momentum of light and single-cell force spectroscopy to quantify the occurring adhesion.

25.2 Materials and methods

25.2.1 RBC preparation and fluorescence microscopy

For experiments using the optical tweezers and microfluidics, fresh blood from healthy donors was obtained by a fingertip needle prick, while for annexin V-FITC fluorescence measurements, human venous blood was drawn from healthy donors. Heparin or EDTA was used as an anticoagulant. The blood was obtained within one day of the experiment from the Institute of Clinical Haematology and Transfusion Medicine of the Saarland University Hospital. The cells were washed three times by centrifugation ($2000 \times g$, 3 min) in a HEPES buffered solution of physiological ionic strength (PIS-solution) containing the following (in mM): 145 NaCl, 7.5 KCl, 10 glucose and 10 HEPES, pH 7.4, at room temperature. The buffy coat and plasma were removed by aspiration. For Ca²⁺ imaging, RBCs were loaded with 5 μM Fluo-4 AM (Molecular Probes, Eugene, USA) from a 1 mM stock solution in dimethyl sulfoxide (DMSO) with 20% Pluronic (F-127, Molecular Probes). Loading was performed in 1 ml of PIS-solution at an RBC haematocrit of approximately 1% for 1 h at 37 °C. The cells were washed by centrifugation once more and equilibrated for de-esterification for 15 min. LPA, prepared from a stock solution of 1 mM in distilled water, and the Ca²⁺ ionophore A23187, prepared from a stock solution of 1 mM in ethanol, were obtained from Sigma Aldrich (St. Louis, USA). To investigate phosphatidylserine (PS) exposure, cells were stained with annexin V-FITC (Molecular Probes). Annexin V-FITC was delivered in a unit size of 500 μl containing 25 mM HEPES, 140 mM NaCl, 1 mM EDTA, pH 7.4, and 0.1% bovine serum albumin. Five hundred microli-

ters of annexin binding buffer (10 mM HEPES, 140 mM NaCl, 2.5 mM CaCl₂, pH 7.4) was added to 1 μl of washed, packed RBCs previously treated with LPA or A23187. Afterwards, 5 μl annexin V-FITC was added, and the cells were mixed gently. The probes were incubated at room temperature in the dark for 15 min and then washed 2 times in 1 mL of annexin binding buffer ($2000 \times g$, 3 min). Finally, the cells were re-suspended in 500 μl of annexin binding buffer and used for microscopy. The measurements were obtained from images taken with a CCD camera (CCD97, Photometrics, USA).

25.2.2 Holographic optical tweezers (HOT)

A schematic drawing of our experimental setup is shown in ❏ Fig. 25.1a. A Nd:YAG infrared laser (Ventus, Laser Quantum, Stockport, UK) with a beam width of 2.5 mm was coupled for coarse alignment with a visible He–Ne laser via a dichroic mirror. The latter was switched off during the experiments. Both beams were expanded five-fold (BM.X, Linos Photonics, Göttingen, Germany) to overfill the 8 mm back aperture of the microscope objective (CFI Plan Fluor 60× oil immersion, Nikon Corp., Tokyo, Japan). The optics were integrated into an inverted fluorescence microscope (TE-2000, Nikon Corp.).

This allowed for combined trapping and fluorescence or differential interference contrast (DIC) measurements. Images were taken with an electron multiplication CCD camera (Cascade 512F, Roper Scientific, Trenton, USA) with a typical frame rate of 100 Hz. To reduce vibrations, the original cooling fan was replaced with water-cooling components. In addition, the optical setup was placed on active vibration isolation elements (Vario Series, Halcyonics, Göttingen, Germany). The phase of the laser's electric field was modified using a spatial light modulator (PPM X8267-15, Hamamatsu Photonics, Hamamatsu City, Japan) to create the desired trap pattern in the focal plane of the microscope objective. In order to test for RBC adhesion after stimulation, a suitable configuration of independently movable optical traps was created

a)

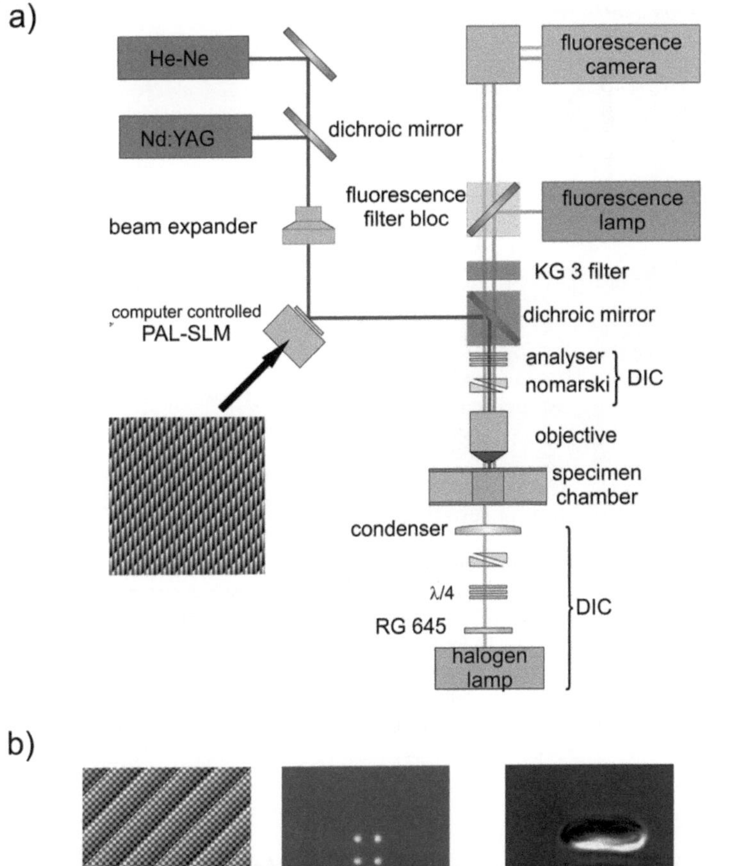

b)

Fig. 25.1 a) A schematic drawing of the experimental setup. For a detailed description see text. b) Left: phase holograms applied to the parallel aligned liquid crystal-spatial light modulator. Middle: trapped 2 μm latex beads. Right: trapped and orientated, discoid RBCs.

(see the schematic sketch in **Fig. 25.2a**). The laser power in each trap was approximately 5 mW. Cells could be moved against one another by replaying a series of kinoforms on the computer-controlled spatial light modulator. As an example, **Fig. 25.1b** shows a quadrotrap in parallel and perpendicular configuration for RBC arrangement along with the corresponding kinoforms.

25.2.3 Microfluidics

In order to provide a controlled yet interchangeable solvent environment, a microfluidic polydimethylsiloxane (PDMS) cell was constructed by standard soft lithography (inset in **Fig. 25.3**). The cell consists of two inlets from which either the cells in the buffer solutions or LPA or the

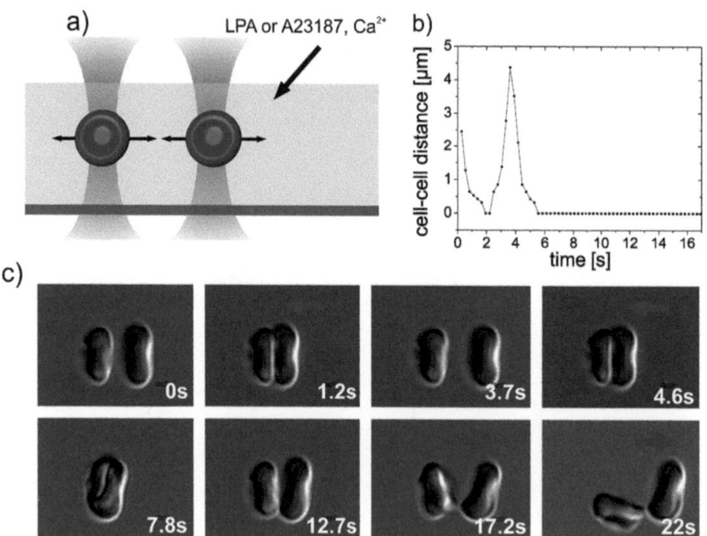

▢ **Fig. 25.2** Probing for adhesion between RBCs after LPA stimulation. Panel (a) represents a sketch of the oscillatory movement of two trapped cells. The actual cell–cell distance over the course of the adhesion test is depicted in panel (b). The graph depicts the separation distance of two red blood cells, as determined using edge detection, over the course of the experiment. Cells were preincubated with 2.5 μM LPA for 5 min. After adhesion occurred at around 6 s, the distance between the cells remained zero. Panel (c) depicts representative images of an adhesion measurement of two RBCs held by 4 optical traps. During a recording period of 26 s, the cells were moved back and forth as indicated by the red arrows. (For interpretation of the references to color in this figure legend, the reader is referred to the web version of the original article.)

ionophore A23187 (final concentration of 40 μM) could be applied. The flow was driven by a hydrostatic pressure difference that allowed for fine tuning. After injection of a new sample, the flow could be brought approximately to rest by eliminating the pressure difference. Behind the y-junction, the flow remained laminar, and the two solutions did not mix, except for very slow diffusion. The RBCs could then be captured by the HOT and transferred to the new environment. Thus, the cells could be transferred from one solvent condition to another in a rapid and controlled manner. The same measurements were also conducted in a Petri dish as follows: RBCs were incubated for 10 min with either A23187 or LPA in the presence of Ca²⁺ (for concentration values, please refer to the figure legends), followed by a fast sequence of adhesion tests between various cells. There were more than 50 cells in each experiment; the total number of tested cells amounted to 250.

25.2.4 Single-cell force spectroscopy

In order to quantify the occurring cell–cell adhesion, an atomic force microscope (AFM) (Cell-Hesion 200 with increased pulling range up to 100 μm, JPK Instruments, Germany) was used to conduct single-cell force spectroscopy measurements. In these measurements, a single RBC was attached to an AFM cantilever with the help of an appropriate adhesive. Cell-Tak™ (BD Biosciences) turned out to be the most efficient adhesive for binding RBCs to the cantilever. A stock solution of 2.4 mg/ml was diluted 1:30 according to instructions from BD. The cantilevers were incubated for at least 30 min at room temperature and used after rinsing with PIS solution. After the cell capture (CellTak™ functionalised cantilevers brought in contact with a 0.2 nN force for 10 s), the approach- and retraction speed were set to 5 μm/s. The pulling range varied between 30 and 50 μm, and the contact time varied from

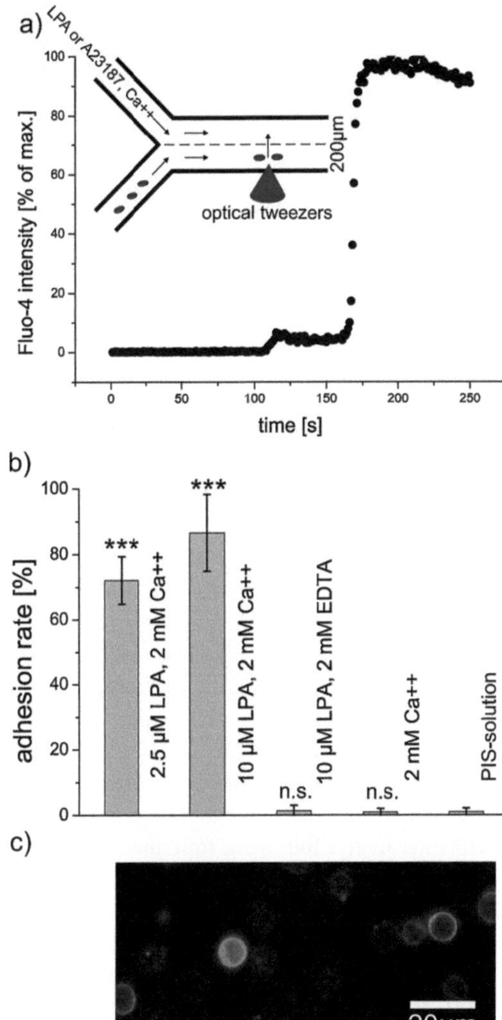

Fig. 25.3 The response of RBCs to LPA. (a) The relative fluorescence signal of a representative RBC in the upper microfluidic channel; t = 0 s is the time when the cell was moved into the LPA solution. The inset shows a schematic picture of the microfluidic chamber used. (b) Results of the LPA measurements conducted in the microfluidic chamber and the Petri dish. The grey bars represent the percentage of cells that showed adhesion. The overall number of cells tested was 60 cells per measurement. In the presence of LPA and Ca^{2+}, a significant number of cells showed adhesion, whereas in the control experiments, only a very small portion of the cells showed an adhesion. The results of the Student's t-test, compared to the control measurement (PIS-solution), are indicated at the top of each bar. (c) A fluorescence image of LPA-treated (2.5 µM) RBCs after annexin V-FITC staining. The annexin V binding indicates the presence of PS on the outer membrane leaflet of the cells.

0 to 120 s. During spectroscopy experiments, the deflection of the cantilever was monitored in real time using a built-in feature of the AFM software. The spring constant of the cantilever was determined by the common thermal noise method. The cantilevers used were tipples TL1 with a nominal spring constant of 0.03 N/m (Nanoworld). The cell morphology was monitored using phase contrast microscopy. In the course of the experiment, the cantilever with the attached cell was lowered onto another cell attached to the substrate, which was coated with 0.005% poly-L-lysine, until a preset defined constant force was reached and kept stationary for a defined contact time. The conditions during contact were determined by the choice of the particular closed-loop mode, specifically at a constant piezo position, after reaching a prescribed maximum pushing force of 400 pN. Subsequently, the cantilever was withdrawn at a constant speed. During approach and retraction, the cantilever deflected as a consequence of the acting forces. This deflection, which is proportional to the acting forces, was recorded in force–distance curves. The retraction curve was typically characterised by the maximum force required to separate the cells from each other, referred to as the maximum unbinding force F_{max}.

25.2.5 Statistical significance

A Student's t-test was used to test the results obtained from the adhesion experiments for statistical significance. Statistical significance of the data was defined as follows: $p > 0.05$ (n.s.), $p \leq 0.05$ (*); $p \leq 0.01$ (**), $p \leq 0.001$ (***).

25.3 Results

25.3.1 Red blood cell stimulation with LPA

As pointed out in Section 25.1, RBCs can be stimulated by LPA, and this has been proposed to contribute to the active participation of RBCs in the later stage of thrombus formation. In order to test for altered intercellular adhesion behavior, we set up holographic optical tweezers (for details, refer to ◘ Fig. 25.1). In the blood flow, cell–cell contact times are rather short when RBCs »bump« into each other. To mimic this physiological condition, two cells were grabbed with the laser foci (compare ◘ Fig. 25.2c) and moved back and forth as depicted in ◘ Fig. 25.2a, in a cyclical manner. Upon stimulation with 2.5 μM LPA, the RBCs adhered to each other as plotted and visualised in an image sequence in ◘ Fig. 25.2b and c, respectively. During the stimulation procedure, most of the RBCs remained in their discocyte shape. The separation force could not be determined by the HOT approach because it exceeds the force of the laser tweezers. In this configuration, 72% of the cells tested showed such an irreversible adhesion (grey bars in ◘ Fig. 25.3b). To exclude any dependencies on the interaction surface due to the anisotropic shape of the cells, we aimed for another condition using spherocytes. This was realised by increasing the LPA concentration to 10 μM, which is still within the physiologically observed range. In parallel to the adhesion experiments, the Ca²⁺ uptake of the cells was followed up by Fluo-4 imaging, as the trace in ◘ Fig. 25.3a shows for the stimulation with LPA. Additionally, a microfluidic system, schematically plotted as an inset in ◘ Fig. 25.3a, allowed a fast change of media by moving the RBCs using the holographic optical tweezers from one »channel« to the other. In this way, we tested cells under five different conditions: (i) PIS-solution, (ii) PIS-solution containing 2 mM Ca²⁺ and 10 μM LPA, (iii) PIS-solution containing 2 mM Ca²⁺ and 2.5 μM LPA, (iv) PIS-solution containing 2 mM Ca²⁺ and no LPA, and finally (v) PIS-solution containing 2 mM EDTA and 10 μM LPA. We used at least 60 cells per condition. The results are summarised in ◘ Fig. 25.3b. The RBC stimulation with LPA (2.5 μM as well as 10 μM) in the presence of extracellular Ca²⁺ led to an immediate qualitative change in the adhesion behavior: cells irreversibly stuck to each other. Additionally, RBCs were stained with an annexin V-FITC conjugate to probe for phosphatidylserine (PS) exposure of the RBCs by means of fluorescence microscopy. In ◘ Fig. 25.3c, we show that the 2.5 μM LPA-induced Ca²⁺ influx indeed was associated with PS transport from the inner leaflet to the outer leaflet of the RBC membrane. This staining was not observed in cells treated with conditions (i), (iv) and (v).

25.3.2 Approaching the signaling entities

The initial stimulation experiments using LPA revealed that an LPA-induced Ca²⁺ influx leads to intercellular RBC adhesion. To test whether this is a pure Ca²⁺ effect or if the presence of LPA is required, experiments were performed using the Ca²⁺ ionophore A23187 as an artificial tool to increase the intracellular Ca²⁺ concentration. After transferring the Fluo-4 loaded RBCs to the A23187 solution, the Fluo-4 fluorescence signal increased almost immediately, i.e., faster than after application with LPA. The Ca²⁺ influx into the cell is depicted in ◘ Fig. 25.4a. As illustrated in ◘ Fig. 25.4b, during the Ca²⁺ increase, the cells undergo a shape transformation from discocytes to spherocytes via an intermediate step of echinocytes. Testing for adhesion was performed using a procedure identical to the LPA experiments performed in the following four media: (i) PIS-solution, (ii) PIS-solution containing 2 mM Ca²⁺ and 40 μM A23187, (iii) PIS-solution containing 2 mM Ca²⁺ and no A23187, and (iv) PIS-solution containing 2 mM EDTA and 40 μM A23187. The results are summarised in ◘ Fig. 25.4c. Intercellular adhesion was significantly increased compared to the controls only when the intracellular Ca²⁺ concentration was increased. There have been controversial reports about PS transition to the outer membrane leaflet that claim merely coincidental increase in Ca²⁺ that is not related to the PS exposure[16,17]. Therefore, RBCs were stained with annexin-V-FITC conjugates in order to visualise PS exposure on the surface of RBCs. As depicted in ◘ Fig. 25.4d, treatment with A23187 in the presence of extracellular Ca²⁺ leads to a clear annexin-V binding (PS staining) on the outer leaflet of the membranes. Under all the other conditions (i), (iii), and (iv), no such staining could be observed.

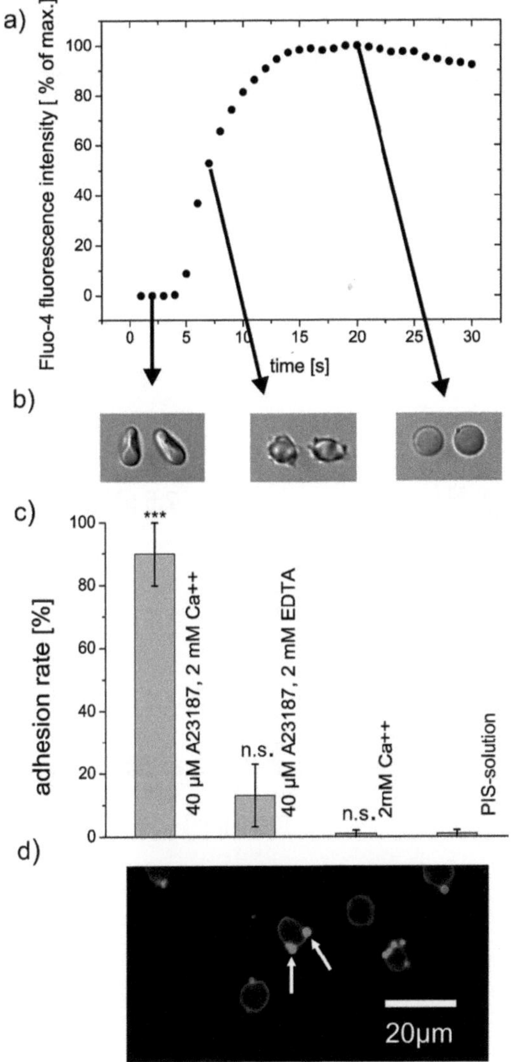

◘ Fig. 25.4 Measurements with the Ca^{2+} ionophore A23187. (a) The relative fluorescence signal of a representative RBC in the upper channel of the microfluidic system; t = 0 s is the time when the cell reached the A23187 solution. The Ca^{2+} increase happens almost instantaneously. The decrease in signal after 15 s is due to photobleaching. (b) The RBCs undergo a shape transformation from discocytes (left) via echinocytes (middle) to spherocytes (right) after transfer into the A23187 buffer solution in the upper channel. (c) The results of the ionophore measurements conducted in the microfluidic chamber and a Petri dish. The black bars represent the percentage of cells that showed adhesion. The number of cells tested was about 60 per measurement. In the presence of A23187 and Ca^{2+}, about 90% of the cells tested adhered, whereas in the control experiments, less than 3% of the cells adhered. The results of the Student's t-test, compared to the control measurement (PIS-solution), are indicated at the top of each bar. (d) A fluorescence image of annexin V-FITC-labeled RBCs. The cells were treated with A23187, and exposure of PS at the cell surface was clearly identified. A vesiculation of the cells was also observed (indicated by arrows).

25.3.3 Quantification of the intracellular adhesion

To allow a discussion of a physiological (or pathophysiological) relevance of the described adhesion process, one needs to determine the separation force. As described above, the separation force exceeds the abilities of the HOT. Therefore, single-cell force spectroscopy[18] was utilised for determination of the force. Two different measurements were conducted: control measurements in which the cells remained untreated and measure-

ments in which the cells were treated with a concentration of 2.5 µM LPA. Example curves for both experiments are depicted in ◘ Fig. 25.5b. It is an inherent part of the single-cell force spectroscopy procedure that cells need to be brought in contact by a certain force application (step 2 in ◘ Fig. 25.5a). In the control measurements, it was only possible to detect a weak interaction between the cells, whereas in the LPA measurements, a pronounced adhesion of the red blood cells could be observed. The step-wise release plotted in ◘ Fig. 25.5b (step 3 in ◘ Fig. 25.5a) was typical for all the cells mea-

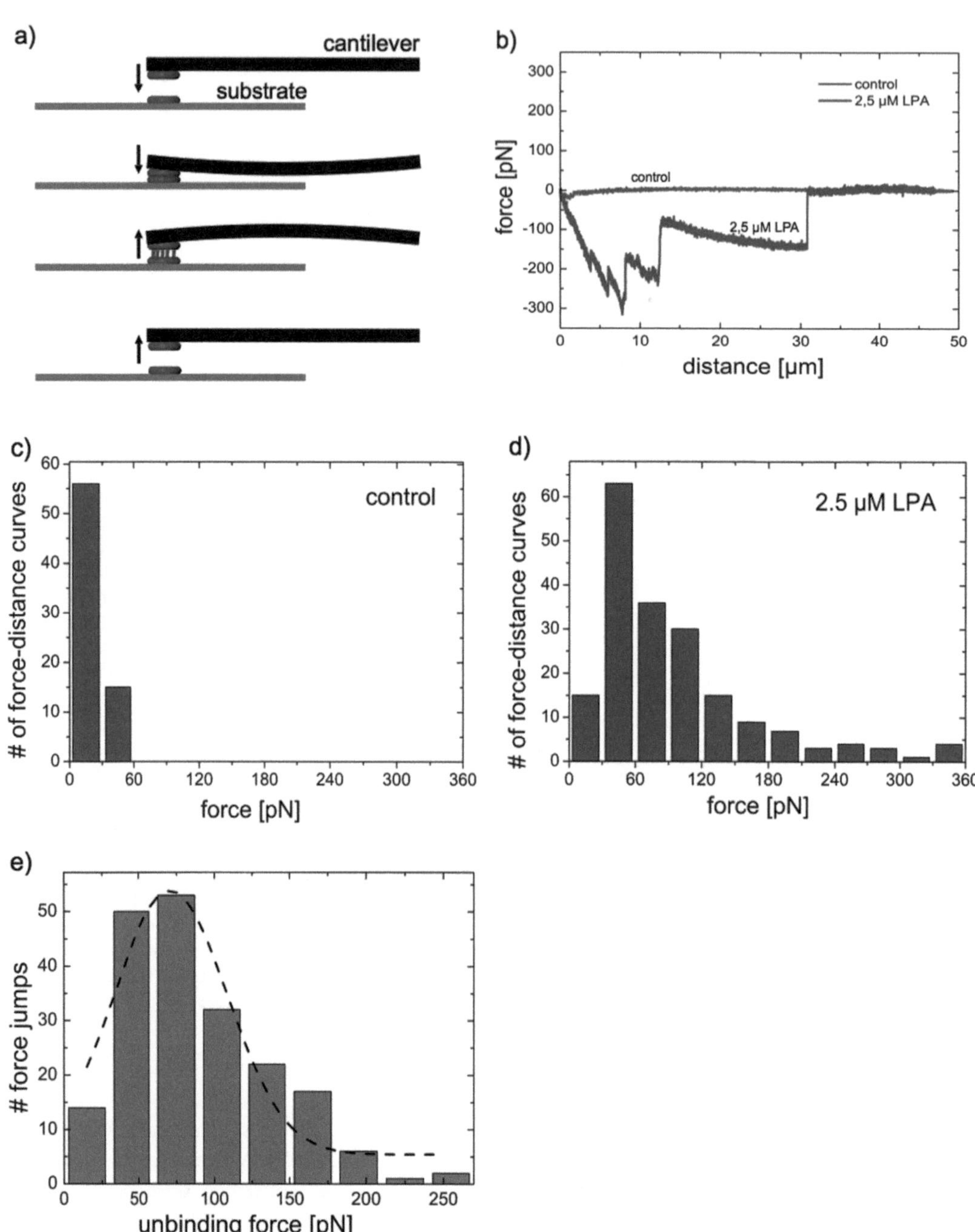

☐ **Fig. 25.5** Force quantification using atomic force spectroscopy (AFS). Panel (a) shows the working principle of the AFS. One cell is attached to a cantilever, while another one is attached to the surface. Over the course of the experiment, the cells are brought into contact and are withdrawn again. The adhesion force of the two cells can be measured by measuring the deflection of the cantilever. Panel (b) shows the combined plot of an example force–distance-curve of a control (green) and an LPA measurement (red). Panel (c) and panel (d) depict the statistics of the measured forces in the controls and LPA measurements, respectively. Panel (e) provides a histogram of the unbinding force of a single tether within the procedure visualised in panels (a) and (b). The dotted line depicts a Gaussian approximation of the bars. (For interpretation of the references to color in this figure legend, the reader is referred to the web version of the original article.)

sured. All results of the measurements are collected into histograms summarised in ◘ Fig. 25.5c and d. The mean value of the maximum unbinding force of untreated red blood cells amounted to 28.8 ± 8.9 pN (s.d.) ($n = 71$), whereas in the LPA experiments, the mean value of the maximum unbinding force amounted to a much higher value of 100 ± 84 pN (s.d.) ($n = 193$, from three different donors) indicating a severe difference in adhesion behavior of untreated and LPA-stimulated RBCs.

25.4 Discussion

25.4.1 LPA stimulation leads to intercellular adhesion

It is an established fact that the stimulation of RBCs with LPA leads to a Ca^{2+} influx through an NSVDC channel[9,12]. It is further known that the increased intracellular Ca^{2+} results in the stimulation of the Ca^{2+}-activated K^+ channel (Gardos channel)[19,20] and the activation of the lipid scramblase[21,22] (see below for further discussion). Based on these mechanisms, an increased aggregation tendency of stimulated RBCs has been hypothesised[9,23]. The development of the HOT provides a tool for testing this hypothesis at the cellular level. Concentrations of 2.5 µM and 10 µM LPA were chosen because they seemed to be within the common range of concentrations used with other cell types[24,25] in addition to RBCs[17]. Moreover, this concentration is comparable to the local LPA concentration in the immediate surroundings of activated platelets, e.g., inside a blood clot[26,27]. While the choice of the LPA concentration did not seem to have any significant effect on the adhesion rate itself, it had an impact on the shape of the RBCs. This relationship is discussed in the next section. After an intracellular Ca^{2+} increase was observed by fluorescence imaging, the setup mode was switched to white light for better HOT operation. Then, the cells were brought into contact and adhered to each other immediately. Because the time for the Ca^{2+} increase varied from cell to cell, which is in agreement with previous investigations[9,28], the time between the initial stimulation of the cell and the final adhesion varied between 10 and 140 s, yielding an average value of 72.75 ± 46.79 s (s.d.). This time range already indicates that under normal physiological conditions, the activation of RBCs is not compatible with an active process contributing to the initiation of a blood clot, but once caught in the fibrin network, the RBCs may actively support clot formation. In addition, to test the necessity of the presence of Ca^{2+} during LPA stimulation, the negative control experiments excluded any interplay between the infrared trapping laser and the adhesion process.

25.4.2 Signaling components

Because LPA is a phospholipid derivative, we examined the extent to which it is directly involved in the adhesion process. Although the concentration used was clearly below the critical micelle concentration (70 µM – 1 mM)[26], which might induce detergent-like effects, LPA is likely to be incorporated into the membrane. From the initial intercellular adhesion after LPA stimulation (◘ Fig. 25.3b), one may propose the following alternative explanations: (i) LPA is directly responsible for the adhesion, (ii) LPA and the Ca^{2+} influx are both necessary to mediate adhesion, and (iii) LPA simply triggers the Ca^{2+} influx, and the Ca^{2+} signaling alone is sufficient to induce adhesion. Option (i) can be excluded immediately because it was tested as a control in the initial set of experiments (◘ Fig. 25.3b). In order to discriminate between options (ii) and (iii), experiments where A23187, a Ca^{2+} ionophore, was added to the RBCs were performed. As shown in ◘ Fig. 25.4, the increased intracellular Ca^{2+} concentration is the dominant signal initiating the adhesion. The Ca^{2+} entry under LPA stimulation is channel-mediated, although the molecular identity of the channel remains unclear[29]. Because LPA is not the major entity in the adhesion process, an alternative molecule or a combination of several entities downstream of the Ca^{2+} signal must control the response. Proteins that are known to be activated in RBCs by an increase in intracellular Ca^{2+} concentration are the Gardos channel, the lipid scramblase, the cysteine protease calpain[30,31] and the Ca^{2+} pump. Although all the proteins are Ca^{2+}-activated, their sensitivity to Ca^{2+} differs. To determine in which order and under what conditions the above mentioned players activate, we refer to the Ca^{2+} concentration with the half maximal effect (EC_{50}). It would therefore be desirable to quantitatively measure the Ca^{2+} concentration in an individual RBC. Unfortunately, this is not possible due to the failure of ratiometric Ca^{2+} sensors in RBCs[28]. Instead, we compare EC_{50} values determined in different studies with the cellular responses we observed in the present study. The smallest EC_{50} for Ca^{2+} is obviously that of the Ca^{2+} pump that keeps the resting Ca^{2+} concentration in RBCs well below 100 nM[32]. With any increase in the intracellular Ca^{2+} concentration, the Ca^{2+} pump will activate. Although the V_{max} of the Ca^{2+} pump was determined in cell populations to be patient- and sample-dependent in a range of 8–20 mmol/(l_{cells} h)[32], the pump activity per cell varies tremendously[33]. This variation explains the broad time range observed between the start of LPA stimulation and the Ca^{2+} increase. While the pump can counterbalance the LPA-induced Ca^{2+} influx for a short time period, during the application of A23187, the amount of Ca^{2+} entering the cell exceeds the V_{max} capacity of the Ca^{2+} pump for all cells. At a Ca^{2+} concentration of 400 nM, the flippase transporting PS actively from the outer membrane leaflet to the inner one is inhibited[34]. Once the Ca^{2+} influx exceeds the transport capacity of the Ca^{2+} pump, the first player that will be activated is the Gardos channel, with an EC_{50} of 4.7 μM[35]. As we can see for the shape transitions upon A23187 stimulation shown in ◻ Fig. 25.4b, cells turn (transiently) into echinocytes as a consequence of KCl loss triggered by K^+ efflux through the Gardos channel. For LPA stimulation, the situation is different: due to the activation of the non-selective cation channel by LPA, a Na^+ influx and, consequently, NaCl uptake counterbalances the KCl loss initiated by the Gardos channel, producing an osmotic equilibrium. For a 2.5 μM LPA stimulation, this equilibrium[36] is reached, whereas for a 10 μM LPA stimulation, the effect of the NaCl uptake overwhelms the KCl loss, resulting in the formation of spherocytes. The next entity to be activated upon Ca^{2+} entry is the scramblase, with an EC_{50} of 29 μM[37]. Scramblase activity was demonstrated by probing for PS in the outer membrane leaflet using annexin V-FITC staining. Staining was

present after both LPA and A23187 stimulation, as depicted in ◻ Fig. 25.4c and d, respectively. Calpain is activated with an EC_{50} of 40 μM, which is very close to the EC_{50} of the scramblase[38]. Under both stimulation conditions (LPA and A23187), we observed vesiculation that has been shown to be associated with the activation of calpain[30], which cleaves spectrin and actin and therefore leads to the breakdown of the cytoskeleton. This is in good agreement with a recent report on exovesiculation by Cueff *et al.*[39]. However, the vesiculation was much more pronounced under A23187 stimulation than under LPA stimulation, as depicted in the representative images in ◻ Fig. 25.4 and ◻ Fig. 25.3, respectively. Therefore, we suggest that the Ca^{2+} concentration after stimulation with 2.5 μM LPA is smaller compared to the A23187 stimulation and might be in the range of EC_{50} of calpain. For A23187 stimulation, a shape change from echinocytes to spherocytes occurs (compare ◻ Fig. 25.4a and b), which is mediated by the encapsulation of microvesicles. However, the occurrence of PS on the outside of the cell makes it a good candidate for initiating the adhesion process. This could be due to simple Ca^{2+}–PS cross-bridging and/or a more complex process involving adhesion proteins. Further evidence for both options comes from aggregation studies of PS vesicles[40,41], where aggregation occurs in solutions of physiological ionic strength containing Ca^{2+} in the mM concentration range. The dependence on high Ca^{2+} concentrations and evidence from further studies reporting enhanced aggregation of PS liposomes in the presence of polymers[42] suggest that additional membrane constituents in the RBC contribute to the aggregation process. This leads to other Ca^{2+}-dependent proteins in RBCs, such as PKC_α[43,44] or the nitric oxide synthase[45,46]. Further research is required to address the question of the molecular identity of the additional components in the adhesion process. Analysing the stepwise unbinding (compare ◻ Fig. 25.5b) when RBCs are separated from each other (corresponding to step 3 in ◻ Fig. 25.5a), we found a Gaussian distribution of the force centered at 71.9 pN (◻ Fig. 25.5e). Such a distribution suggests the formation of tethers and specific bonds between the RBCs that are released one by one during the separation process.

25.4.3 **Relevance to *in vivo* conditions**

The LPA concentration of between 2.5 and 10 µM is a physiologically relevant concentration that is likely to occur locally after platelet activation. Upon stimulation with such an LPA concentration, RBCs adhere irreversibly to each other. The separation force of approximately 100 pN (determined by single cell force spectroscopy) is in a range that is of relevance in the vasculature[47]. As mentioned previously, due to the time course of the Ca^{2+} increase, we regard an initiation of a blood clot based on intercellular RBC adhesion to be irrelevant under physiological conditions. However, once caught in the fibrin network of a blood clot, the adhesion process observed here *in vitro* may support the solidification of the clot. This notion is supported by the aforementioned experimental and clinical investigations reporting a prolongation of bleeding time in subjects with low RBC counts[7-9,48]. Evidence that the adhesion process described in this paper may play a role *in vivo* was recently provided by Chung and coworkers[49]. In this study, an increase in intracellular Ca^{2+} of RBCs associated with a PS exposure was related to prothrombotic activity *in vivo* in a venous thrombosis rat model. Under pathophysiological conditions, intercellular RBC adhesion after Ca^{2+} influx seems to have a more pronounced effect. An example is the vasco-occlusive crisis of sickle cell disease (SCD) patients. Here, the Ca^{2+} influx is mediated by the NMDA-receptor, which has been found to be abundant in RBCs[46]. Our study provides a link between the increased prevalence of the NMDA-receptor in SCD patients[50] and the symptoms of the vasco-occlusive crisis. Further examples where disorders in the ion homeostasis of RBCs are associated with thrombotic events are malaria[51] and thalassemia[52,53]. Therefore, we propose that the Ca^{2+} increase, independent of the entry pathway, followed by PS exposure and RBC aggregation is a general mechanism that may become relevant under pathological conditions.

25.5 **References**

[1] J.L. McWhirter, H. Noguchi and G. Gompper, Flow-induced clustering and alignment of vesicles and red blood cells in microcapillaries. Proc. Natl. Acad. Sci. U.S.A., 106 (2009), pp. 6039–6043.

[2] P. Snabre, G.H. Grossmann and P. Mills, Effects of dextran polydispersity on red blood-cell aggregation. Colloid Polym. Sci., 263 (1985), pp. 478–483.

[3] A. Pribush, D. Zilberman-Kravits and N. Meyerstein, The mechanism of the dextran-induced red blood cell aggregation. Eur. Biophys. J., 36 (2007), pp. 85–94.

[4] E. Evans, Detailed mechanics of membrane-membrane adhesion and separation. 1. Continuum of molecular cross-bridged. 2. Discrete kinetically trapped molecular cross-bridges. Biophys. J., 48 (1985), pp. 175–192.

[5] K.I. Ataga, M.D. Cappellini and E.A. Rachmilewitz, Beta-thalassaemia and sickle cell anaemia as paradigms of hypercoagulability. Br. J. Haematol., 139 (2007), pp. 3–13.

[6] W.W. Duke, The relation of blood platelets to hemorrhagic disease. JAMA, 55 (1910), pp. 1185–1192.

[7] A.J. Hellem, C.F. Borchgrevink and S.B. Ames, The role of red cells in haemostasis: the relation between haematocrit, bleeding time and platelet adhesiveness. Br. J. Haematol., 7 (1961), pp. 42–50.

[8] M. Livio, E. Gotti, D. Marchesi, G. Mecca, G. Remuzzi and G. de Gaetano, Uraemic bleeding: role of anaemia and beneficial effect of red cell transfusions. Lancet, 2 (1982), pp. 1013–1015.

[9] L. Kaestner, W. Tabellion, P. Lipp and I. Bernhardt, Prostaglandin E_2 activates channel-mediated calcium entry in human erythrocytes: an indication for a blood clot formation supporting process. Thromb. Haemost., 92 (2004), pp. 1269–1272.

[10] L. Kaestner, C. Bollensdorff and I. Bernhardt, Non-selective voltage-activated cation channel in the human red blood cell membrane. Biochim. Biophys. Acta, 1417 (1999), pp. 9–15.

[11] L. Kaestner, P. Christophersen, I. Bernhardt and P. Bennekou, The non-selective voltage-activated cation channel in the human red blood cell membrane: reconciliation between two conflicting reports and further characterisation. Bioelectrochemistry, 52 (2000), pp. 117–125.

[12] L. Kaestner and I. Bernhardt, Ion channels in the human red blood cell membrane: their further investigation and physiological relevance. Bioelectrochemistry, 55 (2002), pp. 71–74.

[13] A. Dyrda, U. Cytlak, A. Ciuraszkiewicz, A. Lipinska, A. Cueff, G. Bouyer, S. Egee, P. Bennekou, V.L. Lew and S.L.Y. Thomas, Local membrane deformations activate Ca^{2+}-dependent K^+ and anionic currents in intact human red blood cells. PLoS One, 5 (2010), p. e9447.

[14] T. Oonishi, K. Sakashita, N. Ishioka, N. Suematsu, H. Shio and N. Uyesaka, Production of prostaglandins E_1 and E_2 by adult human red blood cells. Prostaglandins Other Lipid Mediat., 56 (1998), pp. 89–101.

[15] K. Dholakia, G. Spalding and M. MacDonald, Optical tweezers: the next generation. Phys. World, 15 (2002), pp. 31–35.

[16] P.A. Lang, S. Kaiser, S. Myssina, T. Wieder, F. Lang and S.M. Huber, Role of Ca^{2+}-activated K$^+$ channels in human erythrocyte apoptosis. Am. J. Physiol. Cell Physiol., 285 (2003), pp. C1553–C1560.

[17] S.M. Chung, O.N. Bae, K.M. Lim, J.Y. Noh, M.Y. Lee, Y.S. Jung and J.H. Chung, Lysophosphatidic acid induces thrombogenic activity through phosphatidylserine exposure and procoagulant microvesicle generation in human erythrocytes. Arterioscler. Thromb. Vasc. Biol., 27 (2007), pp. 414–421.

[18] J. Friedrichs, J. Helenius and D.J. Muller, Quantifying cellular adhesion to extracellular matrix components by single-cell force spectroscopy. Nat. Protoc., 5 (2010), pp. 1353–1361.

[19] G. Gardos, The function of calcium in the potassium permeability of human erythrocytes. Biochim. Biophys. Acta, 30 (1958), pp. 653–654.

[20] J.F. Hoffman, W. Joiner, K. Nehrke, O. Potapova, K. Foye and A. Wickrema, The hSK4 (KCNN4) isoform is the Ca^{2+}-activated K$^+$ channel (Gardos channel) in human red blood cells. Proc. Natl. Acad. Sci. U.S.A., 100 (2003), pp. 7366–7371.

[21] L.A. Woon, J.W. Holland, E.P. Kable and B.D. Roufogalis, Ca^{2+} sensitivity of phospholipid scrambling in human red cell ghosts. Cell Calcium, 25 (1999), pp. 313–320.

[22] C.W.M. Haest, Distribution and movement of membrane lipids Red Cell Membrane Transport in Health and Disease, Springer Verlag, Berlin (2003), pp. 1–26.

[23] L. Yang, D.A. Andrews and P.S. Low, Lysophosphatidic acid opens a Ca^{2+} channel in human erythrocytes. Blood, 95 (2000), pp. 2420–2425.

[24] R. Dixon, K. Young and N. Brunskill, Lysophosphatidic acid-induced calcium mobilization and proliferation in kidney proximal tubular cells. Am. J. Physiol., 276 (1999), pp. F191–F198.

[25] K. Meerschaert, V. De Corte, Y. De Ville, J. Vandekerckhove and J. Gettemans, Gelsolin and functionally similar actin-binding proteins are regulated by lysophosphatidic acid. EMBO J., 17 (1998), pp. 5923–5932.

[26] T. Eichholtz, K. Jalink, I. Fahrenfort and W.H. Moolenaar, The bioactive phospholipid lysophosphatidic acid is released from activated platelets. Biochem. J., 291 (1993), pp. 677–680.

[27] F. Gaits, O. Fourcade, F. LeBalle, G. Gueguen, A. Gaige, A. GassamaDiagne, J. Fauvel, J. Salles, G. Mauco, M. Simon and H. Chap, Lysophosphatidic acid as a phospholipid mediator: pathways of synthesis. FEBS Lett., 410 (1997), pp. 54–58.

[28] L. Kaestner, W. Tabellion, E. Weiss, I. Bernhardt and P. Lipp, Calcium imaging of individual erythrocytes: problems and approaches. Cell Calcium, 39 (2006), pp. 13–19.

[29] L. Kaestner, Cation channels in erythrocytes – historical and future perspective. Open Biol. J., 4 (2011), pp. 27–34.

[30] C. Berg, I. Engels, A. Rothbart, K. Lauber, A. Renz, S. Schlosser, K. Schulze-Osthoff and S. Wesselborg, Human mature red blood cells express caspase-3 and caspase-8, but are devoid of mitochondrial regulators of apoptosis. Cell Death Differ., 8 (2001), pp. 1197–1206.

[31] H.J. Schatzmann, ATP-dependent Ca^{2+}-extrusion from human red cells. Experientia, 22 (1966), pp. 364–365.

[32] T. Tiffert, R.M. Bookchin and V.L. Lew, Calcium homeostasis in normal and abnormal human red cells Red Cell Membrane Transport in Health and Disease, Springer Verlag, Berlin (2003), pp. 373–405.

[33] V.L. Lew, N. Daw, D. Perdomo, Z. Etzion, R.M. Bookchin and T. Tiffert, Distribution of plasma membrane Ca^{2+} pump activity in normal human red blood cells. Blood, 102 (2003), pp. 4206–4213.

[34] M. Bitbol, P. Fellmann, A. Zachowski and P.F. Devaux, Ion regulation of phosphatidylserine and phosphatidylethanolamine outside inside translocation in human-erythrocytes. Biochim. Biophys. Acta, 904 (1987), pp. 268–282.

[35] T. Leinders, R.G.D.M. Vankleef and H.P.M. Vijverberg, Single Ca(2+)-activated K$^+$ channels in human erythrocytes: Ca^{2+} dependence of opening frequency but not of open lifetimes. Biochim. Biophys. Acta, 1112 (1992), pp. 67–74.

[36] L. Kaestner, A. Juzeniene and J. Moan, Erythrocytes – the »house elves« of photodynamic therapy. Photochem. Photobiol. Sci., 3 (2004), pp. 981–989.

[37] J.G. Stout, Q.S. Zhou, T. Wiedmer and P.J. Sims, Change in conformation of plasma membrane phospholipid scramblase induced by occupancy of its Ca^{2+} building site. Biochemistry, 37 (1998), pp. 14860–14866.

[38] T. Murakami, M. Hatanaka and T. Murachi, The cytosol of human-erythrocytes contains a highly Ca^{2+}-sensitive thiol protease (calpain I) and its specific inhibitor protein (calpastatin). J. Biochem., 90 (1981), pp. 1809–1816.

[39] A. Cueff, R. Seear, A. Dyrda, G. Bouyer, S. Egee, A. Esposito, J. Skepper, T. Tiffert, V.L. Lew and S.L. Thomas, Effects of elevated intracellular calcium on the osmotic fragility of human red blood cells. Cell Calcium, 47 (2010), pp. 29–36.

[40] J. Lansman and D.H. Haynes, Kinetics of a Ca^{2+}-triggered membrane aggregation reaction of phospholipid membranes. Biochim. Biophys. Acta, 394 (1975), pp. 335–347.

[41] S. Ohki, N. Duzgunes and K. Leonards, Phospholipid vesicle aggregation: effect of monovalent and divalent ions. Biochemistry, 21 (1982), pp. 2127–2133.

[42] M. Babincova and E. Machova, Dextran enhances calcium-induced aggregation of phosphatidylserine liposomes: possible implications for exocytosis. Physiol. Res., 48 (1999), pp. 319–321.

[43] R.B. Govekar and S.M. Zingde, Protein kinase C isoforms in human erythrocytes. Ann. Hematol., 80 (2001), pp. 531–534.

[44] B.A. Klarl, P.A. Lang, D.S. Kempe, O.M. Niemoeller, A. Akel, M. Sobiesiak, K. Eisele, M. Podolski, S.M. Huber, T. Wieder and F. Lang, Protein kinase C mediates erythrocyte »programmed cell death« following glucose depletion. Am. J. Physiol. Cell Physiol., 290 (2006), pp. C244–253.

[45] P. Kleinbongard, R. Schulz, M. Muench, T. Rassaf, T. Lauer, A. Goedecke and M. Kelm, Red blood cells express a

functional endothelial nitric oxide synthase. Eur. Heart J., 27 (2006), p. 127.

[46] A. Makhro, J. Wang, J. Vogel, A.A. Boldyrev, M. Gassmann, L. Kaestner and A. Bogdanova, Functional NMDA receptors in rat erythrocytes. Am. J. Physiol. Cell Physiol., 298 (2010).

[47] P. Snabre, M. Bitbol and P. Mills, Cell disaggregation behavior in shear. Biophys. J., 51 (1987), pp. 795–807.

[48] N. Mackman, Triggers, targets and treatments for thrombosis. Nature, 451 (2008), pp. 914–918.

[49] J.Y. Noh, K.M. Lim, O.N. Bae, S.M. Chung, S.W. Lee, K.M. Joo, S.D. Lee and J.H. Chung, Procoagulant and prothrombotic activation of human erythrocytes by phosphatidic acid. Am. J. Physiol. Heart Circ. Physiol., 299 (2010), pp. H347–H355.

[50] A. Bogdanova, A. Makhro, J. Goede, J. Wang, A. Boldyrev, M. Gassmann and L. Kaestner, NMDA receptors in mammalian erythrocytes. Clin. Biochem., 42 (2009), pp. 1858–1859.

[51] V. Luvira, S. Chamnanchanunt, V. Thanachartwet, W. Phumratanaprapin and A. Viriyavejakul, Cerebral venous sinus thrombosis in severe malaria. Southeast Asian J. Trop. Med. Public Health, 40 (2009), pp. 893–897.

[52] A. Eldor and E.A. Rachmilewitz, The hypercoagulable state in thalassemia. Blood, 99 (2002), pp. 36–43.

[53] A.T. Taher, Z.K. Otrock, I. Uthman and M.D. Cappellini, Thalassemia and hypercoagulability. Blood Rev., 22 (2008), pp. 283–292.

Lysophospatidic acid induced red blood cell aggregation in vitro

Lars Kaestner, Patrick Steffen, Duc Bach Nguyen, Jue Wang, Lisa Wagner-Britz, Achim Jung, Christian Wagner, Ingolf Bernhardt

Reprint from J. Bioelectrochem. (2012)[87, 89-95.]

■ Abstract

Under physiological conditions healthy RBCs do not adhere to each other. There are indications that RBCs display an intercellular adhesion under certain (pathophysiological) conditions. Therefore we investigated signalling steps starting with transmembrane calcium transport by means of calcium imaging. We found a lysophosphatidic acid (LPA) concentration dependent calcium influx with an EC_{50} of 5 µM LPA. Downstream signaling was investigated by flow cytometry as well as by video-imaging comparing LPA induced with »pure« calcium mediated phosphatidylserine exposure and concluded the coexistence of two branches of the signaling pathway. Finally we performed force measurements with holographic optical tweezers (HOT): The intercellular adhesion of RBCs (aggregation) exceeds a force of 25 pN. These results support (i) earlier data of a RBC associated component in thrombotic events under certain pathophysiological conditions and (ii) the concept to use RBCs in studies of cellular adhesion behavior, especially in combination with HOT. The latter paves the way to use RBCs as model cells to investigate molecular regulation of cellular adhesion processes.

26.1   Introduction

Adhesion between cells is a vital property that is essential for multi-cell organism. This holds true all the way from primitive cell clusters to mammals that can be regarded as the most complex organisms known. Most organs of the human body form tissues that rely on cell to cell adhesion. The adhesion processes are complex and versatile ranging from direct cell–cell contacts, like occluding junctions, desmosomes or gap junctions, to processes involving extracellular matrix proteins. One of the few organs, where the constitutive cells instead of adhering to each other form a complex liquid, is the blood. The predominant cell type in the blood is the red blood cell (RBC). Although the normal physiological function of RBCs is devoid of intercellular adhesion, there are conceptual reviews proposing the active involvement of RBCs in aggregation processes[1,2].

Already more than 30 years ago, Evan Evans studied intercellular adhesion of RBC on the level of individual cell using micropipettes[3]. Such early studies were based either on hydrodynamic interaction forces or interaction mediated by macromolecules[4] – both of them being reversible. The nature of these effects is based on »physical adhesion« and is therefore different from cross bridging or binding associated »biological adhesion«, although the borderline in-between is blurred.

However, experimental evidence for an involvement of RBC aggregation *in vivo* was published recently[5]. Our own previous work connected intercellular RBC adhesion to an increase in intracellular calcium[6]. Therefore RBC adhesion is both a relevant (patho)physiological process and a model system to study particular aspects of the adhesion process.

Here we investigate several aspects of the RBC aggregation in greater detail ranging from the stimulation of the calcium increase to the cellular signaling cascade.

26.2 Material and methods

26.2.1 RBC preparation and fluorescence microscopy

For experiments fresh blood from healthy donors was obtained by a fingertip needle prick or human venous blood was drawn from healthy donors. Heparin or EDTA was used as an anticoagulant. The obtained blood was used within one day. The cells were washed three times by centrifugation (2000 g, 3–5 min) in a HEPES buffered solution of physiological ionic strength containing the following (in mM): 145 NaCl, 7.5 KCl, 10 glucose and 10 HEPES, pH 7.4, at room temperature. The buffy coat and plasma were removed by aspiration. For Ca^{2+} imaging, RBCs were loaded with 4–5 µM Fluo-4 AM (Molecular Probes, Eugene, USA) from a 1 mM stock solution in dimethyl sulfoxide with 20% Pluronic (F-127, Molecular Probes). Loading was performed in 1 ml of solution of physiological ionic strength at an RBC hematocrit of approximately 1% for 45 min at 37 °C. The cells were washed by centrifugation once more and equilibrated for de-esterification for 15 min. LPA, prepared from a stock solution of 1 mM in distilled water, and the Ca^{2+} ionophore 4-bromo-A23187, prepared from a stock solution of 1 mM in ethanol, were obtained from Sigma-Aldrich (St. Louis, USA). To investigate phosphatidylserine (PS) exposure, cells were stained with annexin V-FITC (Molecular Probes). Annexin V-FITC was delivered in a unit size of 500 µl containing 25 mM HEPES, 140 mM NaCl, 1 mM EDTA, pH 7.4, and 0.1% bovine serum albumin. Five hundred microliters of annexin binding buffer (10 mM HEPES, 140 mM NaCl, 2.5 mM $CaCl_2$, pH 7.4) was added to 1 µl of washed, packed RBCs previously treated with LPA or A23187. Afterwards, 5 µl annexin V-FITC was added, and the cells were mixed gently. The probes were incubated at room temperature in the dark. Finally, the cells were placed on coverslips to perform the microscopy recordings. The measurements were obtained from images taken with CCD cameras, Imago (TILL, Photonics, Gräfelfing, Germany) and CCD97 (Photometrics, Tucson, USA) for Ca^{2+} imaging and annexin V imaging, respectively.

26.2.2 Flow cytometry

Washed RBCs were suspended in a physiological solution containing (mM): NaCl 145, KCl 7.5, glucose 10, HEPES 10, $CaCl_2$ 2 (pH 7.4) at an haematocrit of 0.1%. The Ca^{2+} ionophore 4-bromo-A23187, from a stock solution of 1 mM in absolute ethanol, was added to give a final concentration of 2 µM. The cell suspensions were incubated at 37 °C for different time intervals. The RBCs were washed by quick centrifugation (20 s, 12 000 g) in cold PBS buffer (mM): NaCl 140, KCl 3, Na_2HPO_4 7.5, and KH_2PO_4 1.5 (pH 7.4).

To investigate PS exposure, cells were stained with annexin V-FITC (Molecular Probes). A volume of 500 µl of annexin binding buffer (10 mM HEPES, 140 mM NaCl, 2.5 mM $CaCl_2$, pH 7.4) was added to 1 µl of washed, packed RBCs previously treated with 4-bromo-A23187. 5 µl of annexin V-FITC was then added, and the cells were mixed gently. The probes were incubated at room temperature in the dark for 15 min. Subsequently, the samples were washed once in annexin binding buffer by quick centrifugation (20 s, 12 000 g) to remove unbound annexin V-FITC and resuspended in 500 µl of the same buffer and placed on ice. RBCs showing PS exposed on their outer membrane leaflet were measured in the FL-1 channel using an argon laser at 488 nm of a fluorescence-activated flow cytometer (FACSCalibur, BD Biosciences, Franklin Lakes, New Jersey). The setting of the negative fluorescent gate was obtained in the absence of the Ca^{2+} ionophore 4-bromo-A23187 (negative control). The annexin V-FITC positive RBCs can be calculated in percentage by comparing the number of positive and negative signal events with the control. CellQuest Pro (BD Biosciences) software was used for data acquisition and analysis. The experiment was repeated three times on blood from different donors. For each experiment, 30 000 events were counted. The data are presented as mean values ± SD. For statistical comparison unpaired, two tailed t-test was performed (Prism5, GraphPad Software, La Jolla, USA).

26.2.3 Holographic optical tweezers (HOT)

The set-up of the holographic optical tweezers (HOT) was previously described[6]. In short: A Nd:YAG infrared laser (Ventus, Laser Quantum, Stockport, UK) with a beam width of 2.5 mm was coupled for coarse alignment with a visible He–Ne laser via a dichroic mirror. Both beams were expanded five-fold (BM.X, Linos Photonics, Göttingen, Germany) to overfill the 8 mm back aperture of the microscope objective (CFI Plan Fluor 60× oil immersion, Nikon Corp., Tokyo, Japan). The optics were integrated into an inverted fluorescence microscope (TE-2000, Nikon Corp.). This allowed for combined trapping and fluorescence or differential interference contrast measurements. Images were taken with an electron multiplication CCD camera (Cascade 512F, Roper Scientific, Trenton, USA) with a typical frame rate of 100 Hz. The optical setup was placed on active vibration isolation elements (Vario Series, Halcyonics, Göttingen, Germany). The phase of the laser's electric field was modified using a spatial light modulator (PPM X8267-15, Hamamatsu Photonics, Hamamatsu City, Japan) to create the desired trap pattern in the focal plane of the microscope objective. In order to test for RBC adhesion after stimulation, a suitable configuration of independently movable optical traps was created (cp. ◘ Fig. 26.3A). The laser power in each trap was approximately 5 mW. Cells could be moved against one another by replaying a series of kinoforms on the computer-controlled spatial light modulator.

26.3 Results and discussion

26.3.1 Lysophosphatidic acid (LPA) induced Ca²⁺ entry

Although it is known for more than 10 years that LPA opens a calcium permeable channel in RBCs[7,8],

the molecular identity of the channel is still not resolved[9]. From other cell types it is known that there are in principle five different isoforms of LPA receptors[10]. If the LPA induced Ca^{2+} entry in RBCs is receptor mediated one would expect a clear dose response relationship. The results of concentration dependent experiments are depicted in ◘ Fig. 26.1. The EC_{50} of LPA towards the amplitude of the Ca^{2+} entry is 5 µM, which is compatible with previous studies on cloned LPA receptors[11]. This clear concentration dependent behavior speaks for a receptor mediated signaling and against detergent like effects that should anyway occur only at concentrations above 70 µM[12]. Concerning the present knowledge of LPA receptors, it is fair to assume that their activation translates into the activation of G-proteins[10]. However, the mechanism from G-protein activation to Ca^{2+} entry remains to be elucidated in RBCs.

At a LPA concentration of 10 µM we observe a shape transformation to spherocytes that does not occur at lower LPA concentrations, although there was no difference in the adhesion behavior between 2.5 and 10 µM LPA stimulation[6]. The spherical shape of RBC makes them an ideal object for laser tweezer calibration (see below).

26.3.2 Exposure of phosphatidylserine (PS) to the outer membrane leaflet

Although it is known, that LPA induces PS exposure in RBC, there are conflicting reports about the mechanism. While Chung *et al.*[13] claim it is a totally Ca^{2+}-independent process, we could show that Ca^{2+} alone is sufficient to get a PS exposure in RBCs[6]. Therefore we aimed for a comparison of the temporal course of pure Ca^{2+}-induced and LPA induced PS exposure. The results of this investigation are displayed in ◘ Fig. 26.2. The first difference is that A23187 treated cells could be reliably investigated over a time period of 24 h, while cells treated with LPA beyond 2 h showed such a significant hemolysis that further investigations were impossible. Based on the results presented in ◘ Fig. 26.1 we can assume that the average Ca^{2+} concentration in A23187 treated RBCs is

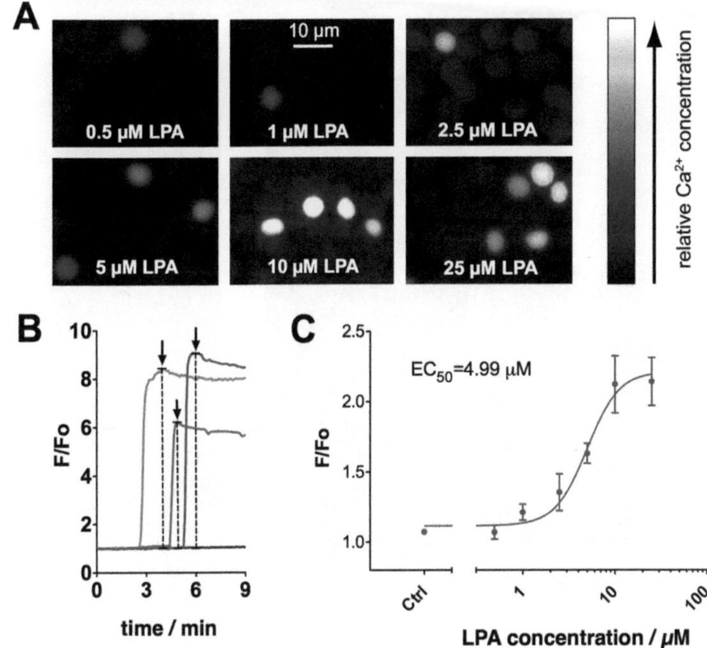

Fig. 26.1 Dose response relationship between lysophosphatidic acid (LPA) concentration and Ca^{2+} influx in RBCs. Panel A shows intensity images of Fluo-4 loaded cells 14 min after stimulation with the given LPA concentration. The time course of LPA stimulation was followed for 15 min. Panel B depicts intensity traces for 3 example RBCs. The amplitude of the Ca^{2+} signal for each cell is highlighted by the arrows and was taken for analysis. To determine the dose–response relationship as illustrated in panel C, the maximum amplitude from at least 158 cells per condition out of three independent experiments were analysed. The self-ratio of the fluorescence F/F_o was plotted against the LPA concentration and the Hill slope of the non-linear fit revealed a mean effective concentration (EC_{50}) of 5 µM.

higher compared to LPA-treated cells. Opposite to that, the image series propose that LPA stimulated cells depict a faster and more intense PS exposure than a »pure« Ca^{2+} increase (compare ◘ Fig. 26.2A and C). This notion is supported by the flow cytometry measurements. Thirty min after stimulation 17.29 ± 1.38% of A23187 treated cells exposed PS in the outer membrane leaflet (◘ Fig. 26.2B), while for 30 min LPA stimulation the responding cells were determined to show a percentage of 35.05 ± 6.66. This is a significant difference (p = 0.01).

These observations suggest the coexistence of two pathways leading to LPA induced PS exposure. One of the pathways is well known involving Ca^{2+}-induced flippase inhibition[14] and scramblase activation[15]. The other one was suggested to involve Ca^{2+}-independent protein kinase C (PKC) isoforms[13]. Since RBCs also contain the Ca^{2+} dependent PKC_α[16], a cross-linking of both signaling pathways seems reasonable. However, further studies need to be performed to refine the signaling pathways and to elucidate their interconnection.

26.3.3 RBC adhesion forces

In a previous study atomic force microscopy (AFM) was used to determine the adhesion force between RBCs after LPA stimulation and was found to be in the range of $F_{AFM} = 100 ± 84$ pN[6]. However control experiments in the same study gave an average separation force of 29 ± 9 pN. This high value is supposed to be an artifact caused by the inherent operation mode of the AFM, which requires an initial pressure to push the cells together. To overcome this potential artifact we aimed on using holographic optical tweezers (HOT) to estimate the adhesion power. Therefore two cells were captured with two independent traps of the HOT and they were brought into contact in an oscillatory manner. Cells that were not treated with LPA did not show any adhesive properties within our experimental force resolution, while LPA treated cells did adhere so strongly to each other that they could not be separated with the HOT. In order to quantify the adhesion forces occurring between adhering RBCs, the HOT needs to be force calibrated. The force F acting on the RBC in the opti-

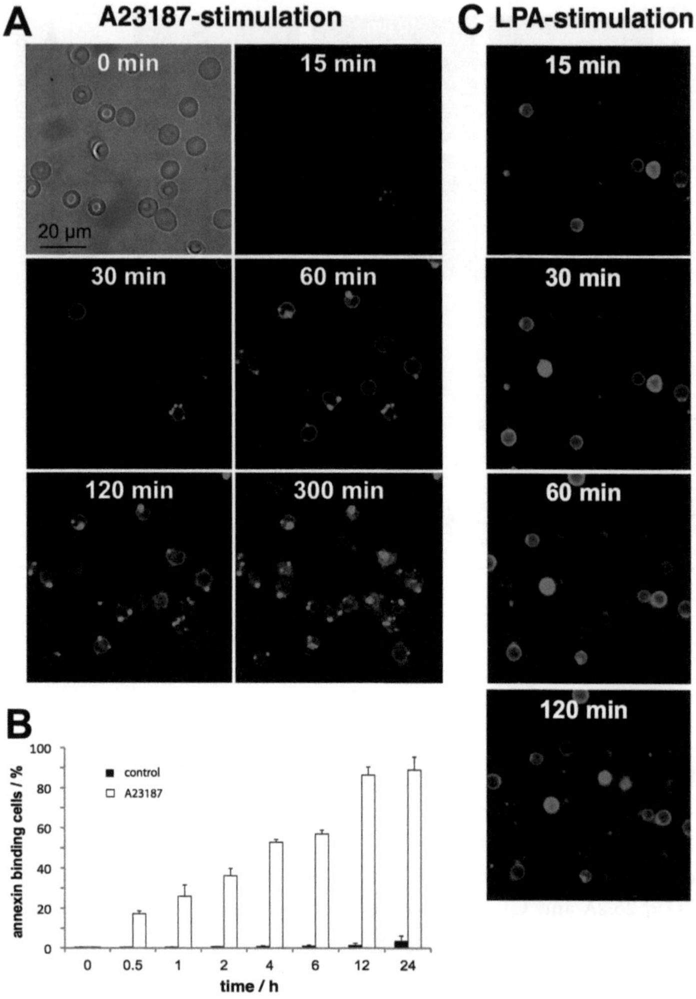

■ **Fig. 26.2** Investigation of phosphatydylserine (PS) exposure to the outer membrane leaflet of RBCs. Panel A depicts an image series of RBCs that ware stimulated with the Ca^{2+}-ionophore bromo-A23187 (2 µM) and simultaneously incubated with FITC labeled annexin V. The first image is a white light image to show the integrity of the cells, while all other images are fluorescence based. The procedure of the PS exposure was followed over 5 h and accompanied with a strong vesicle constriction. Additionally the long term rate (up to 24 h) of PS exposing cells was followed by flow cytometry and is given in panel B. Images of the PS exposure of RBCs stimulated with 2.5 µM LPA are given in panel C. LPA stimulation was restricted to 2 h, since longer stimulations led to significant haemolysis.

cal trap is defined by the gradient intensity. Using a Gaussian shaped laser intensity profile, the force obeys a Hookian spring law:

$$F \propto \kappa x \tag{26.1}$$

where x is the displacement of the particle from the center of the trap. There are many calibration methods commonly used to determine the trap stiffness κ of optical tweezers[17]. Using identical experimental conditions the probe chamber as a whole is brought under oscillatory displacement with amplitude A_0. In that way a hydrodynamic force acts on the blood cells that can be approximated using the Stokes force relation[18]:

$$x(t) = \Re\left(\frac{A_0 v}{\sqrt{v^2 + v_0^2}} e^{-(i2\pi v t + \varphi)} \right) \tag{26.2}$$

with R() denominating the real part and

$$\varphi = -\tan^{\frac{v_0}{v}} \tag{26.3}$$

where

$$v = \frac{\kappa}{2\pi\beta} \tag{26.4}$$

the characteristic frequency of the trap. The latter is generally much higher than the driving fre-

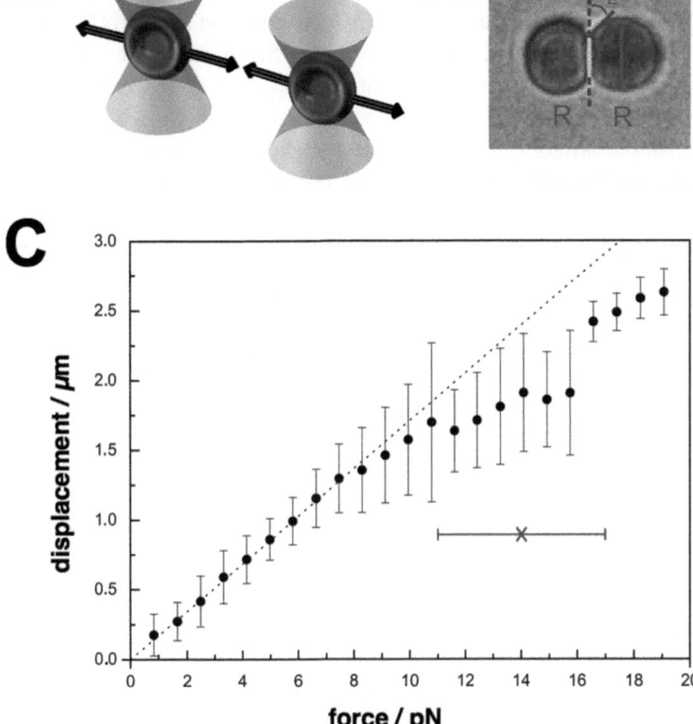

Fig. 26.3 The use of holographic optical tweezers (HOT) to determine the adhesion force. Panel A illustrates the how the holographic optical tweezers were used. The kinoforms on a spatial light modulator generated two foci, which could trap two RBCs. These cells could be independently moved and such brought in contact with each other in an oscillating manner. Panel B shows a picture of two adhering cells (spherocytes), held with two traps. The strength of the traps was weakened so that their influence on the contact angle µ establishment can be neglected. R is given by the diameter of the cell. Panel C depicts the resulting plot of the mean displacement of 12 RBCs. At small displacements the average of all cells is shown, at larger displacements more and more cells did escape from the trap and averaging was only done with the remaining cells. The bar indicates the mean force and standard deviation at escape from the trap. The trap stiffness can be calculated from the initial slope, while in our measurements only the escape force was used.

quency, i.e. the system is over-damped with damping coefficient β leaving Eq. (26.2) to be

$$x(t) \approx \frac{A_0 \nu}{\nu_0} \sin 2\pi\nu\, t \qquad (26.5)$$

Fig. 26.3C shows a plot of the mean displacement of 12 RBCs. Peak hydrodynamic forces have been calculated from the Stokes relation assuming spherical RBCs (see above) with a diameter of 5.6 µm. This corresponds to a sphere with a total volume comparable to that of the native cells (90 µm³). The resulting cell displacements are determined from the center of mass in video microscopy. However, small force displacements are more likely due to membrane deformations and not simple displacements as for the case of rigid beads. This is substantiated by comparing the trap strength $\kappa = 5.85 \pm 0.01$ pN/µm calculated from the initial slope to typical values for RBC elastic moduli[19]. In any case, from these measurements a mean escape force $F_{HOT} = 14 \pm 3$ pN required for cells to leave the trap can be determined. The strength of the observed adhesion of LPA treated cells turned out to be larger than the maximum force applicable using the optical traps. However, we can also conclude that the untreated cells did not show any adhesive force within the measurement resolution of 3 pN. This confirms our original assumption that the adhesion forces for untreated cells, measured with the AFM, are due to instrumental artifacts or due to the strong contact pressure.

We also applied another method to quantify the adhesion forces that is (i) less invasive than the AFM and (ii) based on HOT, but a low trapping strength is sufficient. The method is based on the Young–Dupré relation[20] by taking into account the equilibrium adhesion contact area geometry as shown in **Fig. 26.3B** [21-23]. In the absence of external forces cell–cell adhesion energies W_{adh} are given by the equilibrium angle of contact θ_e and the cell surface tension γ via Young–Dubré relation of capillary forces:

$$W_{adh} = \gamma\left(1 - \cos\theta_e\right) \quad (26.6)$$

in which both cells are considered to be identical in their physical properties[24]. In particular, the surface tension that is given by the shear modulus of the cell must be known *a priori*. Separation forces can be derived from the Derjaguin approximation[25]:

$$F_S = -2\pi W_{adh} R \quad (26.7)$$

where

$$R = \frac{R_1 R_2}{R_1 + R_2} \quad (26.8)$$

is the mean radius on the apex. In these measurements the strength of the traps has been weakened so that their influence on the membrane and contact angle establishment can be neglected, i.e. the zero external force condition is well approximated. The traps serve as weak pinpoints to hold the cells in the focal plane of observation. An equilibrium contact angle of $\theta \approx 45°$ was established where the diameter on apex is identically $D = 6.70 \pm 0.27$ µm for both cells. Values for surface tension may be obtained from Hilbert phase microscopy and are taken from the literature. There is a difference in the values obtained from discocytes compared to spherocytes due to morphological changes in the RBC cytoskeleton[26]. Spherocytes occur to some extend in samples from healthy donors, and for those a value $\gamma = 8.25 \pm 1.6 \times 10^{-6}$ N/m has been reported[27]. The resulting separation force in our study is then $F_{YD} = 25.0 \pm 6.0$ pN. For now, we cannot tell if the literature value for the surface tension holds for the LPA treated spherocytes, but our impression is that they are significantly stiffer than the discocytes and their surface tension and thus the adhesion forces might be even larger. Indeed, an independent measurement of the surface tension would allow for a more accurate determination of the adhesion energy. However, already with a reasonable factor of four for the surface tension one would get adhesion forces with this method in the order of 100 pN which is similar to what we get from the AFM measurements[6].

We can now compare the measured adhesion forces to typical shear stresses in the blood flow. Supposing a RBCs cross section surface of 50 µm² of discocytes of 8 µm we find a maximum sustainable shear strain of $\tau = F_{HOT}/c \approx 2.8 \pm 0.6$ N/m² from the Stokes force calibration and $\tau \approx 5.0 \pm 1.2$ N/m² from the contact angle, respectively. c is a characteristic constant taking into account the random orientation of cell doublets when tumbling in the vascular flow[28]. The values mentioned are above those from measurements on depletion mediated rouleaux formation and mostly above shear strain typically found in veins ($\tau = 1$–6 N/m²) or aortas ($\tau = 0.5$–2 N/m²). Thus the RBC adhesion should resist most shear forces in the vascular system.

26.4 Conclusion

The present study allowed us to refine and extend previous studies about the action of LPA on RBCs[6-8,13], although the molecular mechanism from LPA-receptor activation in RBC to Ca^{2+} entry (most probably by a non-selective ion channel) still needs to be elucidated. Here we are now presenting a hypothesis of a cellular signaling scheme reaching from LPA stimulation to RBC aggregation as depicted in ◘ Fig. 26.4. Although not all steps are completely resolved, we believe this is a general scheme, where the LPA receptor mediated Ca^{2+} entry can be substituted by other Ca^{2+} permeable channels, such as NMDA receptors[29] or prostaglandin E_2 activated pathways[30,31]. This points to a relevance of this pathway in pathophysiological situations like sickle cell anemia, thalassemia or complications after blood transfusion.

Furthermore we show that the RBCs and holographic optical tweezers are complementary to investigate intercellular adhesion. This system can be utilised to explore the molecular mechanism of adhesion processes. Such investigations are very promising, especially for the low force range since HOT is a very sensitive tool — providing a force resolution of approximately 3 pN in the range up to 25 pN. Estimates on the adhesion forces based on a Young Dupré approach were within in the range of the results of the HOT and AFM mea-

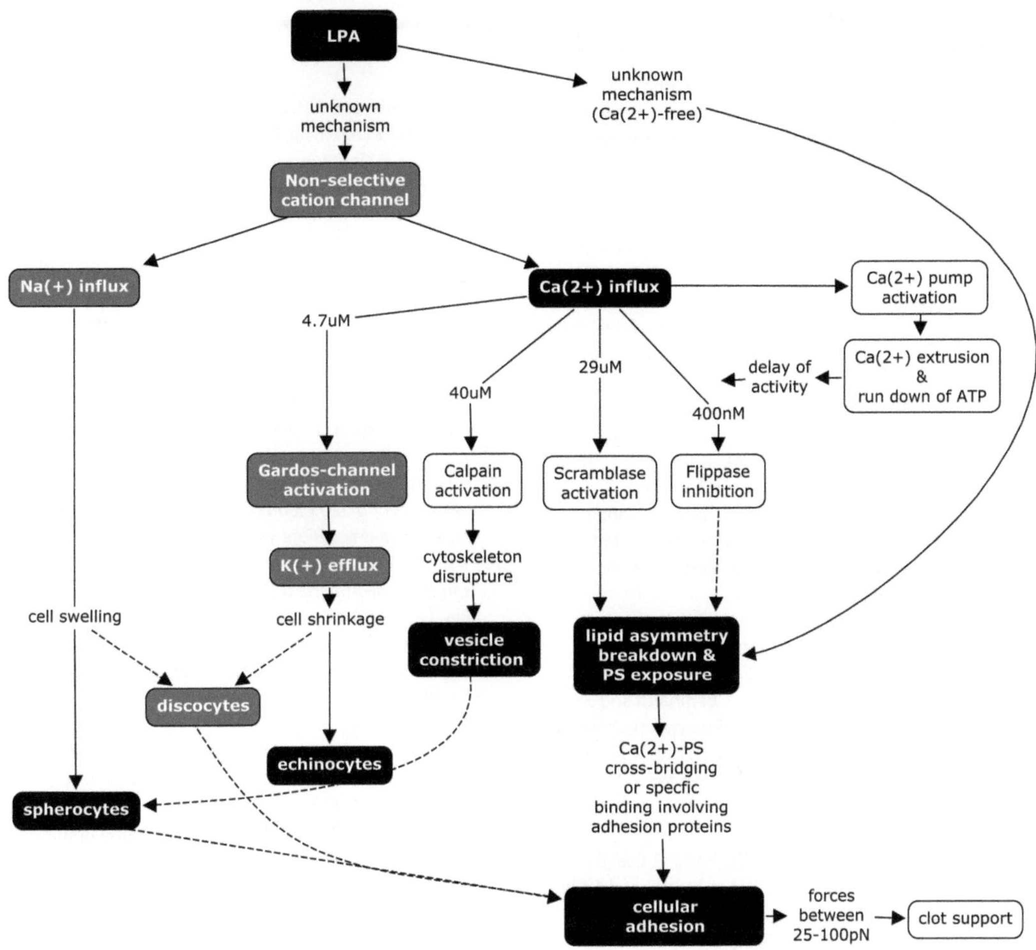

❏ **Fig. 26.4** Hypothesis of the signalling cascade in human red blood cells upon LPA stimulation. The black boxes depict observations and results from the current paper. The gray boxes represent previously published results by the authors, whereas the information in the white boxes is retrieved from scientific literature. LPA activates a non-selective cation channel[7,8]. This channel is permeable to Na[+] as well as to calcium[32,33]. The Ca^{2+} influx is presented in Fig. 26.1. This influx leads to a concentration dependent switching process of the Ca^{2+} pump[34], the flippase inhibition[14], the scramblase activation[15], the calpain activation[35] and the Gardos channel activation[36]. The shape changes RBCs undergo after stimulation with LPA and the consecutive Ca^{2+} influx are well described[2,13,29]. The cell changes are accompanied by the encapsulation of microvesicles, which can be seen in Fig. 26.2. The mechanism behind the shape transformation is a Ca^{2+} influx upon the activation of the non-selective cation channel, what activates the Gardos channel. A K[+] loss-induced shrinkage of the cells leads to the formation of echinocytes. Caused by the Ca^{2+}-activated calpain the RBC cytoskeleton is breaking down. This triggers the microvesicle constriction and the consecutive formation of spherocytes. The lipid asymmetry breakdown and phosphatidylserine exposure is mediated by the inhibition of the flippase and the activation of the scramblase. There is a second Ca^{2+}-independent pathway involved, which is LPA mediated and leads to PS exposure. For a discussion about the relation of these two pathways see main text. Finally the signaling results in adhesion properties occurring between RBCs that were determined to result in forces greater than 25 pN (Fig. 26.3) but can reach values around 100 pN[6].

surements. HOT in combination with microfluidics[6] even allows probing two different cell populations, e.g., differently treated RBCs. Additionally, HOT can be combined with fluorescence microscopy and such a multi parameter read-out can be performed to refine results.

26.5   **References**

[1] D.A. Andrews and P.S. Low, Role of red blood cells in thrombosis. Curr. Opin. Hematol., 6 (1999), pp. 76–82.

[2] L. Kaestner, A. Juzeniene and J. Moan, Erythrocytes-the »house elves« of photodynamic therapy. Photochem. Photobiol. Sci., 3 (2004), pp. 981–989.

[3] E.A. Evans, Minimum energy analysis of membrane deformation applied to pipet aspiration and surface adhesion of red blood cells. Biophys. J., 30 (1980), pp. 265–284.

[4] P. Snabre, G. Grossmann and P. Mills, Effects of dextran polydispersity on red blood-cell aggregation. Colloid Polym. Sci., 263 (1985), pp. 478–483.

[5] J. Noh, K. Lim, O. Bae, S. Chung, S. Lee, K. Joo, S. Lee and J. Chung, Procoagulant and prothrombotic activation of human erythrocytes by phosphatidic acid. Am. J. Physiol. Heart Circ. Physiol., 299 (2010), pp. H347–H355.

[6] P. Steffen, A. Jung, D.B. Nguyen, T. Müller, I. Bernhardt, L. Kaestner and C. Wagner, Stimulation of human red blood cells leads to Ca^{2+}-mediated intercellular adhesion. Cell Calcium, 50 (2011), pp. 54–61.

[7] L. Yang, D.A. Andrews and P.S. Low, Lysophosphatidic acid opens a Ca(++) channel in human erythrocytes. Blood, 95 (2000), pp. 2420–2425.

[8] L. Kaestner, W. Tabellion, E. Weiss, I. Bernhardt and P. Lipp, Calcium imaging of individual erythrocytes: problems and approaches. Cell Calcium, 39 (2006), pp. 13–19.

[9] L. Kaestner, Cation channels in erythrocytes — historical and future perspective. Open Biol. J., 4 (2011), pp. 27–34.

[10] J.W. Choi, D.R. Herr, K. Noguchi, Y.C. Yung, C. Lee, T. Mutoh, M. Lin, S.T. Teo, K.E. Park, A.N. Mosley and J. Chun, LPA receptors: subtypes and biological actions. Annu. Rev. Pharmacol. Toxicol., 50 (2010), pp. 157–186.

[11] K. Bandoh, J. Aoki, A. Taira, M. Tsujimoto and H. Arai, Lysophosphatidic acid (LPA) receptors of the EDG family are differentially activated by LPA species: structure–activity relationship of cloned LPA receptors. FEBS Lett., 478 (2000), pp. 159–165.

[12] T. Eichholtz, K. Jalink and I. Fahrenfort, The bioactive phospholipid lysophosphatidic acid is released from activated platelets. Biochem. J., 291 (1993), pp. 677–680.

[13] S.M. Chung, O.N. Bae, K.M. Lim, J.Y. Noh, M.Y. Lee, Y.S. Jung and J.H. Chung, Lysophosphatidic acid induces thrombogenic activity through phosphatidylserine exposure and procoagulant microvesicle generation in human erythrocytes. Arterioscler. Thromb. Vasc. Biol., 27 (2007), pp. 414–421.

[14] M. Bitbol, P. Fellmann, A. Zachowski and P.F. Devaux, Ion regulation of phosphatidylserine and phosphatidylethanolamine outside–inside translocation in human erythrocytes. Biochim. Biophys. Acta, 904 (1987), pp. 268–282.

[15] L.A. Woon, J.W. Holland, E.P. Kable and B.D. Roufogalis, Ca^{2+} sensitivity of phospholipid scrambling in human red cell ghosts. Cell Calcium, 25 (1999), pp. 313–320.

[16] R.B. Govekar and S.M. Zingde, Protein kinase C isoforms in human erythrocytes. Ann. Hematol., 80 (2001), pp. 531–534.

[17] K.C. Neuman and S.M. Block, Optical trapping. Rev. Sci. Instrum., 75 (2004), pp. 2787–2809.

[18] S. Grover, R. Gauthier and A. Skirtach, Analysis of the behaviour of erythrocytes in an optical trapping system. Opt. Express, 7 (2000), pp. 533–539.

[19] E.A. Evans, Structure and deformation properties of red blood cells: concepts and quantitative methods. Methods Enzymol., 173 (1989), pp. 3–35.

[20] A. Pribush, D. Zilberman-Kravits and N. Meyerstein, The mechanism of the dextran-induced red blood cell aggregation. Eur. Biophys. J., 36 (2007), pp. 85–94.

[21] Z.W. Zhang and B. Neu, Role of macromolecular depletion in red blood cell adhesion. Biophys. J., 97 (2009), pp. 1031–1037.

[22] S. Pierrat, F. Brochard-Wyart and P. Nassoy, Enforced detachment of red blood cells adhering to surfaces: statics and dynamics. Biophys. J., 87 (2004), pp. 2855–2869.

[23] Y. Chu, S. Dufour, J.P. Thiery, E. Perez and F. Pincet, Johnson–Kendall–Roberts theory applied to living cells. Phys. Rev. Lett., 94 (2005), p. 028102.

[24] S. Bailey, S. Chiruvolu and J. Israelachvili, Measurements of forces involved in vesicle adhesion using freeze-fracture electron microscopy. Langmuir, 6 (1990), pp. 1326–1329.

[25] B. Derjaguin and V. Muller, Effect of contact deformations on the adhesion of particles. J. Colloid Interface Sci., 53 (1975), pp. 314–326.

[26] G. Lim H.W., M. Wortis and R. Mukhopadhyay, Stomatocyte–discocyte–echinocyte sequence of the human red blood cell: evidence for the bilayer — couple hypothesis from membrane mechanics. Proc. Nat. Acad. Sci. U.S.A., 99 (2002), pp. 16766–16769.

[27] G. Popescu, T. Ikeda, K. Goda and C. Best-Popescu, Optical measurement of cell membrane tension. Phys. Rev. Lett., 97 (2006), pp. 218101–218104.

[28] P. Snabre, M. Bitbol and P. Mills, Cell disaggregation behavior in shear flow. Biophys. J., 51 (1987), pp. 795–807.

[29] A. Makhro, J. Wang, J. Vogel, A.A. Boldyrev, M. Gassmann, L. Kaestner and A. Bogdanova, Functional NMDA receptors in rat erythrocytes. Am. J. Physiol. Cell Physiol., 298 (2010), pp. C1315–C1325.

[30] L. Kaestner and I. Bernhardt, Ion channels in the human red blood cell membrane: their further investigation and physiological relevance. Bioelectrochemistry, 55 (2002), pp. 71–74.

[31] L. Kaestner, W. Tabellion, P. Lipp and I. Bernhardt, Prostaglandin E_2 activates channel-mediated calcium entry in human erythrocytes: an indication for a blood clot formation supporting process. Thromb. Haemost., 92 (2004), pp. 1269–1272.

[32] L. Kaestner, C. Bollensdorff and I. Bernhardt, Non-selective voltage-activated cation channel in the human red blood cell membrane. Biochim. Biophys. Acta, 1417 (1999), pp. 9–15.

[33] L. Kaestner, P. Christophersen, I. Bernhardt and P. Bennekou, The non-selective voltage-activated cation channel in the human red blood cell membrane: reconciliation between two conflicting reports and further characterisation. Bioelectrochemistry, 52 (2000), pp. 117–125.

[34] H. Schatzmann, Calcium movements across the membrane of human red cells. J. Physiol., 201 (1969), pp. 369–395.

[35] C.P. Berg, I.H. Engels, A. Rothbart, K. Lauber, A. Renz, S.F. Schlosser, K. Schulze-Osthoff and S. Wesselborg, Human mature red blood cells express caspase-3 and caspase-8, but are devoid of mitochondrial regulators of apoptosis. Cell Death Differ., 8 (2001), pp. 1197–1206.

[36] G. Gardos, The permeability of human erythrocytes to potassium. Acta Physiol. Hung., 10 (1956), pp. 185–189.

Regulation of phosphatidylserine exposure in red blood cells

Duc Bach Nguyen, Lisa Wagner-Britz, Sara Maia, Patrick Steffen, Christian Wagner, Lars Kaestner, Ingolf Bernhardt

Reprint from Cellular Physiology and Biochemistry (2011) **28**: 847-856.

■ Abstract

The exposure of phosphatidylserine (PS) on the outer membrane leaflet of red blood cells (RBCs) serves as a signal for eryptosis, a mechanism for the RBC clearance from blood circulation. The process of PS exposure was investigated as function of the intracellular Ca^{2+} content and the activation of PKCα in human and sheep RBCs. Cells were treated with lysophosphatidic acid (LPA), 4-bromo-A23187, or phorbol-12 myristate-13 acetate (PMA) and analysed by flow cytometry, single cell fluorescence video imaging, or confocal microscopy. For human RBCs, no clear correlation existed between the number of cells with an elevated Ca^{2+} content and PS exposure. Results are explained by three different mechanisms responsible for the PS exposure in human RBCs: (i) Ca^{2+}-stimulated scramblase activation (and flippase inhibition) by LPA, 4-bromo-A23187, and PMA; (ii) PKC activation by LPA and PMA; and (iii) enhanced lipid flop caused by LPA. In sheep RBCs, only the latter mechanism occurs suggesting absence of scramblase activity.

27.1   Introduction

Phospholipids are asymmetrically distributed in the plasma membrane of most, if not all, biological cells. Sphingomyelin (SM) and phosphatidylcholine (PC) are found predominantly in the outer leaflet of the membrane bilayer, while phosphatidylserine (PS) and phosphatidylethanolamine (PE) are located mostly in the inner leaflet[1]. The distribution of the membrane phospholipids is regulated by three proteins: flippase[2], floppase, and scramblase[3-5]. PS exposure on the outer leaflet of the membrane has been described as a marker for apoptosis in nucleated cells[5]. Although the apoptosis of RBCs is still under discussion, they undergo suicidal death with signs of apoptosis such as PS exposure, membrane blebbing and vesicle formation[6]. This process was termed eryptosis by Lang *et al.*[7].

Based on a correlation between decreased haematocrit and longer bleeding times[8] and experiments of Andrews and Low[9], an active role of RBCs in thrombus formation has been proposed[9]. Kaestner *et al.* published a more detailed signalling cascade based on Ca^{2+} uptake via a non-selective cation channel that could be activated by prostaglandin E_2 (PGE_2)[10,11]. This channel was believed to be a voltage-activated non-selective cation channel[12,13]. Recent considerations[14] propose the PGE_2 receptor-activated non-selective cation channel[10,11] and the voltage activated cation channel[12,13] are of different identity. Ca^{2+} entry can also be caused by a mechanical deformation of RBCs as described by Dyrda *et al.*[15].

PGE_2 and LPA are local mediators released by platelets after their activation within the coagulation cascade and, in case of PGE_2, are released by RBCs under mechanical stress[16]. Moreover, in recent reports, we showed that PS exposure induced by LPA was associated with cell-cell adhesion of human RBCs[17,18]. PS exposure at the outer leaflet of the RBC membrane is of importance for the adhesion of RBCs to endothelium in some diseases such as sickle cell anaemia, malaria, and diabetes[19].

Although it is known that in RBCs LPA induces PS exposure at the outer membrane leaflet, there are conflicting reports about its mechanism. While Chung *et al.*[20] claim that it is a totally Ca^{2+}-independent process, we showed that Ca^{2+} alone is sufficient to induce PS exposure in human RBCs[17]. Woon *et al.* found that an increase of the intracellular Ca^{2+} level in RBCs results in the exposure of PS to the outer membrane leaflet due to activation of the scramblase and inhibition of the flippase[21]. Protein kinase Cα (PKCα) has been also described to be involved in the PS exposure on RBCs[20,22,23]. In human RBCs, in addition to PKCα, only a limited number of PKC isoforms (ζ, λ/ι, and μ) have been detected[24]. The molecular mechanisms for the participation of these proteins in the Ca^{2+} uptake and/ or in the PS exposure remain to be elucidated.

Here, we investigated the parameters regulating PS exposure at the outer membrane leaflet of human RBCs by addressing Ca^{2+} content manipulated by an ionophore and PKC activation by phorbol-12 myristate-13 acetate (PMA). For comparison, sheep RBCs were studied because these cells have a completely different membrane phospholipid distribution. Like in human RBCs, PS and PE are present at the inner membrane leaflet, however, sheep RBC membranes lack PC and the outer layer consists exclusively of SM[25,26].

27.2 Materials and Methods

27.2.1 Blood and solutions

Human venous blood from healthy donors was obtained from the Institute of Clinical Haematology and Transfusion Medicine of Saarland University Hospital. Sheep blood samples were obtained from the sheep farm Ernst in Blieskastel, Germany. Citrate or EDTA (for human blood) or heparin (for sheep blood) were used as anticoagulants. Freshly drawn blood samples were stored at 4 °C and used within one day.

Blood was centrifuged at 2,000 g for 5 min at room temperature (RT) and the plasma and buffy coat was removed by aspiration. Subsequently, RBCs were washed 3 times in HEPES-buffered physiological solution (HPS) containing (mM): NaCl 145, KCl 7.5, glucose 10, HEPES 10, pH 7.4, under the same conditions. Finally, RBCs were re-suspended in HPS and kept at 4 °C. To assess the effect of cell volume, the last wash was done in a high-K^+ solution (mM): NaCl 2.5, KCl 150, glucose 10, HEPES 10, pH 7.4.

27.2.2 Intracellular Ca^{2+} content

The application of fluo-4 for Ca^{2+} measurements in intact RBCs has been reported elsewhere[27]. It has been also shown that fura-2 or indo-1 cannot be applied for those measurements due to quenching by haemoglobin[27].

RBCs were loaded with 5 μM fluo-4 AM from a 1 mM stock solution in dimethyl sulfoxide (DMSO) with 20% Pluronic F-127 in 1 ml HPS at a haematocrit of 1% for 45 min at 37 °C in the dark. Cells were washed 3 times by centrifugation (20 s, 12,000 g) in HPS and finally equilibrated for de-esterification for 15 min at a haematocrit of 0.5% at RT.

Intracellular free Ca^{2+} levels of single RBCs were monitored using an inverted fluorescence microscope (Eclipse TE2000-E, Nikon, Tokyo, Japan). A diluted RBC suspension (approximately 0.025% haematocrit), prepared in HPS with 2 mM $CaCl_2$, was placed on a cover slip coated with poly-L-lysine 5 min before the start of the imaging procedure. Ca^{2+} ionophore 4-bromo-A23187 (the non-fluorescent variant of A23187, thereafter named A23187), LPA, and PMA were added immediately before the start of the imaging procedure. The experiments were performed at RT in a dark room. Control conditions consisted of the same procedure but cells were re-suspended in the presence of respective amounts of reagent solvents (ethanol, water, or DMSO).

Our experimental approach was designed to provide a quantitative assessment of fluorescence intensity while maintaining identical imaging parameters under all conditions. This was accomplished by taking images with an electron multiplication CCD camera (CCD97, Photometrics, Tucson, USA) using a 100x 1.4 (NA) oil immersion lens with infinity corrected optics. An image was taken every 20 s (exposure time 500 ms) for a 30 min experiment using the imaging software Metavue (Universal Imaging Corp., Marlow, UK).

Fluo-4 was excited with a xenon lamp-based monochromator (Visitron Systems, Puchheim, Germany) at a centre wavelength of 488 nm. Emission was recorded at 520/15 nm. The background corrected fluorescence signals were normalised to the initial fluorescence (F_o). For each experiment 20 - 30 cells were analysed from each sample and at least three different blood samples were used.

Additionally, the intracellular free Ca^{2+} content of RBCs was measured by flow cytometry (FACS Calibur and Cell Quest Pro software, Becton Dickinson Biosciences, Franklin Lakes, USA). Preparation of cells, solutions and experimental conditions were the same as described above (except that samples were analysed after 12 min). Ca^{2+} content was measured in the FL-1 channel and analysed using the mean value of the relative fluorescence of 30,000 cells from each blood sample. For each experiment at least three different blood samples were used.

27.2.3 PS exposure

PS exposure on the outer membrane leaflet of RBCs was evaluated by annexinV-FITC binding to this phospholipid. RBCs prepared as described above were treated with A23187, LPA, or PMA or with ethanol, water, or DMSO (for controls) in HPS with 2 mM $CaCl_2$ at 37 °C for 30 min (0.1% haematocrit). Cells were washed in the ice-cold HPS with 2 mM $CaCl_2$ by centrifugation (20 s, 12,000 g). Approximately 10^6 cells were incubated with 5 µl of annexin V-FITC in 500 µl of annexin binding buffer containing (mM): NaCl 145, HEPES 10, $CaCl_2$ 2.5, pH 7.4, for 15 min at RT. Annexin V-FITC was delivered in unit size of 500 µl containing 25 mM HEPES, 140 mM NaCl, 1 mM EDTA, pH 7.4, and 0.1% bovine serum albumin. After incubation, samples were kept on ice and analysed immediately by flow cytometry (FACS Calibur and Cell Quest Pro software). PS exposure was measured in the FL-1 channel excited by an argon laser at 488 nm. Control conditions for the negative fluorescent gate were obtained in the absence of A23187, LPA, or PMA. Annexin V positive RBCs were determined as percentage by relating positive and negative signal events to the control condition. For each experiment at least three different blood samples were used and 30,000 events for each sample were counted.

The kinetics of PS exposure was recorded using a fluorescence microscope (the same as described above for Ca^{2+} measurements). A diluted RBC suspension (approximately 0.025% haematocrit) was prepared in the HPS with 2 mM $CaCl_2$ and 5 µl annexin V-FITC and transferred to a cover slip (coated with poly-L-lysine) 5 min before the start of the imaging procedure. A23187, LPA, and PMA (or the solvents ethanol, water, or DMSO, for controls) were added immediately before the start of the imaging procedure. Experiments were performed at RT in a dark room. An image was taken every 30 s (exposure time 500 ms).

27.2.4 Double labelling experiments

To evaluate the relationship between the intracellular Ca^{2+} content and the PS exposure, RBCs were labelled with both fluo-4 AM and annexin V-alexa 568. Briefly, cells loaded with fluo-4 AM as described before were induced for PS exposure using A23187, LPA, or PMA. After washing in the HPS at RT, cells were incubated with annexin V-alexa 568 for 15 min at RT. The double labelling experiments were carried out using the confocal laser scanning fluorescence microscope LSM 510 META (Carl Zeiss AG, Jena, Germany). Samples were scanned with an argon laser (488 nm) for fluo-4 and HeNe laser (543 nm) for annexin V-alexa 568.

27.2.5 Reagents

All chemicals used (except fluorescent dyes and Pluronic F-127) were purchased from Sigma-Aldrich (Munich, Germany). Fluo-4 AM, annexin V-FITC, and Pluronic F-127 were obtained from Molecular Probes (Eugene, USA) and annexin V-alexa from Roche Diagnostics GmbH (Mannheim, Germany). Stock solutions for A23187 (1 mM), LPA (1 mM), and PMA (10 mM) were dissolved in ethanol, water, and DMSO, respectively.

27.2.6 **Statistical significance**

Data are presented as mean values ± SD of at least 3 independent experiments. The significance of differences was tested by Student's t-test. Statistical significance of the data was defined as follows: $p > 0.05$ (n.s.), $p = 0.05$ (*); $p = 0.01$ (**), $p = 0.001$ (***).

27.3 **Results**

An increased intracellular Ca^{2+} content of RBCs results in the activation of several processes, important for PS exposure: (i) activation of the scramblase and inhibition of flippase[4], (ii) activation of the Ca^{2+}-activated K^+ channel leading to loss of KCl and water, causing cell shrinkage[28-30], (iii) activation of PKCα[22], and (iv) activation of calpain resulting in cytoskeleton destruction, membrane blebbing, and micro-vesiculation[31,32]. In addition, elevated intracellular Ca^{2+} activates the Ca^{2+} pump.

In order to correlate between increased intracellular Ca^{2+} content and PS exposure of human RBCs, we focussed on LPA treatment shown to increase the intracellular Ca^{2+} content[17,18,20,22,27]. For comparison, we used the Ca^{2+} ionophore A23187. In addition, we treated the RBCs with PMA, a known activator of the Ca^{2+}-dependent conventional PKC (cPKC) isoforms (α, βI, βII, γ) and the Ca^{2+}-independent novel PKC (nPKC) isoforms (δ, ε, η, θ) but not the Ca^{2+}-independent, non phorbol ester-binding atypical PKC (aPKC) isoforms (ζ, λ/ι, μ)[33,34]. As mentioned before, PKCα[22] as well as PKCζ[24], and PKCλ/ι[24] are present in human RBCs and may play a role in PS exposure. So far no nPKC isoforms were detected in RBCs.

◘ Fig. 27.1 shows the kinetics of Ca^{2+} uptake after treating RBCs with LPA, A23187, or PMA measured by fluorescence video imaging. Addition of 2.5 μM LPA immediately led to an increase of the intracellular Ca^{2+} content that remains relatively constant up to 30 min. At higher LPA concentrations (3.5 - 10 μM) the rate of haemolysis continuously increased (data not shown). Treatment with 2 μM A23187 significantly increased

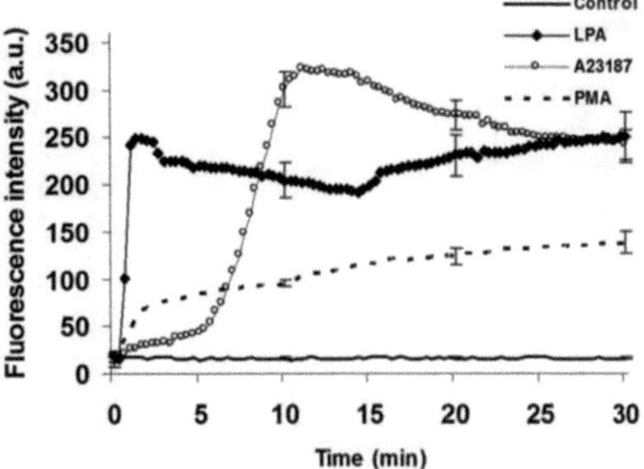

◘ **Fig. 27.1** Fluo-4 AM fluorescence intensity (arbitrary units) of human RBC for Ca^{2+} content after treatment with 2.5 μM LPA, 2 μM A23187 or 6 μM PMA in the presence of 2 mM extracellular Ca^{2+} measured with single cell fluorescence microscopy over a time interval of 30 min. Curves show the mean values of at least 60 RBCs from 3 different blood samples. Error bars represent 25% of SD.

the Ca^{2+} content but only after a delay of about 5 min. The final fluorescence intensity, i.e. the intracellular Ca^{2+} content, was the same under both conditions after 30 min. At higher concentrations (up to 10 μM) the lag phase was reduced without affecting the maximal fluorescence intensity (data not shown). Similar to the treatment with LPA, exposure to 6 μM PMA, a recommended concentration to activate cPKCs and nPKCs[22], increased the Ca^{2+} content without a lagphase, but failed to reach comparably high fluorescence intensities.

Comparable experiments were carried out using flow cytometry analysis and results are represented in ◘ Fig. 27.2. ◘ Fig. 27.2A depicts the percentage of cells with an elevated Ca^{2+} content after LPA, A23187, or PMA treatment in comparison to control conditions after 12 min when the fluorescence intensity after A23187 treatment had reached maximal values (see ◘ Fig. 27.1). ◘ Fig. 27.2B shows the corresponding fluorescence intensities of the reacting cells, i.e. a significantly enhanced Ca^{2+} content after stimulation by the three substances. ◘ Fig. 27.2A shows that in 99.2% of the cells there was a significant increase of Ca^{2+} content in response to A23187. In the presence of LPA about 80% of the cells raised their Ca^{2+} content, comparable to the Ca^{2+} ionophore treatment. With PMA

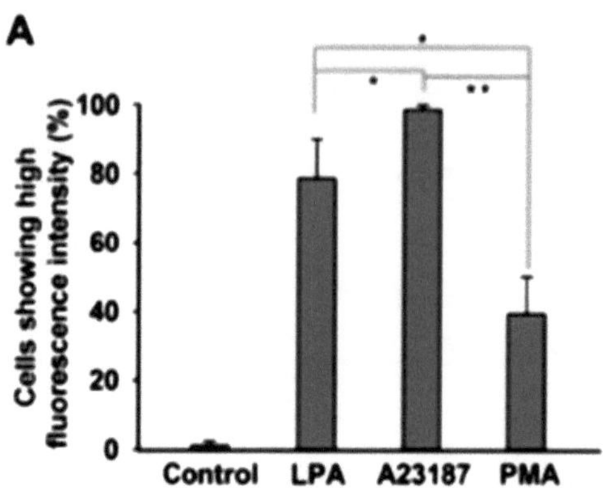

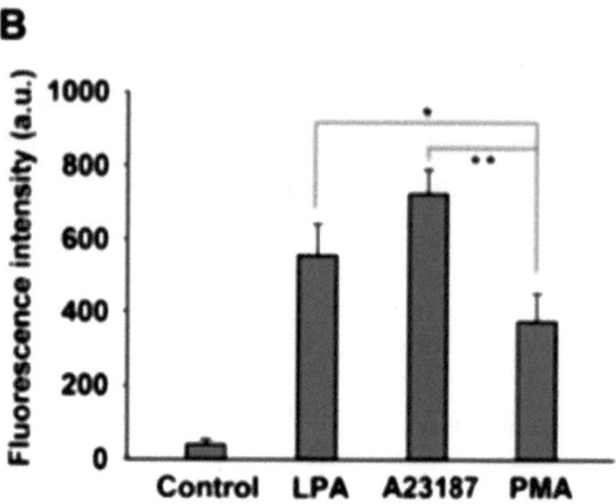

◘ **Fig. 27.2** Flow cytometric analysis of human RBC Ca2+ content after treatment with 2.5 µM LPA, 2 µM A23187 or 6 µM PMA in the presence of 2 mM extracellular Ca^{2+} after 12 min of incubation. Mean values of 3 different blood samples (with 30,000 cells per sample) are shown. Error bars represent SD, p = 0.05 (*); p = 0.01 (**), p = 0.001 (***). Control measurements were done in the presence of solvents (ethanol, water, or DMSO, see Material and Methods). All data measured in the presence of the reagents are significantly different from control, p = 0.001. A: Percentage of RBCs showing a significant higher Ca^{2+} content compared to control after treatment with LPA, A23187 or PMA. B: Increase of Ca^{2+} content of RBCs in response to LPA, A23187 or PMA treatment shown as arbitrary units (a.u.) of fluorescence intensity analysed in cells with higher Ca^{2+} content.

only about 40% of the cells increased their Ca^{2+} content reaching only half the fluorescence intensity as compared to A23187 or LPA (◘ Fig. 27.2B). With 10 µM PMA, the number of cells reacting by exposing PS increased significantly, reaching a value of about 70%, although the Ca^{2+} content was not enhanced under these conditions (not shown in ◘ Fig. 27.2, but cf. with fluorescence microscopy results presented in ◘ Fig. 27.1). It should be mentioned that on average only 1.3% of control cells had about 5-fold higher fluo-4 fluorescence intensity compared to the nonreacting control cells (for non-reacting cells data not shown).

The PS exposure was studied in LPA- and A23187-treated cells over a time period of 120 min using fluorescence microscopy (◘ Fig. 27.3). At times longer than 120 min in LPA treated RBCs haemolysis occurred. In contrast, no significant haemolysis was seen with A23187-treated cells at longer incubation times (up to 24 h). PMA treatment also did not result in a significant haemolysis after long time incubation (up to 6 h, data not shown). The images of treatments with LPA and A23187 (◘ Fig. 27.3A) as well as PMA (not shown) reveal that vesiculation occurred. The kinetic of PS exposure of human RBCs treated with LPA

or A23187 is presented in ◘ Fig. 27.3B. It can be seen that the process of PS exposure is faster and more pronounced in case of LPA compared to A23187. Although the kinetics is a convolution of the annexin V binding to PS and the PS exposure rate, we can deduct from the comparison of ◘ Fig. 27.1 and ◘ Fig. 27.3B that the PS exposure is a slower process than the Ca^{2+} entry.

PS exposure was additionally measured by flow cytometry 30 min after incubation of RBCs with LPA, A23187, or PMA and results are presented in ◘ Fig. 27.4. The number of annexin V-positive cells after LPA treatment in the presence of Ca^{2+} was about 35% and significantly higher as compared to the A23187 treatment (20%). Also, PMA induced the highest PS exposure in human RBCs. It should be stated that these measurements using fluorescence microscopy (◘ Fig. 27.3B) represent integrated values of the fluorescence intensity whereas the data obtained from the flow cytometry experiments depict the numbers of PS exposing cells (in %).

Based on the data obtained from flow cytometry experiments (quadrant analysis), we estimated the number of vesicles detached from the RBC membrane after 30 min. In case of LPA, A23187,

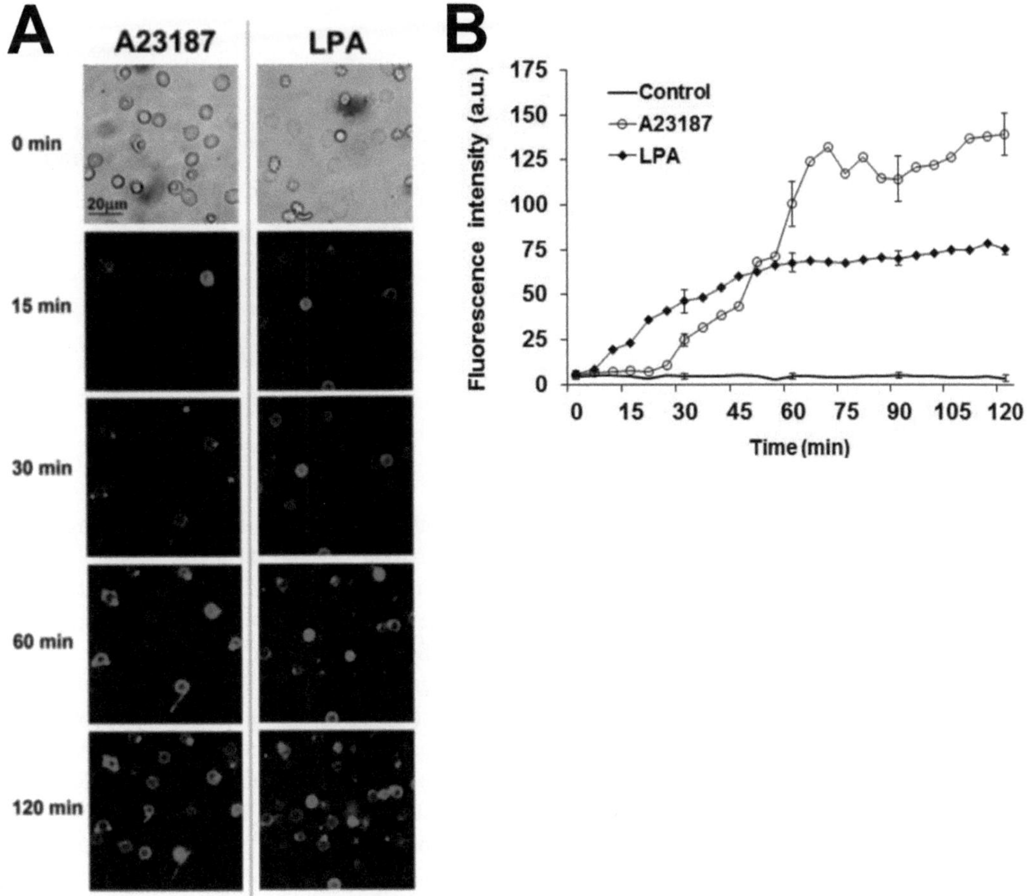

Fig. 27.3 PS exposure evaluation of human RBCs treated with 2 µM A23187 or 2.5 µM LPA in the presence of 2 mM extracellular Ca^{2+} up to 120 min and labelled with annexin V-FITC using fluorescence microscopy. A: Representative images out of 3 independent experiments. Note that blood samples from different donors are presented. B: Kinetics of fluorescence intensity shown as arbitrary units (a.u.); curves show the mean values of at least 30 RBCs from 3 different blood samples. Error bars represent SD.

or PMA treatment the number of vesicles per 30,000 RBCs was estimated as 10,698 ± 2,363 (n = 3), 6,626 ± 2,898 (n = 4), and 15,509 ± 6,382 (n = 4), respectively.

Experiments were also carried out after treating human RBCs with LPA, A23187, or PMA in the absence of extracellular Ca^{2+} (presence of 1 mM EGTA). In case of LPA and A23187, there were no significant differences in the PS exposure between treatment and control conditions (Fig. 27.4, (-) Ca^{2+}). Only in the case of PMA treatment, about 50% of RBCs showed PS exposure (compared to about 80% in the presence of Ca^{2+}, see Fig. 27.4, (+) Ca^{2+}). The data obtained for control conditions in the presence of Ca^{2+} shown in Fig. 27.4 were not significantly different from the control conditions in the absence of Ca^{2+} (data not shown).

Comparing the results of Fig. 27.2 and Fig. 27.4 it is evident that there is no correlation between the number of cells showing an elevated Ca^{2+} content and increased PS exposure. It should be noted that the threshold for the forward scatter in the flow cytometry measurements was set at a value so that only events based on the size

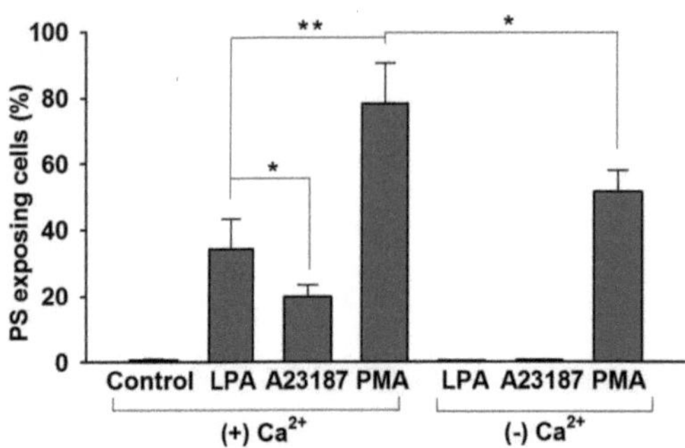

Fig. 27.4 Flow cytometry analysis of human RBCs showing a significant higher percentage PS exposure compared to control after a 30 min treatment with 2.5 μM LPA, 2 μM A23187 or 6 μM PMA in the presence of 2 mM extracellular Ca^{2+} [(+) Ca^{2+}]. For comparison cells were treated in the absence of extracellular Ca^{2+} (presence of 1 mM EGTA) [(-) Ca^{2+}]. Bars show mean value of 3 different blood samples (30,000 cells of each blood sample were analysed). Error bars represent SD, p ≤ 0.05 (*); p ≤ 0.01 (**). Control measurements were done in the presence of solvents (ethanol, water, or DMSO) and extracellular Ca^{2+} (see Material and Methods). LPA or A23187 treatments in the absence of extracellular Ca^{2+} were not significantly different from control. All data measured in the presence of reagents were significantly different from control, p ≤ 0.001.

Tab. 27.1 FACS analysis of RBCs showing a significant higher PS exposure compared to control conditions (in %) after treatment with 2.5 μM LPA, 2 μM A23187 or 6 μM PMA in the presence of 2 mM extracellular Ca^{2+} in HPS (7.5 mM KCl) or in a high K$^+$ solution (150 mM KCl). Mean values ± SD of 3 different blood samples (30,000 cells of each blood sample were analysed).

	7.5 mM KCl [%]	150 mM KCl [%]
LPA	36.3 ± 7.3	23.1 ± 2.9
4-bromo-A23187	27.7 ± 7.7	10.8 ± 5.5
PMA	78.5 ± 6.0	76.9 ± 8.8

of intact cells were counted, since micro-vesicle formation occurred after treatment with all 3 substances (**Fig. 27.3**).

A lack of correlation between Ca^{2+} content and PS exposure was further supported by confocal fluorescence microscopy with double labelling experiments (for both, intracellular Ca^{2+} and PS in the outer membrane leaflet). The upper panels of **Fig. 27.5** show the effect of LPA treatment. Most of the cells exhibited both, an elevated Ca^{2+} content and external PS exposure (**Fig. 27.5**, upper panels, white arrows). However, some cells displayed an increased Ca^{2+} level, but no PS exposure (**Fig. 27.5**, upper panels, green arrows), whereas other cells externalise PS without a higher Ca^{2+} content (**Fig. 27.5**, upper panels, red arrows). The same observation was found for A23187-treated cells (middle panels of **Fig. 27.5**) although it should be noted that PS exposure without high Ca^{2+} content is relatively rare in both cases. The lower panels of **Fig. 27.5** show that after PMA treatment PS exposure was more prevalent at low

Ca^{2+} content in agreement with the findings presented in **Fig. 27.2** and **Fig. 27.4**.

Experiments after PMA treatment, but in the complete absence of external Ca^{2+} (presence of 1 mM EGTA), were carried out using confocal fluorescence microscopy and double labelling of the cells (for Ca^{2+} and PS). In agreement with the flow cytometry measurements, many cells revealed PS exposure but none increased Ca^{2+} (data not shown). In addition, the presence of 1 mM o-vanadate to inhibit the Ca^{2+} pump failed to produce further significant changes (data not shown).

To consider a possible effect of cell volume alteration on PS exposure, experiments were carried out with the HPS replaced by a solution containing 150 mM KCl plus 2.5 mM NaCl instead of 145 mM NaCl plus 7.5 mM KCl. Under these conditions no net K$^+$ efflux occurs through the Ca^{2+}-dependent K$^+$ channel and hence no KCl loss and accompanying cell shrinkage. **Tab. 27.1** reveals that a substantial part of the PS exposure in the presence of LPA or A23187 was due to volume decrease which was confirmed by analysing the forward scatter (FSC) by flow cytometry recordings (data not shown). Note that the PS exposure data in HPS (7.5 mM KCl) presented in **Tab. 27.1** were not identical to the data shown in **Fig. 27.4** since different blood samples were analysed for both investigations.

In comparison, in sheep RBCs (**Fig. 27.6**) LPA (upper panels) also increased the Ca^{2+} content. However, in comparison to human RBCs only very

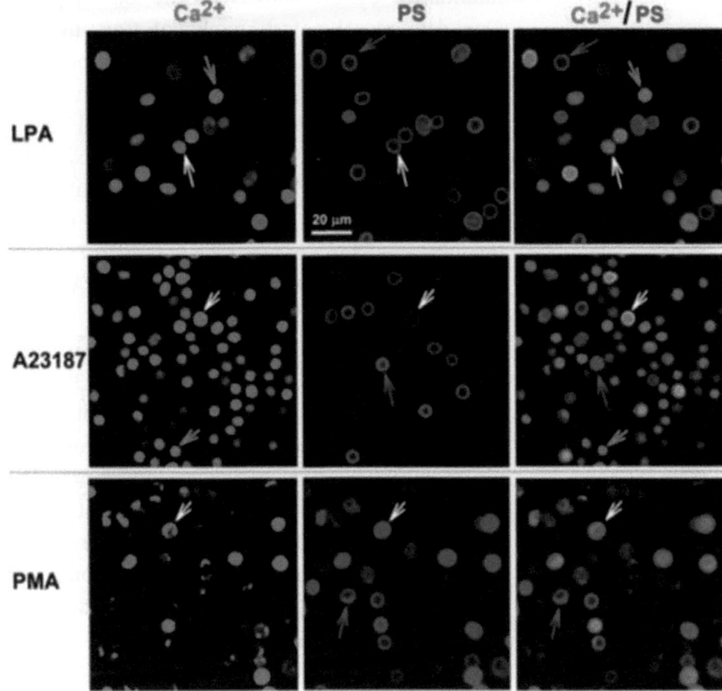

Fig. 27.5 Human RBCs double labelled with fluo-4 and annexin V-alexa 568 (Ca²⁺/PS double scan, green and red, right column) after stimulated to increase Ca²⁺ content (green, left column) and PS exposure (red, middle column), respectively. Upper panels - 2.5 µM LPA, middle panels - 2 µM 23187, lower panels - 6 µM PMA. White arrows indicate cells with both, elevated Ca²⁺ content and an increased PS exposure, green arrows cells with increased Ca²⁺ content only, and red arrows cells with externalised PS only. These are representative images out of 3 independent experiments. (For interpretation of the references to color in this figure legend, the reader is referred to the web version of the original article.)

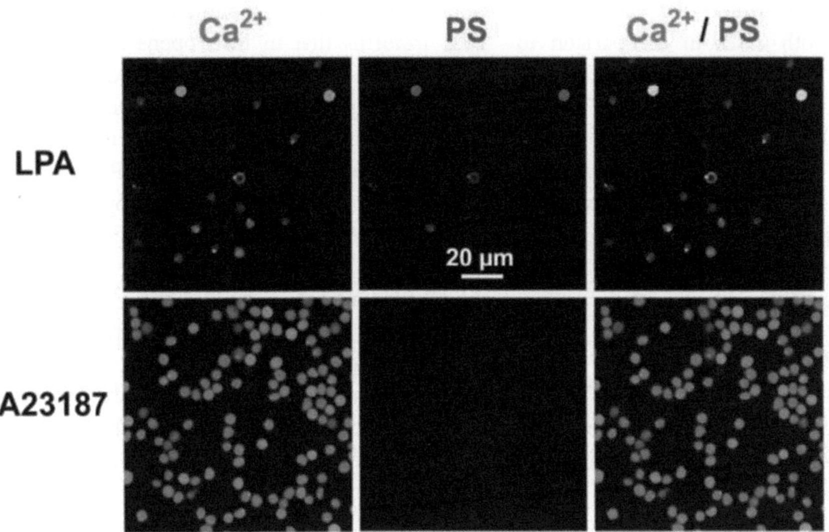

Fig. 27.6 Sheep RBCs double labelled with fluo-4 and annexin V-alexa568 (Ca²⁺/PS double scan, green and red, right column) after stimulation to increase Ca²⁺ content (green, left column) and PS exposure (red, middle column), respectively. Upper panels - 2.5 µM LPA, lower panels - 2 µM A23187. There are two cells with elevated Ca²⁺ content and increased PS exposure. Representative images out of 3 independent experiments. (For interpretation of the references to color in this figure legend, the reader is referred to the web version of the original article.)

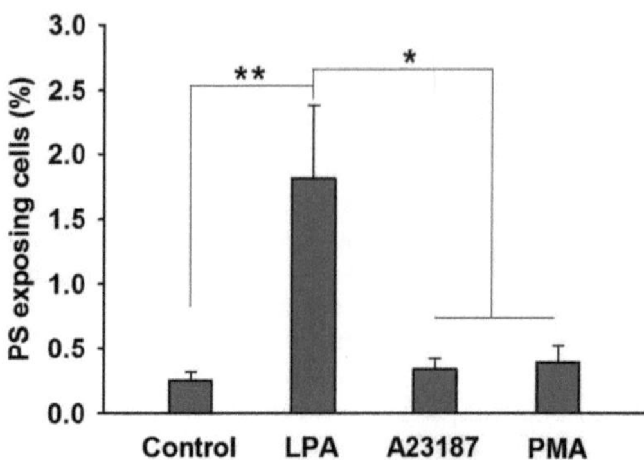

◘ Fig. 27.7 Flow cytometry analysis of sheep RBCs showing a significant higher PS exposure compared to control (in %) after treatment with 2.5 μM LPA, 2 μM A23187 or 6 μM PMA in the presence of 2 mM extracellular Ca^{2+} (incubation time 30 min). Columns show mean value of 3 different blood samples (30,000 cells of each blood sample were analysed). Error bars represent SD, $p \leq 0.05$ (*); $p \leq 0.01$ (**). Control measurements were done in the presence of solvents only (ethanol, water, or DMSO) and extracellular Ca^{2+} (see Material and Methods). Data for treatments with A23187 or PMA are statistically not significantly different from control.

few cells showed PS exposure (◘ Fig. 27.5, upper panels). For A23187 (◘ Fig. 27.6, lower panels) and PMA treatment (data not shown) there was no PS exposure although the intracellular Ca^{2+} content was increased in both cases in comparison to control conditions. These results were confirmed by flow cytometry and are depicted in ◘ Fig. 27.7. Only 2% of sheep RBCs treated with LPA expose PS on the outer membrane leaflet whilst A23187 and PMA treatment failed to induce PS exposure (◘ Fig. 27.7), thus differing significantly from those obtained with human RBCs (◘ Fig. 27.4).

27.4 Discussion

The main idea was to investigate the process of PS exposure as a function of the intracellular Ca^{2+} content and the activation of PKCα in human and sheep RBCs. Flow cytometry, single cell fluorescence microscopy, and confocal fluorescence microscopy were used.

In HPS, the Ca^{2+} permeability of the RBC membrane is very low[35] and the basal level of free cytoplasmic Ca^{2+} is in the range of 60 nM

maintained by the activity of the Ca^{2+} pump[36]. Our data reveal a fast uptake of Ca^{2+} by LPA stimulation (◘ Fig. 27.1). This effect is probably mediated by a specific pathway rather than a non-specific leak since it has been proposed that a non-selective cation channel is activated by a still unclear mechanism[17,27]. Other Ca^{2+} permeable cation channels in the RBC membrane also may be involved[7,32,37] (for a recent review see [14]). Furthermore the Ca^{2+} entry showed a clear LPA dose-dependent response[18].

Ca^{2+} uptake mediated by A23187 reached the same levels as LPA treatment after 30 min, although a delay in the response of about 5 min occurred, probably because A23187 partitions into the membrane before acting as ionophore. The effect of PMA (even at 10 μM) was fast at the beginning of the experiment without ever inducing high levels of intracellular Ca^{2+}, as compared to LPA and A23187 treatments. Flow cytometry experiments (◘ Fig. 27.2B) performed at 12 min (when the Ca^{2+} content reached the maximum values, cf. ◘ Fig. 27.1) are consistent with data obtained by fluorescence microscopy. The mechanism by which PMA increases intracellular Ca^{2+} is different from the LPA and A23187 stimulations. Andrews et al.[38] assume that PMA activates PKC isoforms that in turn opens a ω-agatoxin-TK-sensitive, $Ca_v2.1$-like (P/Qtype) Ca^{2+} channel. It has been proposed that this channel may be identical with the voltage-activated non-selective cation channel[14]. PKC activation by PMA is Ca^{2+}-independent, whereas the channel opening leads to an increase of the intracellular Ca^{2+} content. In addition, PMA may activate the Ca^{2+} pump[34] and inhibit the Ca^{2+}-activated K^+ channel[34,39].

From the images depicting PS exposure at different time points presented in ◘ Fig. 27.3, one can see that the increased Ca^{2+} content also leads to a vesiculation of the cells, in agreement with findings reported in literature[31,32]. Calpain, a calcium-dependent proteolytic enzyme, cleaves spectrin and actin, leading to cytoskeleton breakdown[40-42], and therefore may be involved in vesiculation.

PMA induced a greater PS exposure in comparison to LPA and A23187 (◘ Fig. 27.4, (+) Ca^{2+}). PS exposure was not directly correlated with the Ca^{2+} content since RBCs stimulated with PMA, at the same time point (30 min), displayed two-fold lower Ca^{2+}-dependent fluorescence when compared to LPA and A23187 treatment (cf. ◘ Fig. 27.1 and ◘ Fig. 27.4).

These findings were further supported by the results of the double labelling experiments presented in ◘ Fig. 27.5. After treatment with LPA or A23187 many cells (in case of LPA most of the cells) showed an increased Ca^{2+} content and external PS exposure. However, in both cases some cells had an increased Ca^{2+} level but no PS exposure and, vice versa, some cells depicted externalised PS in the absence of an increased Ca^{2+} content. This situation was rare in both cases and it cannot be ruled out that, after 30 min of incubation, a reduction of the intracellular Ca^{2+} content due to the activation of the Ca^{2+} pump occurred in some cells, in agreement with the findings shown in ◘ Fig. 27.1 (i.e. a sub-population of cells behaved different from others). In case of the PMA treatment, some cells displayed an enhanced Ca^{2+} content and PS exposure. Interestingly, more cells as compared to LPA or A23187 treated cells, show PS exposure without having an elevated Ca^{2+} content. In a control experiment, i.e. in the absence of external Ca^{2+} (◘ Fig. 27.4), some cells showed PS exposure. Thus the primary cause of PS exposure stimulated by PMA is different from the LPA or A23187 stimulation. Finally, the cell volume seems to be important for PS exposure in agreement with findings of Lang et al.[43,44] but only in case of LPA or A23187 treatment. However, we cannot rule out the possibility that the extracellular K^+ concentration might contribute to the observed volume effect on the PS exposure in case of LPA or A23187 treatment, which would be in agreement with the results of Wolfs et al.[45].

Surprisingly, comparable experiments with sheep RBCs did not show PS exposure after treatment with the Ca^{2+} ionophore or PMA although the intracellular Ca^{2+} content was enhanced. However, treatment with LPA resulted in a very small number of cells externalising PS. These effects can be explained by assuming absence of scramblase activity. In turn, LPA had an additional but minor effect on PS exposure (see below). The presumed absence of the scramblase activity in

sheep RBCs is supported by findings of the group of A. Herrmann (Humboldt University, Berlin) applying spin-labelled lipids and investigating the PS movement across the RBC membrane of a great variety of animal species by electron spin resonance (A. Herrmann, personal communication). Indeed, data mining using the scramblase amino acid sequences against protein databases failed to show any match with proteins from sheep. Furthermore, the external leaflet of the sheep RBC membrane contains 100% SM and no PC[25,26]. SM is transported across the membrane very slowly (even slower than PC)[46] and therefore no need for the presence of the scramblase exists in sheep RBCs.

Based on the findings of this paper, it seems reasonable to assume at least 3 different mechanisms being responsible for the PS exposure in human RBCs:

First, a Ca^{2+}-stimulated scramblase activation (and flippase inhibition)[4,6,7,47,48] brought about by LPA, A23187, or PMA treatment inducing Ca^{2+} uptake by different pathways. LPA activates the non-selective cation channel[14] whereas PMA opens the ω-agatoxin-TKsensitive, $Ca_v2.1$-like (P/Q-type) Ca^{2+} channel[38]. This first mechanism is a rather general one that can occur under various pathological conditions, e.g., by activation of NMDA receptors in sickle cell disease as discussed by Makhro et al.[49].

Second, PMA directly activates PKCα involved in PS exposure in human RBCs[22,23]. In addition, LPA activates in human RBCs both the PKCα (Ca^{2+}-dependent) and the PKCζ (Ca^{2+}-independent)[20]. Whether there is a direct activation of the scramblase in human RBCs by PKCα and/or PKCζ remains to be shown. The more pronounced effect of PS exposure after treatment of RBCs with PMA as compared to LPA can be explained by assuming that PMA activates all available PKCα whereas LPA stimulation triggers a signalling cascade[50] resulting only in partial activation of the PKCα pool.

Third, an additional mechanism has to be taken into consideration in the case of sheep RBCs. LPA may insert into the membrane (more pronounced in the presence of Ca^{2+} since it becomes more hydrophobic) stimulating the flop of PS from the inner to the outer membrane leaflet. Such an event would explain the higher rate of haemolysis after LPA treatment as compared to A23187 or PMA treatment. In comparison to the first and second mechanisms, this third effect only plays a minor role in the PS exposure which may be due to the fact that the LPA concentration used was far below the micelle forming concentration[51].

How the 3 pathways are interconnected remains to be elucidated in further investigations. The situation is even more challenging since other parameters, not being addressed here, are also known to affect the PS exposure in human RBCs. For instance, it has been shown that activation of caspases, production of ceramides, oxidative stress as well as ATP depletion increase the PS exposure[7,31,32,52]. ATP depletion is a Ca^{2+}-dependent process because at higher intracellular Ca^{2+} levels the Ca^{2+} pump is activated and will consume more ATP. It has also been demonstrated that the intracellular K^+ content and the cell shape can affect the PS exposure in RBCs[45,53]. Furthermore, it has been demonstrated that RBCs show an increased PS exposure after long-term storage and induced by hyperosmotic shock[54].

27.5 Conclusion

PS exposure in RBCs is a complex process involving at least 3 mechanisms. A variety of Ca^{2+}-dependent as well as Ca^{2+}-independent parameters play a substantial role for PS exposure (see e.g., [52,55,56]). Since the Ca^{2+}-dependent and Ca^{2+}-independent mechanisms are not additive, an interaction between those pathways can be assumed. Cell volume and probably the Ca^{2+} pump modulate the signalling cascade leading to PS exposure. Furthermore, the majority of PS is transported to the outer membrane leaflet mediated by scramblase activity but an additional minor effect based on lipid flop has to be taken into consideration.

Externalisation of PS to the outer membrane leaflet is important for the adherence of RBCs to endothelium[19,57,58] as well as for the RBC-RBC adhesion[9,17,18]. Knowledge of the underlying mechanisms is important for clincial prevention of thrombus formation.

27.6 References

[1] Haest CWM: Distribution and movement of membrane lipids; in Bernhardt I, Elorry JC (eds): Red cell membrane transport in health and disease. Berlin, Springer, 2003, pp 1-25.

[2] Devaux PF, Herrmann A, Ohlwein N, Kozlov MM: How lipid flippases can modulate membrane structure. Biochim Biophys Acta 2008;1778:1591-1600.

[3] Bevers EM, Comfurius P, Dekkers DW, Zwaal RF: Lipid translocation across the plasma membrane of mammalian cells. Biochim Biophys Acta 1999;1439:317-330.

[4] Daleke DL: Regulation of phospholipid asymmetry in the erythrocyte membrane. Curr Opin Hematol 2008;15:191-195.

[5] Zwaal RF, Comfurius P and Bevers EM: Surface exposure of phosphatidylserine in pathological cells. Cell Mol Life Sci 2005;62:971-988.

[6] Lang F, Gulbins E. Lerche H, Huber SM, Kempe DS, Foller M: Eryptosis, a window to systemic disease. Cell Physiol Biochem 2008;22:373–380.

[7] Lang F, Lang KS, Lang PA, Huber SM, Wieder T: Mechanisms and significance of eryptosis. Antioxid Redox Signal 2006;8:1183-1192.

[8] Hellem AJ, Borchgrevink CF, Ames SB: The role of red cells in haemostasis: the relation between haematocrit, bleeding time and platelet adhesiveness. Br J Haematol 1961;7:42-50.

[9] Andrews DA, Low PS: Role of red blood cells in thrombosis. Curr Opin Hematol 1999;6:76-82.

[10] Kaestner L, Bernhardt I: Ion channels in the human red blood cell membrane: their further investigation and physiological relevance. Bioelectrochemistry 2002;55:71-74.

[11] Kaestner L, Tabellion W, Lipp P, Bernhardt I: Prostaglandin E2 activates channel-mediated calcium entry in human erythrocytes: an indication for a blood clot formation supporting process. Thromb Haemost 2004;92: 1269-1272.

[12] Kaestner L, Bollensdorff C, Bernhardt I: Non-selective voltage-activated cation channel in the human red blood cell membrane. Biochim Biophys Acta 1999;1417:9-15.

[13] Kaestner L, Christophersen P, Bernhardt I, Bennekou P: The non-selective voltage-activated cation channel in the human red blood cell membrane: reconciliation between two conflicting reports and further characterisation. Bioelectrochemistry 2000;52:117-125.

[14] Kaestner L: Cation channels in erythrocytes - historical and future perspective. Open Biol J 2011;4:27-34.

[15] Dyrda A, Cytlak U, Ciuraszkiewicz A, Lipinska A, Cueff A, Bouyer G, Egee S, Bennekou P, Lew VL, Thomas SLY: Local membrane deformations activate Ca^{2+}-dependent K^+ and anionic currents in intact human red blood cells. PLoS One 2010;5:e9447.

[16] Oonishi T, Sakashita K, Ishioka N, Suematsu N, Shio H, Uyesaka N: Production of prostaglandins E_1 and E_2 by

adult human red blood cells. Prostaglandins Other Lipid Mediat 1998;56:89-101.

[17] Steffen P, Jung A, Nguyen DB, Mueller T, Bernhardt I, Kaestner L, Wagner C: Stimulation of human red blood cell leads to Ca^{2+}-mediated intercellular adhesion. Cell Calcium 2011;50:54-61.

[18] Kaestner L, Steffen P, Nguyen DB, Wang J, Wagner-Britz L, Jung A, Wagner C, Bernhardt I: Lysophosphatidic acid induced red blood cell aggregation in vitro. Bioelectrochemistry 2012;87:89-95.

[19] Closse C, Dachary-Prigent J, Boisseau MR: Phosphatidylserine-related adhesion of human erythrocytes to vascular endothelium. Br J Haematol 1999;107:300-302.

[20] Chung SM, Bae ON, Lim KM, Noh JY, Lee MY, Jung YS, Chung JH: Lysophosphatidic acid induces thrombogenic activity through phosphatidylserine exposure and procoagulant microvesicle generation in human erythrocytes. Arterioscler Thromb Vasc Biol 2007;27:414-421.

[21] Woon LA, Holland JW, Kable EP, Roufogalis BD: Ca^{2+} sensitivity of phospholipid scrambling in human red cell ghosts. Cell Calcium 1999;25:313-320.

[22] De Jong K, Rettig MP, Low PS, Kuypers FA: Protein kinase C activation induces phosphatidylserine exposure on red blood cells. Biochemistry 2002;41:12562-12567.

[23] Klarl BA, Lang PA, Kempe DS, Niemoeller OM, Akel A, Sobiesiak M, Eisele K, Podolski M, Huber SM, Wieder T, Lang F: Protein kinase C mediates erythrocyte »programmed cell death« following glucose depletion. Am J Physiol 2006;290:C244-C253.

[24] Govekar RB, Zingde SM: Protein kinase C isoforms in human erythrocytes. Ann Hematol 2001;80:531-534.

[25] Van Dijck PW, De Kruijff B, Van Deenen LL, De Gier J, Demel RA: The preference of cholesterol for phosphatidylcholine in mixed phosphatidylcholine phosphatidylethanolamine bilayers. Biochim Biophys Acta 1976;455:576-587.

[26] Saito M, Tanaka Y, Ando S: Thin-layer chromatography-densitometry of minor acidic phospholipids: application to lipids from erythrocytes, liver, and kidney. Anal Biochem 1983;132:376-383.

[27] Kaestner L, Tabellion W, Weiss E, Bernhardt I, Lipp P: Calcium imaging of individual erythrocytes: problems and approaches. Cell Calcium 2006;39:13-19.

[28] Gardos G: The permeability of human erythrocytes to potassium. Acta Physiol Hung 1956;10:185-189.

[29] Hoffman JF, Joiner W, Nehrke K, Potapova O, Foye K, Wickrema A: The hSK4 (KCNN4) isoform is the Ca^{2+}-activated K^+ channel (Gardos channel) in human red blood cells. Proc Natl Acad Sci USA 2003;100:7366-7371.

[30] Lang PA, Kaiser S, Myssina S, Wieder T, Lang F, Huber SM: Role of Ca^{2+}-activated K^+ channels in human erythrocyte apoptosis. Am J Physiol 2003;285:C1553-C1560.

[31] Lang KS, Lang PA, Bauer C, Duranton C, Wieder T, Huber SM, Lang F: Mechanisms of suicidal erythrocyte death. Cell Physiol Biochem 2005;15:195-202.

[32] Foeller M, Huber SM, Lang F: Erythrocyte programmed cell death. IUBMB Life 2008;60:661-668.

[33] Lord JM, Pongracz J: Protein kinase C: a family of isoenzymes with distinct roles in pathogenesis. Clin Mol Pathol 1995;48:M57-M64.

[34] Fathallah H, Sauvage M, Romero JR, Canessa M, Giraud F: Effects of PKC alpha activation on Ca^{2+} pump and K(Ca) channel in deoxygenated sickle cells. Am J Physiol 1997;273:C1206-C1214.

[35] Tiffert T, Bookchin RM, Lew VL: Calcium homeostasis in normal and abnormal human red cells; in Bernhardt I, Elory JC (eds): Red cell membrane transport in health and disease. Berlin, Springer, 2003, pp 373-405.

[36] Gilboa H, Chapman BE, Kuchel PW: 19F NMR magnetization transfer between 5-FBAPTA and its complexes. An alternative means for measuring free Ca^{2+} concentration, and detection of complexes with protein in erythrocytes. NMR Biomed 1994;7:330-338.

[37] Duranton C, Huber S, Tanneur V, Lang K, Brand V, Sandu C, Lang F: Electrophysiological properties of the plasmodium falciparum-induced cation conductance of human erythrocytes. Cell Physiol Biochem 2003;13:189-198.

[38] Andrews DA, Yang L, Low PS: Phorbol ester stimulates a protein kinase C-mediated agatoxin-TK-sensitive calcium permeability pathway in human red blood cells. Blood 2002;100:3392-3399.

[39] Del Carlo B, Pellegrini BM, Pellegrino M: Modulation of Ca^{2+}-activated K^+ channels of human erythrocytes by endogenous protein kinase C. Biochim Biophys Acta 2003;1612:107-116.

[40] Anderson DR, Davis JL, Carraway KL: Calcium-promoted changes of the human erythrocyte membrane. Involvement of spectrin, transglutaminase, and a membrane-bound protease. J Biol Chem 1977;252:6617-6623.

[41] Hatanaka M, Yoshimura N, Murakami T, Kannagi R, Murachi T: Evidence for membrane-associated calpain I in human erythrocytes. Detection by an immunoelectrophoretic blotting method using monospecific antibody. Biochemistry 1984;23:3272-3276.

[42] Lofvenberg L, Backman L: Calpaininduced proteolysis of beta-spectrins. FEBS Lett 1999;443:89-92.

[43] Lang F, Lang KS, Wieder T, Myssina S, Birka C, Lang PA, Kaiser S, Kempe D, Duranton C, Huber SM: Cation channels, cell volume and the death of an erythrocyte. Pflügers Arch 2003;447:121-125.

[44] Lang F, Gulbins E, Szabo I, Lepple-Wienhues A, Huber SM, Duranton, C, Lang KS, Lang PA, Wieder T: Cell volume and the regulation of apoptotic cell death. J Mol Recognit 2004;17:473-480.

[45] Wolfs JLN, Comfurius P, Bekers O, Zwaal RFA, Balasubramanian K, Schroit AJ, Lindhout T, Bevers EM: Direct inhibition of phospholipid scrambling activity in erythrocytes by potassium ions. Cell Mol Life Sci 2009;66:314-323.

[46] Van Meer G, Holthuis JC: Sphingolipid transport in eukaryotic cells. Biochim Biophys Acta 2000;1486:145-170.

[47] Williamson P, Kulick A, Zachowski A. Schlegel RA, Devaux PF: Ca^{2+} induces transbilayer redistribution of all major phospholipids in human erythrocytes. Biochemistry 1992;31:6355-6360.

[48] Daleke DL: Regulation of phospholipid asymmetry in the erythrocyte membrane. Curr Opin Hematol 2008;15: 191-195.

[49] Makhro A, Wang J, Vogel J, Boldyrev AA, Gassmann M, Kaestner L, Bogdanova A: Functional NMDA receptors in rat erythrocytes. Am J Physiol Cell Physiol 2010;298:C1315-C1325.

[50] Choi JW, Herr DR, Noguchi K, Yung YC, Lee C, Mutoh T, Lin M, Teo ST, Park KE, Mosley AN, Chun J: LPA receptors: subtypes and biological actions. Annu Rev Pharmacol Toxicol 2010;50:157-186.

[51] Eichholtz T, Jalink K, Fahrenfort I: The bioactive phospholipid lysophosphatidic acid is released from activated platelets. Biochem J 1993;291:677-680.

[52] Mandal D, Moitra PK, Saha S, Basu J: Caspase 3 regulates phosphatidylserine externalization and phagocytosis of oxidatively stressed erythrocytes. FEBS Lett 2002;513:184-188.

[53] Wolfs JLN, Comfurius P, Bevers EM, Zwaal RFA: Influence of erythrocyte shape on the rate of Ca^{2+}-induced scrambling of phosphatidylsrine. Mol Membr Biol 2003;20:83-91.

[54] Bosman GJCGM, Cluitmans JCA, Groenen YAM, Were JM, Willekens FLA, Novotny VMJ: Susceptibility to hyperosmotic stress-induced phosphatidylserine exposure increases during red blood cell storage. Transfusion 2011;51:1072-1087.

[55] Lang PA, Kempe DS, Tanneur V, Eisele K, Klarl BA, Myssina S, Jendrossek V, Ishii S, Shimizu T, Waidmann M, Hessler G, Huber SM, Lang F, Wieder T: Stimulation of erythrocyte ceramide formation by platelet-activating factor. J Cell Sci 2005;118:1233-1243.

[56] Kiedaisch V, Akel A, Niemoeller OM, Wieder T, Lang F: Zinc-induced suicidal erythrocyte death. Am J Clin Nutr 2008;87:1530-1534.

[57] Telen MJ: Red blood cell surface adhesion molecules: their possible roles in normal human physiology and disease. Semin Hematol 2000;37:130-142.

[58] Setty BN, Kulkarni S, Stuart MJ: Role of erythrocyte phosphatidylserine in sickle red cell-endothelial adhesion. Blood 2002;99:1564-1571.

Cation Channels in Erythrocytes - Historical and Future Perspective

Lars Kaestner

Reprint from The Open Biology Journal (2011) **4**, 27-34.

■ **Abstract**

Compared to ion pumps and carriers, ion channels have been quite late recognised as individual transport identities in red blood cells. Here the transition from cell population based flux experiments to single cell based patch-clamp investigations are described in detail. The present knowledge of cation channels in red blood cells from their molecular identity to their electrophysiological properties are summarised. Tendencies and novel concepts for future research concerning ion-transport across the red blood cell membrane are discussed.

28.1 Introduction

Cation channels are versatile transporters, responsible for the adjustment of the resting membrane potential in most cells, for the induction of the action potential in excitable cells and the transmembrane transduction of many signalling cascades and ion-exchange in general. Historically, methods of tracer flux experiments using red blood cells have been extremely popular for investigating transmembrane transport processes (e.g., [1,2]). The major reasons for this popularity are summarised in ■ Tab. 28.1.

Even nowadays the availability (reason (i) ■ Tab. 28.1), especially the aspect of using human cells, holds a strong argument. Nevertheless, a decent number of studies have been performed on animal red blood cells. In the following the reasons for using animal cells are discussed:

a. the size of the cells - it is not by chance that the first microscopically observed cells at all (probably around 1660) have been frog red blood cells[3]. For an illustration of differing cell sizes refer to ■ Fig. 28.1A;

b. to avoid the special condition of lacking cell organelles and sub-cellular structures (argument (iii) in ■ Tab. 28.1) by investigating red blood cells of e.g., birds or amphibians, that contain a few organelles like a nucleus or mitochondria and are therefore more representative for cells in general;

c. disease models might be more advantageous on certain animals, for instance the malaria *Plasmodium gallinaceum* in chicken red blood cells (cp. ■ Fig. 28.1B and Thomas *et al.* 2001[4]) is not infectious to humans and therefore eases cell handling;

■ **Tab. 28.1** Reasons for popularity of red blood cells in investigating transmembrane transport processes.

(i)	availability (also in large numbers) without seriously affecting the donor - such they are the easiest available living cells from humans
(ii)	individual cells (not connected as in tissue), so if the surrounding medium is changed they are immediately and completely covered by the new medium
(iii)	especially mammalian red blood cells do not contain cell organelles or subcellular compartments (cp. ■ Fig. 28.1A) that could be used as ion stores
(iv)	the cellular population seems very homogeneous, cell to cell differences are minor since growth and development are almost stagnating in red blood cells.

d. animal red blood cells may display certain properties that are absent or not inducible in human red blood cells - examples are transgenic animals or lamprey erythrocytes having an inactive band 3 protein[5,6].

Within this review the focus is set to human red blood cells, however, animal red blood cells are covered if their investigation made a significant contribution to the cation channel identification or investigation.

With the advent of methods addressing membrane properties more directly than flux experiments, namely the patch-clamp technique, but also single molecule imaging techniques or cloning and over-expressions of channels e.g., in oocytes, on the one hand a lot of open questions in the field of red blood cell transmembrane transport could be resolved (see below). On the other hand, other cell types, not meeting the conditions of being individual cells *per se* and not containing cell organelles (reasons (ii) and (iii) in ◙ Tab. 28.1, respectively), became easier to investigate and ousted red blood cells somewhat from their central role of transport investigations. However, in the recent years it became evident, that the homogeneous cell population (argument (iv) in ◙ Tab. 28.1) is of limited validity. There are subpopulations behaving different from the rest of the cells indicating that red blood cells are much more versatile than one would expect from membrane-bags just holding the haemoglobin.

28.2 The discovery of cation channels in erythrocytes: from flux-measurements to electrophysiology

The transport mediated by the cation channels in the red blood cell membrane was in the major fraction observed before ion channels were established facts of life[7]. However, the discovery of cation transport across a biological membrane in general and across red blood cell membranes was a long way. The earliest observations considering cation concentration alterations under certain conditions date back more than 100 years e.g., Abderhalden 1898[8] and Bang 1909[9].

The investigations measuring real transport across a membrane were boosted by the discovery of radioactivity, allowing a precise measurement of the movement of radio- isotopes across red blood cell membranes. Among the first papers proving a cation permeability of the membrane under physiological conditions were the experiments by Elma and Waldo Cohn published in 1939[10]. ^{24}NaCl was intravenously injected into dogs and blood samples were examined after several hours *in vitro* or *in vivo*. Considerably amounts of ^{24}Na were found inside erythrocytes proving a Na-permeability. The radioactive tracer flux measurements became state of the art and the predominant technique in investigating ion transport for almost 50 years. Such measurements based on observations in cell populations have detected general principles. However, such approaches are hardly be able to identify special transport properties of a minor subpopulation of the cells. The state of knowledge concerning passive cation transport in the red blood cell membrane at the advent of the cellular electrophysiology is comprehensively summarised by Lew and Beaugé[11].

The investigations »predicting« the Gardos channel date back to the 1930s when Ørskov measured a K^+ loss caused by Pb^{2+} ref [12] and Wilbrandt, who detectet a K^+ efflux upon inhibition of glycolysis[13]. Both investigations did not even hypothesise the Ca^{2+}-activated K^+-channel, but the K^+-fluxes they measured were caused by exactly that channel. However, the transport process as a Ca^{2+}-activated K^+-pathway was demonstrated by Gardos[14] and the effect named after him - later on for consistence the channel was also called Gardos channel. However, in the 1970s novel electrophysiological methods appeared, revolutionising the investigations and the understanding of membrane transport in general. As one of the precurser of the patch-clamp technique the micro-electrode methodology appeared. In the 1970s this was applied on Amphiuma red blood cells because of their enormous size exceeding even the size of frog red blood cells (cp. ◙ Fig. 28.1A). It was shown that the hyperpolarisation occuring with the micropuncture of the cells was Ca^{2+}-dependent and determined by the membranes K^+ permeability[15].

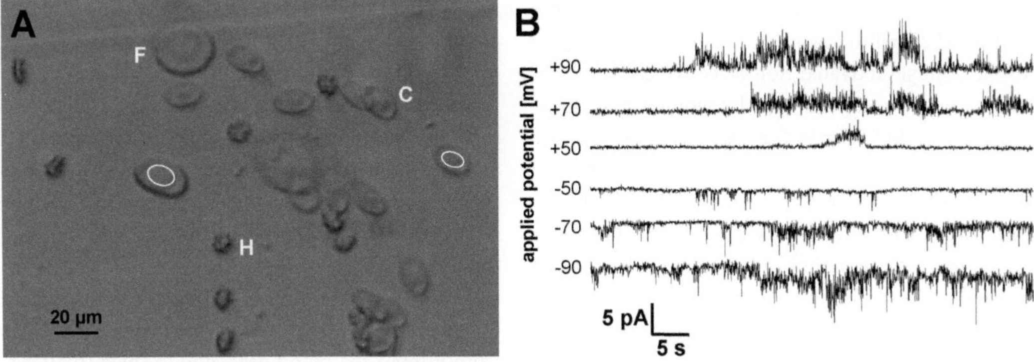

◻ Fig. 28.1 Red blood cells of several species. The image A provides a white light image comparing red blood cells from frogs (F) chicken (C) and human (H). B depicts one of the first current-traces measured in malaria-infected (Plasmodium gallinaceum) chicken red blood cells. Recordings in cell-attached configuration, identical pipette and bath solution containing in mM: 150 NaCl, 5 KCl, 1.4 CaCl$_2$, 1 MgCl$_2$, 5 glucose, 10 HEPES, pH adjusted to 7.4. Fig. 28.1B modified from Kaestner 2001[71].

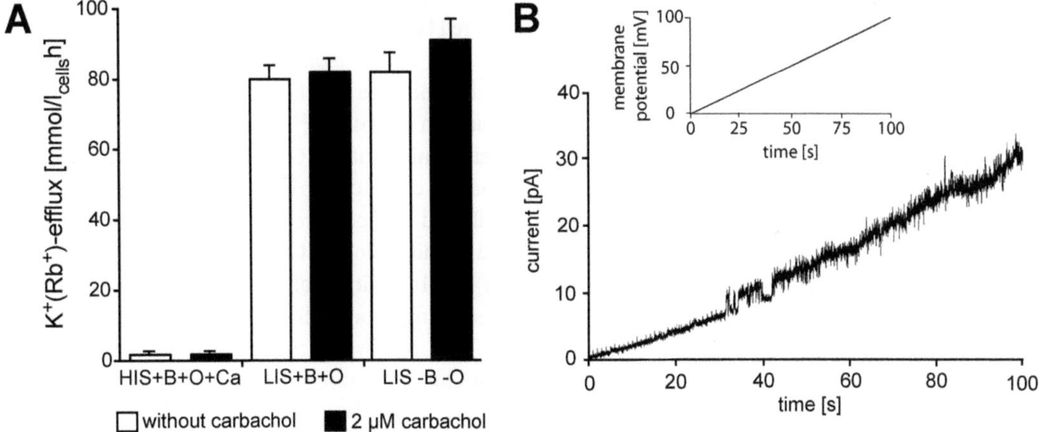

◻ Fig. 28.2 Pannel A displays a radioactive tracer flux measurement. Each bar is the mean (±SEM) of measurements from 3 independent blood samples from healthy donors. To address channel activity both major K$^+$-transport entities, namely the Na$^+$/K$^+$/2Cl$^-$-cotransporter and the Na$^+$/K$^+$-pump were inhibited with 100 µM bumetanide (B) and 100 µM ouabain (O), respectively. To be able to monitor an indirect channel activity (Gardos channel mediated K$^+$-efflux induced by Ca^{2+}-influx), 2 mM Ca^{2+} was added to the external solution in the experiments performed in solutions of high ionic strength (HIS: 145 mM NaCl, 10 mM MesTRIS, 10 mM glucose). To depolarise the cells, measurements were performed in solution of low ionic strength (LIS: 10 mM MesTRIS, 10 mM glucose, 250 mM sucrose) leading to a measured membrane potential of approximately 45 mV[72]. Since carbachol activates the non-selective voltage activated cation channel[34], a significant difference should be observed in LIS-solutions in the presence and in the absence of carbachol. This change could not be measured. To ensure that bumetanide and ouabain do not interfere with the non-selective voltage activated cation channel, the experiments in LIS-solution were repeated without the presence of the mentioned inhibitors. Again, a significant difference between the presence and absence of carbachol was not detectable. Pannel B is a plot of a current trace recorded from an excised patch showing a number of biophysical properties of the non-selective voltage activated cation channel. The pipette and bath solution contained 20 mM Na-tartrate and 70 mM Na-tartrate, respectively. Additionally, both solutions contained 2.5 mM BaCl$_2$, 10 mM glucose, 100 µM bumetanide, 100 µM ouabain and 10 mM MOPS. The tonicity of the solutions was adjusted to 300 mOsM with sucrose and the pH-value was adjusted to 7.4 with NaOH. The inset shows the applied voltage ramp in the voltage-clamp mode, while the current trace nicely reflects the open probability of the channel being zero up to a membrane potential of approximately 25 mV and being about 1 from 75 mV upwards. The conductance is approximately 21 pS and the current-voltage relationship is within the range of the open probability roughly linear.

In similarity to the first observations of the Gardos channel the first results nowadays assigned to the voltage-activated non-selective cation channel, namely the experiments by Donlon and Rothstein from 1969[16] were flux based. They found a three-phasic increase in the salt efflux from red blood cells with decreasing extracellular NaCl concentration and already discussed the hypothetical chance of a channel involvement. However, a voltage-activated cation channel was assigned to red blood cells by the work of Halperin *et al.* two decades later[17]. Remarkably it is a flux based paper giving evidence of a voltage-activated cation channel even after the establishment of the patch-clamp technique. The membrane potential was modulated by either varying the K^+ concentration gradient in the presence of valinomycin or by varying the concentration gradient of the permeant anion nitrate in the presence of 4,4'-diisothio-cyanostilbene-2,2'-disulfonic acid (DIDS). Interestingly, this results could not be verified when the Rb^+-flux was measured in solutions of low ionic strength (LIS) activating the channel by carbachol (❏ Fig. 28.2A and P. Bennekou, personal communication).

It should be mentioned that parallel to the electrophysiological characterisation of the voltage-activated non-selective cation channel (see section 28.3, below) a $K^+(Na^+)/H^+$ exchanger was proposed and characterised[18,19]. Both transport systems were assigned to be responsible for the residual K^+ and Na^+ fluxes in LIS solutions, but as pointed out before (❏ Fig. 28.2A) under widely used LIS conditions the channel could not be observed in radioactive tracer flux experiments and most of the LIS effect is mediated by the $K^+(Na^+)/H^+$ exchanger. Only a portion of the initial flux measured by Donlon and Rothstein, namely the second phase of the three-phasic increase is channel mediated[20,21].

28.3 Classification and properties of channels achieved by the patch-clamp technique

The patch-clamp technique as it was introduced by Neher, Sakmann and colleagues provided a tool for direct ion transport measurements through single channels and such provided a direct functional assay for a biophysical investigation of ion channels even allowing for versatile manipulation options. During the initial 20 years of patch-clamp measurements on human red blood cells, investigations were limited to the cell-attached configuration and excised patches, where channels in the piece of membrane touched by the micropipette could be recorded, either with the rest of the cell between the two electrodes or under direct voltage control of the membrane, respectively. This is caused by the difficulty to achieve whole cell recordings from human red blood cells, due to their small size and the special shape, which is determined by the cytoskeleton[7]. In the early days of patch-clamping red blood cells, the possibility to achieve whole cell recordings on human red blood cells was thought to be impossible[22]. However, with the establishment of automatised pipette pullers, whole cell recordings on red blood cells became a routine method.

The first person investigating ion channels in red blood cells was Hamill, who published current traces in an abstract[23] clearly depicting single channel openings of the Gardos channel giving a single channel conductance of 18 pS. This initial report was followed by a comprehensive chapter about »Potassium and Chloride Channels in Red Blood Cells« in the 1st edition of the classic »Single Channel Recording« book edited by Sakmann and Neher[24]. This chapter was focusing on frog red blood cells, but contained all classical patch-clamp approaches including whole cell recordings. The first comprehensive analysis of the Gardos channel in human red blood cells was performed by Grygorczyk and Schwarz[25], revealing basic properties such as the K^+ to Na^+ selectivity of 15:1, inward rectification and an estimation of (in average) 10 channels per cell. After this initial report a number of follow-up paper appeared correlating the patch-clamp data to the flux measurements[26] and further characterising the Gardos channel (❏ Tab. 28.2). The following years more detailed reports were published, e.g., [27,28] culminating in two papers by Leinders *et al.* [29,30] summarising and complementing the knowledge of the opening behaviour of the Gardos channel.

Single channel currents of the voltage-activated non-selective cation channel in human red blood cells were firstly measured by Christophersen and Bennekou[33]. Measurements have been performed

◘ Tab. 28.2 Overview of all Cation Channels and their Properties Described in Red Blood Cells

property	reference	remark
I. The Gardos Channel		
flux measurement based evidence	Gardos 1958[14]	short, but precise report published as »preliminary notes«
microelectrode measurements of Ca^{2+} induced hyperpolarisation	Lassen et al. 1974[15], 1976[73], Gardos et al. 1976[74]	performed on giant red blood cells from Amphiuma means
elelctrophysiological identification as a 18 pS channel by patch-clamp	Hamill 1981[23], 1983[23]	major part of characterisation performed on frog red blood cells
elelctrophysiological characterisation such as modulation of single channel conductance, selectivity of K^+ over Na^+ being 17:1, temperature dependence	Grygorczyk and Schwarz 1983[25], Grygorczyk et al. 1984[26], Grygorczyk and Schwarz 1985[69], Grygorczyk 1987[70]	all measurements performed on human red blood cells - inside out patches
Ca^{2+} dependence of opening behaviour and the concentration dependent action of various metal ions	Leinders et al. 1992a[29], b[30]	2 side by side articles, all measurements performed on human red blood cells - inside out patches
molecular identity (hSK4)	Hoffman et al. 2003[43]	Western blots from ghost membranes
whole cell conductance was noninactivating and inward rectifying for K^+; selectivity of $K^+,Rb^+:Cs^+:Na^+,Li^+$, $NMDG:NH_4^+$ was 51:5:>1:0; inhibition (EC_{50}) by charybdotoxin (28 nM), clotrimazole (153 nM), nitrendipine (27 nM), Stychodactyla toxin (291 nM), margatoxin (459 nM), miconazole (785 nM)econazole (2.4 μM), cetiedil (79 μM); activation (EC_{50}) by 1-ethyl-2-benzimidazolinone (74 μM)	Jensen et al. 1998[42]	channel properties not investigated in red blood cells, although the characterisation was before the molecular identification of hSK4 in red blood cells it was already proposed to be the Gardos channel
II. Low Conductance Cation Channel		
15 pS caused by inward current	Grygorczyk and Schwarz 1983[25]	cell attached configuration with the pipette solution in mM: 140 KCl, 1 EGTA, 1 $MgCl_2$, 10 MOPS
17 pS, K^+ selectivity	Bennekou and Christophersen 1988[31]	abstract
8.4 pS in physiological ionic strength, 17 pS at approximatly bisected ionic strength	Kaestner and Bernhardt 2002[32]	when in inside out patches (symmetrical 150 mM KCl solution) the bath solution was exchanged for 75 mM KCl no shift in the reversal potential could be observed
III. The Voltage-activated Non-selective Cation Channel		
flux measurement based evidence for voltage activation < 40 mV, permeability for Na^+, K^+ and Ca^{2+} inhibition by ruthenium red ▼	Halperin et al. 1989[17]	potential was adjusted by either varying the K^+ concentration gradient in the presence of valinomycin or by varying the concentration gradient of anion nitrate in the presence of DIDS

◘ Tab. 28.2 *Continued*

property	reference	remark
elelctrophysiological identification, zero current conductance of 35 pS, open probability increases between -30 mV and +30 mV from 0 to 100%	Christophersen and Bennekou 1991[33]	all measurements were performed in symmetrical 500 mM salt solutions
acetylcholine sensitivity	Bennekou 1993[34]	measurements were performed in symmetrical 500 mM salt solutions
elelctrophysiological characterisation under physiological conditions, permeability for divalent cations	Kaestner *et al.* 1999[20], 2000[44]	the 2000 paper reconciles the discrepancy between the report of Chistophersen and Bennekou and the 1999 paper
hysteretic behaviour of the open probability	Kaestner *et al.* 2000[44], Bennekou *et al.* 2004[57]	the property was initially found in patch-clamp experiments, but confirmed in cell suspension
whole-cell investigations	Rodighiero *et al.* 2004[58]	confirmation of previous single channel recordings
activation by clotrimazole and analogues; inhibition: 30% by 100 μM La^{3+}, max. 70% by ruthenium red (IC$_{50}$ of 3.7 μM); inactivation by N-ethyl-maleimide (IC$_{50}$ of 660 μM), iodoacetamide (IC$_{50}$ of 480 μM), 2,4'-dibromoacetophenone and declined pH values down to 6.0	Barksmann *et al.* 2004[68], Bennekou *et al.* 2004[59], 2006[60]	paper trilogy based on flux experiments; CCCP-estimated membrane potentials
indications for identity Ca$_V$2.1	Andrews *et al.* 2002[48]	Western blots from ghost membranes and ω-agatoxin pharmacology in flow cytometry
for Ca$_V$2.1 no activators are known, but is blocked by ω-conotoxin MVIIC, piperidines, substituted diphenylpiperidines, piperazines, volatile anesthetics, gabapentin, mibefradil, peptide toxins DW13.3 and ω-conotoxin SVIB	Catterall *et al.* 2005[6]	review, Ca$_V$2.1 channel properties not investigated in red blood cells
IV. The Receptor-activated Non-selective Cation Channel		
flux measurement based evidence	Li *et al.* 1996[67], Yang *et al.* 2000[61]	indirect measurement by Gardos channel activity in the 1996 paper; ^{45}Ca^{2+} flux and Fluo-3 based FACS in the 2000 paper
activation by osmotic shock and oxidative stress	Huber *et al.* 2001[37], Duranton *et al.* 2002[38]	performed in whole cell configuration on red blood cell ghosts
activation by prostaglandine E$_2$ and lysophospatidic acid	Kaestner and Bernhardt 2002[32], Kaestner *et al.* 2004[35]	at the time of publication believed to be a property of the voltage-activated non-selective cation channel; still not completely resolved
inhibition by ethylisopropylamiloride and erythropoietin ▼	Lang *et al.* 2003[62], Myssina *et al.* 2003[63]	channel inhibition protects against eryptosis

◘ Tab. 28.2 *Continued*

property	reference	remark
V. TRP C6 Channel		
identification (Western blot)	Föller *et al.* 2008[46]	the particular Western blot was provided by V. Flockerzi, M. Meissner and M. Freichel (Saarland University, Germany)
Ca^{2+} entry changes into human ghosts upon antibody incubation	Föller *et al.* 2008[46]	performed by FACS analysis
no selective or high affinity ligands for activation or inhibition; activation by phospholipase C signalling or directly by diacylglycerol	Abramowitz and Birnbaumer 2009[64]	review, channel properties not exclusively investigated in red blood cells
VI. NMDA Channel		
identification (Western blot, imunohistochemistry)	Makhro *et al.* 2010[47]	performed on rat red cells, predominantly present in reticulocytes and young erythrocytes
identification in human red blood cells	Bogdanova *et al.* 2009[65]	increased occurrence in sickle-cell patients
agonists: L-glutamate, glycine, D-2-amino-5-phosphonovalerate, 7-chlorokynurenate; sensitivity and affinity depends on heteromeric subunit composition	Yamakura and Shimoji 1999[66]	channel properties not investigated in red blood cells

in 500 mM salt solutions and as turned-out later on, showed a frequent appearance because it is coupled to an acetylcholine receptor of nicotinic type and a contamination of utensils with nicotine due to tobacco smoking in the lab took place[34]. However, the channel could be verified in physiological solutions under agonist free conditions, with an extremely low open probability[20]. The biophysical properties of the channel are nicely displayed by a single current trace depicted in ◘ Fig. 28.2B: The ramp reflects the I-V relationship and the channel follows the typical open probability - for a detailed description refer to the figure legend of ◘ Fig. 28.2B.

In 2002 the first electrophysiological report of a receptor-like channel not showing voltage dependent activation behaviour appeared[32]. At the time of publication it was in similarity to the follow-up reports[35,36] believed that this channel is identical with the non-selective voltage-activated cation channel. Whether this is the case or not could still not be resolved completely. As a working hypothesis these reports were grouped in ◘ Tab. 28.2 with other observations by the group of Florian Lang depicting channel activation by osmotic shock and oxidative stress[37,38].

Once the channels are identified on a molecular level (see section 28.4, below), a huge number of reports about the channels from other cell types become available. A full coverage of these sources would blast the frame of the present review. Therefore it is just pinpointed to comprehensive reviews (◘ Tab. 28.2, last line for each channel, if applicable).

28.4 Towards the molecular idendity of the channels

Identifying the amino-acid blueprint of ion-channels in red blood cells needs a different strategy compared to other cell types, where cloning the protein is the method of choice. Looking at the recent work of proteomics in red blood cells approximately 150 membrane proteins were identified[39,40]. However, there are basically no obvious candidates for ion channels, although their existence is clearly proved by the patch-clamp technique (compare section 28.3, above). All identification concepts so far mainly rely on antibody recog- nition, either in Western blots or in immunohistochemistry.

As for the Gardos channel at the end of the 1990s a Ca^{2+}-activated K^+-channel (hSK4) was cloned from, among other tissues, human T-lymphocytes[41] and was assigned to the KCNN4-gene. Shortly afterwards this cloned channel was characterised[42] and from that time the hSK4-channel was the hottest candidate for the Gardos channel, because electrophysiological characterisation as well as the pharmacology of the channel was virtually identical. In 2003 Joe Hoffman and co-workers presented a paper[43] providing further evidence: The Western blots of erythrocyte membranes displayed a clear band for SK4. Additionally, the developmental route was followed from Northern and Western blots of human erythroid progenitor cells to RT-PCR of RNA from reticulocytes also discriminating all other isoforms of the SK-family. These results whipped away all doubts for the hSK4 being identical with the Gardos channel.

For the non-selective cation conductance the situation is slightly more complex. In conjunction with the Ca^{2+}-permeability a report described Western blots against the α1A-subunit of the voltage activated $Ca_V2.1$-channel in human red blood cells[48]. From the electrophysiological characterisation it is very hard to compare the data of the voltage-activated non-selective cation channel in red blood cells to $Ca_V2.1$-channel recordings in other cells, since these properties are differentially affected by the coexpression level of particular β-subunits as well as by alternative splicing of the α1A-subunit[49]. To clarify this, further investigations are compulsory.

For the low-conductance cation channel it is not even clear whether these are substates of channels covered above or if they have their own identity. No molecular data are available.

It seems to be clear that the TRPC6 and the NMDA channels are present in red blood cells. For the TRPC6 and the NMDA-channel, the groups of Flockerzi/Freichel and Bogdanowa provided the Western blots, respectively (for references refer to ◼ Tab. 28.2). However, both channels cannot completely account for Ca^{2+}-entry measured by patch-clamp, flow cytometry and fluorescence imaging before[37,44,45]. The TRPC6-channel covers just a fraction of the non-selective cation-influx ([46] and own unpublished results). The NMDA-channel is just found in a fraction (subpopulation) of the cells[47].

28.5 Research perspective

The future basic research of red blood cells relies - as for most other cell types - on genetic manipulation. This manipulation might be more difficult in red blood cells, since the lack of protein transcription and translation machinery in mature mammalian red blood cells provokes more sophisticated methods. Presently there are two major strategies: (i) manipulation of stem cells and consecutive differentiation into red blood cells as described by Douay and coworkers[50]. A proof of principle that the expression of fluorescent fusion proteins works out in red blood cells was shown by Dickinson and coworkers when using the GFP expression to follow blood flow properties[51], (ii) the use of transgenic animals. Animals having a protein knocked-out or protein over expression, can be investigated independent of their initial purpose of breeding if the method to breed the transgene animal leads to a global and tissue-unspecific gene manipulation. Since the generation of a transgenic mouse is very labour intensive such synergic effects with other research fields can be used. However, within the recent years it became more popular to breed tissue specific and inducible transgenic animals. This may prevent the re-use of transgenic animals in red cell research that have been originally dedicated to other research areas.

Furthermore, the future mode of investigation will be single cell based such as patch-clamp, flow cytometry or live cell imaging, because it enables to identify also smaller subpopulations of cells that can hardly be addressed by measurements of huge populations of cells such as tracer flux analysis.

In that respect it has to be mentioned that these detection methods need to be further exploited.

Presently it is e.g., still not possible to measure intracellular calcium concentrations in red blood cells in a quantitative manner. It has been proposed but never performed to solve that problem by fluorescence lifetime imaging[45].

Further concepts imply the expression of genetically encoded biosensors (GEBs) that have been proved to be useful tools in many different cell types, either as biomarker[52,53] or as a functional sensor[54,55]. As for other protein manipulation, the challenges described above lead to the concept of GEB expression in erythroid progenitor cells and consecutive differentiation into red blood cells still containing the sensor.

Independent of the direction of future technological developments, investigations of red blood cell cation channels in health and disease will foster the understanding of blood related pathologies and may provide deeper insight in general cellular physiology[56].

28.6 References

[1] Schatzmann HJ. Why red cells? In: Raess BU, Tunnicliff G, Eds. The Red Cell Membrane A Model for solute Transport. Clifton: Humana Press 1989: pp. 3-17.

[2] Lew VL, Glynn IM, Ellory JC. Net synthesis of ATP by reversal of the sodium pump. Nature 1970; 225: 865-6.

[3] Swammerdam J. Bybel der Natuur. In: Hill J, Ed. The book of nature. London: Bookseller C. G. Seyffert 1737.

[4] Thomas SL, Egee S, Lapaix F, Kaestner L, Staines HM, Ellory JC. Malaria parasite Plasmodium gallinaceum up-regulates host red blood cell channels. FEBS Lett 2001; 500: 45-51.

[5] Lapaix F, Egee S, Gibert L, Decherf G, Thomas SL. ATP-sensitive K^+ and Ca^{2+}-activated K^+ channels in lamprey (Petromyzon marinus) red blood cell membrane. Pflugers Arch 2002; 445: 152- 60.

[6] Virkki LV, Nikinmaa M. Two distinct K^+ channels in lamprey (Lampetra fluviatilis) erythrocyte membrane characterized by single channel patch clamp. J Membr Biol 1998; 163: 47-53.

[7] Bennekou P, Christophersen P. Ion Channels. In: Bernhardt I, Ellory JC, Eds. Red cell membrane transport in health and disease. Berlin: Springer Verlag 2003; pp. 139-52.

[8] Abderhalden E. Über den Blut Kalium Gehalt verschiedener Säugetiere. Hoppe-Seyler's Zeitschrift für Physiologische Chemie 1889; 25: 65-115.

[9] Bang I. Physiko-chemische Verhältnisse derBlutkörperchen. Biochemische Zeitschrift 1909; 16: 255-76.

[10] Cohn WE, Cohn ET. Permeability of red corpuscles of the dog to sodium ion. Proc Soc Exp Biol Med 1939; 41: 445-9.

[11] Lew VL, Beaugé L. Passive Cation Fluxes in Red Cell Membranes. In: Giebisch G, Tosteson DC, Ussing HH, Eds. Transport across biological membranes. Vol. II. Berlin: Springer-Verlag 1979; pp. 81-115.

[12] Ørskov SL. Untersuchungen über den Elnfluß von Kohlensäure und Blei auf die Permeabilität der Blutkörperchen für Kalium und Rubidium. Biochem Z 1935; 279: 250-61.

[13] Wilbrandt W. A relation between the permeability of the red cell and its metabolism. Trans Faraday Soc 1937; 33: 956-9.

[14] Gardos G. The function of calcium in the potassium permeability of human erythrocytes. Biochim Biophys Acta 1958; 30: 653-4.

[15] Lassen UV, Pape L, Vestergaard-Bogind B, Bengtson O. Calcium- related hyperpolarization of the Amphiuma red cell membrane following micropuncture. J Membr Biol 1974; 18: 125-44.

[16] Donlon JA, Rothstein A. The cation permeability of erythrocytes in low ionic strength media of various tinicities. J Membr Biol 1969; 1: 37-52.

[17] Halperin JA, Brugnara C, Tosteson MT, Van Ha T, Tosteson DC. Voltage-activated cation transport in human erythrocytes. Am J Physiol 1989; 257: C986-96.

[18] Richter S, Hamann J, Kummerow D, Bernhardt I. The monovalent cation »leak« transport in human erythrocytes: an electroneutral exchange process. Biophys J 1997; 73: 733-45.

[19] Bernhardt I, Bogdanova AY, Kummerow D, Kiessling K, Hamann J, Ellory JC. Characterization of the K^+ (Na^+)/H^+ monovalent cation exchanger in the human red blood cell membrane: effects of transport inhibitors. Gen Physiol Biophys 1999; 18: 119-37.

[20] Kaestner L, Bollensdorff C, Bernhardt I. Non-selective voltage- activated cation channel in the human red blood cell membrane. Biochim Biophys Acta 1999; 1417: 9-15.

[21] Bernhardt I, Weiss E, Robinson HC, Wilkins R, Bennekou P. Differential effect of HOE642 on two separate monovalent cation transporters in the human red cell membrane. Cell Physiol Biochem 2007; 20: 601-6.

[22] Schwarz W, Grygorczyk R, Hof D. Recording single-channel currents from human red cells. Methods Enzymol 1989; 173: 112- 21.

[23] Hamill OP. Potassium channel currents in human red blood cells. J Physiol 1981; 319: 97P-8P.

[24] Hamill OP. Potassium and chloride channels in red blood cells. In: Sakmann B, Neher E, Eds. Single-Channel Recording. New York, London: Plenum Press 1983; pp. 451-71.

[25] Grygorczyk R, Schwarz W. Properties of the Ca^{2+}-activated K^+ conductance of human red cells as revealed by the patch-clamp technique. Cell Calcium 1983; 4: 499-510.

[26] Grygorczyk R, Schwarz W, Passow H. Ca^{2+}-activated K^+ channels in human red cells. Comparison of single-channel currents with ion fluxes. Biophys J 1984; 45: 693-8.

[27] Shields M, Grygorczyk R, Fuhrmann GF, Schwarz W, Passow H. Lead-induced activation and inhibition of potassium-selective channels in the human red blood cell. Biochim Biophys Acta 1985; 815: 223-32.

[28] Fuhrmann GF, Schwarz W, Kersten R, Sdun H. Effects of vanadate, menadione and menadione analogs on the Ca^{2+}-activated K^+ channels in human red cells. Possible relations to membrane- bound oxidoreductase activity. Biochim Biophys Acta 1985; 820: 223-34.

[29] Leinders T, van Kleef RG, Vijverberg HP. Distinct metal ion binding sites on Ca(2+)-activated K^+ channels in inside-out patches of human erythrocytes. Biochim Biophys Acta 1992; 1112: 75-82.

[30] Leinders T, van Kleef RG, Vijverberg HP. Single Ca(2+)-activated K^+ channels in human erythrocytes: Ca^{2+} dependence of opening frequency but not of open lifetimes. Biochim Biophys Acta 1992; 1112: 67-74.

[31] Bennekou P, Christophersen P. The occurance of a low conduc- tance channel in the human red cell. Acta Physiol Scand 1988; 13:

[32] Kaestner L, Bernhardt I. Ion channels in the human red blood cell membrane: their further investigation and physiological relevance. Bioelectrochemistry 2002; 55: 71-4.

[33] Christophersen P, Bennekou P. Evidence for a voltage-gated, non- selective cation channel in the human red cell membrane. Biochim Biophys Acta 1991; 1065: 103-6.

[34] Bennekou P. The voltage-gated non-selective cation channel from human red cells is sensitive to acetylcholine. Biochim Biophys Acta 1993; 1147: 165-7.

[35] Kaestner L, Tabellion W, Lipp P, Bernhardt I. Prostaglandin E_2 activates channel-mediated calcium entry in human erythrocytes: an indication for a blood clot formation supporting process. Thromb Haemost 2004; 92: 1269-72.

[36] Kaestner L, Tabellion W, Weiss E, Bernhardt I, Lipp P. Calcium imaging of individual erythrocytes: Problems and approaches. Cell Calcium 2006; 39: 13-19.

[37] Huber SM, Gamper N, Lang F. Chloride conductance and volume- regulatory nonselective cation conductance in human red blood cell ghosts. Pflugers Arch 2001; 441: 551-8.

[38] Duranton C, Huber SM, Lang F. Oxidation induces a Cl(-)-dependent cation conductance in human red blood cells. J Physiol 2002; 539: 847-55.

[39] Low TY, Seow TK, Chung MC. Separation of human erythrocyte membrane associated proteins with one-dimensional and two- dimensional gel electrophoresis followed by identification with matrix-assisted laser desorption/ionization-time of flight mass spectrometry. Proteomics 2002; 2: 1229-39.

[40] Bruschi M, Seppi C, Arena S, et al. Proteomic analysis of erythrocyte membranes by soft Immobiline gels combined with differential protein extraction. J Proteome Res 2005; 4: 1304-9.

[41] Logsdon NJ, Kang J, Togo JA, Christian EP, Aiyar J. A novel gene, hKCa4, encodes the calcium-activated potassium channel in human T lymphocytes. J Biol Chem 1997; 272: 32723-6.

[42] Jensen BS, Strobaek D, Christophersen P, et al. Characterization of the cloned human intermediate-conductance Ca^{2+}-activated K^+ channel. Am J Physiol 1998; 275: C848-56.

[43] Hoffman JF, Joiner W, Nehrke K, Potapova O, Foye K, Wickrema A. The hSK4 (KCNN4) isoform is the Ca^{2+}-activated K^+ channel (Gardos channel) in human red blood cells. Proc Natl Acad Sci USA 2003; 100: 7366-71.

[44] Kaestner L, Christophersen P, Bernhardt I, Bennekou P. The non- selective voltage-activated cation channel in the human red blood cell membrane: reconciliation between two conflicting reports and further characterisation. Bioelectrochemistry 2000; 52: 117-25.

[45] Kaestner L, Tabellion W, Weiss E, Bernhardt I, Lipp P. Calcium imaging of individual erythrocytes: problems and approaches. Cell Calcium 2006; 39: 13-19.

[46] Foller M, Kasinathan RS, Koka S, et al. TRPC6 contributes to the Ca(2+) leak of human erythrocytes. Cell Physiol Biochem 2008; 21: 183-92.

[47] Makhro A, Wang J, Vogel J, et al. Functional NMDA receptors in mammalian erythrocytes. Am J Physiol Cell Physiol 2010; 298: C1315-25.

[48] Andrews DA, Yang L, Low PS. Phorbol ester stimulates a protein kinase C-mediated agatoxin-TK-sensitive calcium permeability pathway in human red blood cells. Blood 2002; 100: 3392-9.

[49] Catterall WA, Perez-Reyes E, Snutch TP, Striessnig J. International Union of Pharmacology. XLVIII. Nomenclature and structure- function relationships of voltage-gated calcium channels. Pharmacol Rev 2005; 57: 411-25.

[50] Giarratana MC, Kobari L, Lapillonne H, et al. Ex vivo generation of fully mature human red blood cells from hematopoietic stem cells. Nat Biotechnol 2005; 23: 69-74.

[51] Jones EA, Baron MH, Fraser SE, Dickinson ME. Measuring hemodynamic changes during mammalian development. Am J Physiol Heart Circ Physiol 2004; 287: H1561-9.

[52] Scheller N, Resa-Infante P, de la Luna S, et al. Identification of PatL1, a human homolog to yeast P body component Pat1. Biochim Biophys Acta 2007; 1773: 1786-92.

[53] Kaestner L, Scholz A, Hammer K, Vecerdea A, Ruppenthal S, Lipp P. Isolation and genetic manipulation of adult cardiac myocytes for confocal imaging. J Vis Exp 2009; 31: http://www.jove.com/index/ Details.stp?ID=1433.

[54] Viero C, Kraushaar U, Ruppenthal S, Kaestner L, Lipp P. A primary culture system for sustained expression of a calcium sensor in preserved adult rat ventricular myocytes. Cell Calcium 2008; 43: 59-71.

[55] Kaestner L, Ruppenthal S, Schwarz S, Scholz A, Lipp P. Concepts for optical high content screens of excitable primary isolated cells for molecular imaging. In: Licha K, Lin CP, Eds. SPIE biomedical optics. Munich: SPIE 2009; pp. 737-45.

[56] Mohandas N, Gallagher PG. Red cell membrane: past, present, and future. Blood 2008; 112: 3939-48.

[57] Bennekou P, Barksmann TL, Jensen LR, Kristensen BI, Christophersen P. Voltage activation and hysteresis of the non-selective voltage-dependent channel in the intact human red cell. Bioelectrochemistry 2004; 62: 181-5.

[58] Rodighiero S, De Simoni A, Formenti A. The voltage-dependent nonselective cation current in human red blood cells studied by means of whole-cell and nystatin-perforated patch-clamp techniques. Biochim Biophys Acta 2004; 1660: 164-70.

[59] Bennekou P, Barksmann TL, Kristensen BI, Jensen LR, Christophersen P. Pharmacology of the human red cell voltage- dependent cation channel. Part II: inactivation and blocking. Blood Cells Mol Dis 2004; 33: 356-61.

[60] Bennekou P, Barksmann TL, Christophersen P, Kristensen BI. The human red cell voltage-dependent cation channel. Part III: Distribution homogeneity and pH dependence. Blood Cells Mol Dis 2006; 36: 10-14.

[61] Yang L, Andrews DA, Low PS. Lysophosphatidic acid opens a Ca(++) channel in human erythrocytes. Blood 2000; 95: 2420-5.

[62] Lang KS, Myssina S, Tanneur V, Wieder T, Huber SM, Lang F, Duranton C. Inhibition of erythrocyte cation channels and apoptosis by ethylisopropylamiloride. Naunyn Schmiedebergs Arch Pharmacol 2003; 367: 391-6.

[63] Myssina S, Huber SM, Birka C, Lang PA, Lang KS, Friedrich B, Risler T, Wieder T, Lang F. Inhibition of erythrocyte cation channels by erythropoietin. J Am Soc Nephrol 2003; 14: 2750-7.

[64] Abramowitz J, Birnbaumer L. Physiology and pathophysiology of canonical transient receptor potential channels. FASEB J 2009; 23: 297-328.

[65] Bogdanova A, Makhro A, Goede J, et al. NMDA receptors in mammalien erythrocytes. Clin Biochem 2009; 42: 1858-9.

[66] Yamakura T, Shimoji K. Subunit- and site-specific pharmacology of the NMDA receptor channel. Prog Neurobiol 1999; 59: 279-98.

[67] Li Q, Jungmann V, Kiyatkin A, Low PS. Prostaglandin E_2 stimulates a Ca^{2+}-dependent K^+ channel in human erythrocytes and alters cell volume and filterability. J Biol Chem 1996; 271: 18651- 6.

[68] Barksmann TL, Kristensen BI, Christophersen P, Bennekou P. Pharmacology of the human red cell voltage-dependent cation channel; Part I. Activation by clotrimazole and analogues. Blood Cells Mol Dis 2004; 32: 384-8.

[69] Grygorczyk R, Schwarz W. Ca^{2+}-activated K^+ permeability in human erythrocytes: modulation of single-channel events. Eur Biophys J 1985; 12: 57-65.

[70] Grygorczyk R. Temperature dependence of Ca^{2+}-activated K^+ currents in the membrane of human erythrocytes. Biochim Biophys Acta 1987; 902: 159-68.

[71] Kaestner L. Passive elektrogene Transportprozesse durch die Membran von Humanerythrozyten und Erythrozyten von gesunden und Malaria-infizierten Hühnern. Berlin: Logos Verlag 2001.

[72] Bernhardt I, Erdmann A, Glaser R, Reichmann G, Bleiber R. Influence of lipid composition on passive ion transport of erythrocytes. In: Klein RA, Schmitz B, Eds. Topics in lipid research fram structural elucidation to biological function. London: The Royal Society of Chemistry 1986; pp. 243-8.

[73] Lassen UV, Pape L, Vestergaard-Bogind B. Effect of calcium on the membrane potential of Amphiuma red cells. J Membr Biol 1976; 26: 51-70.

[74] Gardos G, Lassen UV, Pape L. Effect of antihistamines and chlorpromazine on the calcium-induced hyperpolarization of the Amphiuma red cell membrane. Biochim Biophys Acta 1976; 448: 599-606.

Subject Index

A

AB0-blood group system 22, 26
absorption 4, 9, 83
acousto-optical deflector 6, 38, 41
acquisition speed 5
actin 25, 26, 130, 135, 136, 137, 140, 142, 143, 147, 196, 218
action potential 43, 94, 135, 136, 137, 146, 147, 223
adeno-associated virus 18
adenosine-5'-triphosphate 108, 122, 132, 220
adenovirus 18, 60, 63, 67, 70, 136
adhesion force 23, 203, 205, 206
Aequorea victoria 14
aequorin 14
agglutination 22, 26
aggregability 22
aggregation index 22
Alzheimer's disease 171
angiogenesis 96
apoaequorin 14
apoptosis 26, 29, 42, 169, 210
Asante Calcium Red 4
atomic absorption spectroscopy 15
atomic force microscopy 22, 190, 203
atrioventricular node 40
autofluorescence 14, 39, 42, 43, 52, 83, 85, 88
avalanche photo diode 47

B

back-thinned camera 7, 38, 43, 47
bilirubin 86, 90
biocompatibility 5, 49, 53, 59
blips 45
blood vessel 24, 94, 95, 98, 101, 104

bumetanide 153, 154, 156
butanedione monoxime 42, 64, 135

C

calcium-induced calcium release 25, 135
calmodulin 5, 65, 68, 70, 111
calpain 26, 195, 196, 213, 218
calreticulin 104, 122
calsequestrin 24, 107
capillary sprouting 99
capillary vessel 39
carbachol 153, 154, 156, 157, 159, 226
carbon electrode 49, 50, 54, 69
carbon fibre 21
cardiac valve 42
caspase 220
cellular contraction 21, 24
ceramide 220
channelrhodopsin 19
Chlamydomonas reinhardtii 19
chromophore 9, 17
circularly permuted fluorescent protein 5
clot formation 162, 163, 166, 187, 195
clotrimazole 174, 175, 178, 184, 227, 228
coagulation 21, 166, 169, 187, 210
coelenterazine 14
collagen 39, 42, 43, 60, 61, 64
competent cells 60, 70
computer tomography 36
confocal principle 5
contamination 3, 229
contractility 2, 55, 104
contraction 105
coronary vessel 42
Coulter counter 15
cross striation 121, 124
cytochalasin D 42, 135
cytochrome C oxidase 122
cytokinesis 135

D

deformability 22
dendrimer 18
detubulation 25
diabetes 210
diabetic cardiomyopathy 25, 148
diffraction-limited spot 4, 12, 19
dihydropyridine 99, 105
dyadic junction 130
dynamic-link library 51
dyssynchrony index 146

E

echinocyte 192, 196
ejection fraction 21
elastic coating 57
elastin 42
electron multiplying camera 38
electron spin resonance 219
electroporation 18, 63
embryonic death 24
embryonic development 103
endoscopic imaging 36
entactin 64
enzyme-linked immunosorbent assay 17
epicard 39
eryptosis 26, 210
erythropoietin 184
evanescent wave 12
excitationcontraction coupling 3, 81, 107, 117, 135, 146
extracellular matrix 42, 43, 46, 55, 60, 68
extracorporal circulation 59

F

fibrin 197
fibroblast 101, 126
fibronectin 60

U

V

W

Y

FSC
www.fsc.org
MIX
Papier aus verantwortungsvollen Quellen
Paper from responsible sources
FSC® C105338